Essentials of MATLAB® Programming

Stephen J. Chapman

BAE SYSTEMS Australia

Australia • Canada • Mexico • Singapore • Spain • United Kingdom • United States

THOMSON
NELSON

Essentials of MATLAB® Programming
by Stephen J. Chapman

Associate Vice-President and Editorial Director:
Evelyn Veitch

Publisher:
Bill Stenquist

Sales and Marketing Manager:
John More

Developmental Editor:
Kamilah Reid Burrell

Permissions Coordinator:
Vicki Gould

Production Services:
RPK Editorial Services

Copy Editor:
Harlan James

Proofreader:
Erin Wagner

Indexer:
RPK Editorial Services

Production Manager:
Renate McCloy

Creative Director:
Angela Cluer

Interior Design:
Carmela Pereira

Cover Design:
Andrew Adams

Compositor:
Interactive Composition Corporation

Printer:
Webcom Limited

Printed and bound in United States of America
1 2 3 4 07 06 05

For more information contact Nelson, 1120 Birchmount Road, Toronto, Ontario, Canada, M1K 5G4.
Or you can visit our Internet site at http://www.nelson.com

Library of Congress Control Number: 2005931495

ISBN: 0-495-07300-8

North America
Nelson
1120 Birchmount Road
Toronto, Ontario M1K 5G4
Canada

Asia
Thomson Learning
5 Shenton Way #01-01
UIC Building
Singapore 068808

Australia/New Zealand
Thomson Learning
102 Dodds Street
Southbank, Victoria
Australia 3006

Europe/Middle East/Africa
Thomson Learning
High Holborn House
50/51 Bedford Row
London WC1R 4LR
United Kingdom

Latin America
Thomson Learning
Seneca, 53
Colonia Polanco
11560 Mexico D.F.
Mexico

Spain
Paraninfo
Calle/Magallanes, 25
28015 Madrid, Spain

This book is dedicated to my son,
MIDN David S. Chapman, of the
Royal Australian Navy.

Preface

MATLAB (short for MATrix LABoratory) is a special-purpose computer program optimized to perform engineering and scientific calculations. It started life as a program designed to perform matrix mathematics, but over the years it has grown into a flexible computing system capable of solving essentially any technical problem.

The MATLAB program implements the MATLAB language and provides a very extensive library of predefined functions to make technical programming tasks easier and more efficient. This extremely wide variety of functions makes it much easier to solve technical problems in MATLAB than in other languages such as Fortran or C. This book introduces the MATLAB language and shows how to use it to solve typical technical problems.

This book is *not* primarily a "how to use MATLAB" text (although students will learn how to use MATLAB to solve problems while using the text). Instead, the book teaches MATLAB as a technical programming language, in place of other languages such as Basic, Fortran, or C++. Most engineering curricula now require MATLAB and use it as an essential tool throughout the program. At the same time, most engineering curricula require students to become familiar with at least the basics of computer programming. The intention of this book is to satisfy both requirements simultaneously in a single course, freeing up precious time in engineering degree programs.

This book makes no pretense at being a complete description of all of MATLAB's hundreds of functions. Instead, it teaches the student how to use MATLAB as a language to solve problems, and how to locate any desired function with MATLAB's extensive on-line help facilities.

Essentials of MATLAB Programming is designed to serve as the text for an "Introduction to Programming/Problem Solving" course for freshman engineering students. This material should fit comfortably into a 9-week, 3-hour course.

The Advantages of MATLAB for Technical Programming

MATLAB has many advantages compared to conventional computer languages for technical problem solving. Among them are:

1. **Ease of Use**
   MATLAB is an interpreted language, like many versions of Basic. Like Basic, it is very easy to use. The program can be used as a scratch pad to evaluate expressions typed at the command line, or it can be used to execute large prewritten programs. Programs may be easily written and modified with the built-in integrated development environment and debugged with the MATLAB debugger. Because the language is so easy to use, it is ideal for educational use and for the rapid prototyping of new programs.

   Many program development tools are provided to make the program easy to use. They include an integrated editor/debugger, on-line documentation and manuals, a workspace browser, and extensive demos.

2. **Platform Independence**
   MATLAB is supported on many different computer systems, providing a large measure of platform independence. At the time of this writing, the language is supported on Windows NT/2000/XP, Linux, Unix, and the Macintosh. Programs written on any platform will run on all of the other platforms, and data files written on any platform may be read transparently on any other platform. As a result, programs written in MATLAB can migrate to new platforms when the needs of the user change.

3. **Predefined Functions**
   MATLAB comes complete with an extensive library of predefined functions that provide tested and prepackaged solutions to many basic technical tasks. For example, suppose that you are writing a program that must calculate the statistics associated with an input data set. In most languages, you would need to write your own subroutines or functions to implement calculations such as the arithmetic mean, standard deviation, median, etc. These and hundreds of other functions are built right into the MATLAB language, making your job much easier.

   In addition to the large library of functions built into the basic MATLAB language, there are many special-purpose toolboxes available to help solve complex problems in specific areas. For example, a user can buy standard toolboxes to solve problems in signal processing, control systems, communications, image processing, and neural networks, among many others.

4. **Device-Independent Plotting**
   Unlike other computer languages, MATLAB has many integral plotting and imaging commands. The plots and images can be displayed on any graphical output device supported by the computer on which MATLAB is running. This capability makes MATLAB an outstanding tool for visualizing technical data.

5. **Graphical User Interface**
   MATLAB includes tools that allow a program to interactively construct a Graphical User Interface (GUI) for his or her program. With this capability, the programmer can design sophisticated data analysis programs that can be operated by relatively inexperienced users.

Features of This Book

Many features of this book are designed to emphasize the proper way to write reliable MATLAB programs. These features should serve a student well as he or she is first learning MATLAB and should also be useful to the practitioner on the job. They include:

1. **Emphasis on Top-Down Design Methodology**
   The book introduces a top-down design methodology in Chapter 3, and uses it consistently throughout the rest of the book. This methodology encourages a student to think about the proper design of a program *before* beginning to code. It emphasizes the importance of clearly defining the problem to be solved and the required inputs and outputs before any other work is begun. Once the problem is properly defined, it teaches the student to employ stepwise refinement to break the task down into successively smaller subtasks and to implement the subtasks as separate subroutines or functions. Finally, it teaches the importance of testing at all stages of the process, both unit testing of the component routines and exhaustive testing of the final product.

   The formal design process taught by the book may be summarized as follows:

   1. *Clearly state the problem that you are trying to solve.*
   2. *Define the inputs required by the program and the outputs to be produced by the program.*
   3. *Describe the algorithm that you intend to implement in the program.* This step involves top-down design and stepwise decomposition, using pseudocode or flow charts.
   4. *Turn the algorithm into MATLAB statements.*
   5. *Test the MATLAB program.* This step includes unit testing of specific functions, and also exhaustive testing of the final program with many different data sets.

2. **Emphasis on Functions**
   The book emphasizes the use of functions to logically decompose tasks into smaller subtasks. It teaches the advantages of functions for data hiding. It also emphasizes the importance of unit-testing functions before they are combined into the final program. In addition, the book teaches about the common mistakes made with functions, and how to avoid them.

3. **Emphasis on MATLAB Tools**
   The book teaches the proper use of MATLAB's built-in tools to make programming and debugging easier. The tools covered include the Editor/Debugger, Workspace Browser, Help Browser, and GUI design tools.

4. **Good Programming Practice Boxes**
   These boxes highlight good programming practices when they are introduced for the convenience of the student. In addition, the good programming practices introduced in a chapter are summarized at the end of the chapter. An example Good Programming Practice Box is shown below.

Good Programming Practice

Always indent the body of an `if` construct by two or more spaces to improve the readability of the code.

5. **Programming Pitfalls Boxes**
   These boxes highlight common errors so that they can be avoided. An example Programming Pitfalls Box is shown below.

Programming Pitfalls

Make sure that your variable names are unique in the first 63 characters. Otherwise, MATLAB will not be able to tell the difference between them.

Pedagogical Features

This book is specifically designed to be used in a freshman "Introduction to Program/Problem Solving" course. It should be possible to cover this material comfortably in a 9-week, 3-hour-per-week course. If there is insufficient time to cover all of the material in a particular Engineering program, Chapters 6 and 7 may be deleted, and the remaining material will still teach the fundamentals of programming and using MATLAB to solve problems. This feature should appeal to harassed engineering educators trying to cram ever more material into a finite curriculum.

The book includes several features designed to aid student comprehension. A total of 12 quizzes appear scattered throughout the chapters, with answers to all questions included in Appendix C. These quizzes can serve as a useful self-test of comprehension. In addition, there are approximately 130 end-of-chapter exercises. Answers to all exercises are included in the Instructor's Manual. Good programming practices are highlighted in all chapters with special Good Programming Practice boxes, and common errors are highlighted in Programming Pitfalls boxes. End-of-chapter materials include Summaries of Good Programming Practice and Summaries of MATLAB Commands and Functions.

The book is accompanied by an Instructor's Manual, containing the solutions to all end-of-chapter exercises. The source code for all examples in the book is available from the book's Web site, and the source code for all solutions in the Instructor's Manual is available separately to instructors.

A Final Note to the User

No matter how hard I try to proofread a document like this book, it is inevitable that some typographical errors will slip through and appear in print. If you should spot any such errors, please drop me a note via the publisher, and I will do my best to get them eliminated from subsequent printings and editions. Thank you very much for your help in this matter.

I will maintain a complete list of errata and corrections at the book's World Wide Web site, which is http://engineering.thomsonlearning.com. Please check that site for any updates and/or corrections.

STEPHEN J. CHAPMAN
Melbourne, Australia
April 25, 2005 (Anzac Day)

Contents

Chapter 2 MATLAB Basics 21

CHAPTER 1

Introduction to MATLAB

MATLAB (short for MATrix LABoratory) is a special-purpose computer program optimized to perform engineering and scientific calculations. It started life as a program designed to perform matrix mathematics, but over the years it has grown into a flexible computing system capable of solving essentially any technical problem.

The MATLAB program implements the MATLAB programming language, and provides a very extensive library of predefined functions to make technical programming tasks easier and more efficient. This book introduces the MATLAB language as it is implemented in MATLAB Version 7, and shows how to use it to solve typical technical problems.

MATLAB is a huge program, with an incredibly rich variety of functions. Even the basic version of MATLAB without any toolkits is much richer than other technical programming languages. There are more than 1,000 functions in the basic MATLAB product alone, and the toolkits extend this capability with many more functions in various specialties. This book makes no attempt to introduce the user to all of MATLAB's functions. Instead, it teaches a user the basics of how to use MATLAB, and how to write, debug, and optimize good MATLAB programs, as well as a subset of the most important functions. Just as importantly, it teaches the programmer how to use MATLAB's own tools to locate the right function for a specific purpose from the enormous number of choices available.

1.1 The Advantages of MATLAB

MATLAB has many advantages compared with conventional computer languages for technical problem solving. Among them are:

1. **Ease of Use**
   MATLAB is an interpreted language, like many versions of Basic. Like Basic, it is very easy to use. The program can be used as a scratch pad to evaluate expressions typed at the command line, or it can be used to execute large pre-written programs. Programs may be easily written and modified with the built-in integrated development environment, and debugged with the MATLAB debugger. Because the language is so easy to use, it is ideal for the rapid prototyping of new programs.

   Many program development tools are provided to make the program easy to use. They include an integrated editor/debugger, on-line documentation and manuals, a workspace browser, and extensive demos.

2. **Platform Independence**
   MATLAB is supported on many different computer systems, providing a large measure of platform independence. At the time of this writing, the language is supported on Windows NT/2000/XP, Linux, several versions of Unix, and the Macintosh. Programs written on any platform will run on all of the other platforms, and data files written on any platform may be read transparently on any other platform. As a result, programs written in MATLAB can migrate to new platforms when the needs of the user change.

3. **Predefined Functions**
   MATLAB comes complete with an extensive library of predefined functions that provide tested and prepackaged solutions to many basic technical tasks. For example, suppose that you are writing a program that must calculate the statistics associated with an input data set. In most languages, you would need to write your own subroutines or functions to implement calculations such as the arithmetic mean, standard deviation, median, etc. These and hundreds of other functions are built right into the MATLAB language, making your job much easier.

   In addition to the large library of functions built into the basic MATLAB language, there are many special-purpose toolboxes available to help solve complex problems in specific areas. For example, a user can buy standard toolboxes to solve problems in signal processing, control systems, communications, image processing, and neural networks, among many others. There is also an extensive collection of free user-contributed MATLAB programs that are shared through the MATLAB Web site.

4. **Device-Independent Plotting**
   Unlike most other computer languages, MATLAB has many integral plotting and imaging commands. The plots and images can be displayed

on any graphical output device supported by the computer on which MATLAB is running. This capability makes MATLAB an outstanding tool for visualizing technical data.

5. **Graphical User Interface**
   MATLAB includes tools that allow a programmer to interactively construct a graphical user interface (GUI) for his or her program. With this capability, the programmer can design sophisticated data analysis programs that can be operated by relatively inexperienced users.

6. **MATLAB Compiler**
   MATLAB's flexibility and platform independence is achieved by compiling MATLAB programs into a device-independent p-code, and then interpreting the p-code instructions at run-time. This approach is similar to that used by Microsoft's Visual Basic language. Unfortunately, the resulting programs can sometimes execute slowly, because the MATLAB code is interpreted rather than compiled. We will point out features that tend to slow program execution when we encounter them. The latest versions of MATLAB have partially overcome this problem by introducing just-in-time (JIT) compiler technology. The JIT compiler compiles portions of the MATLAB code as it is executed to increase overall speed.

   A separate MATLAB compiler is available. This compiler can compile a MATLAB program into a stand-alone executable that can run without a MATLAB license. It is a great way to convert a prototype MATLAB program into an executable suitable for sale and distribution to users.

1.2 Disadvantages of MATLAB

MATLAB has two principal disadvantages. The first is that it is an interpreted language and therefore can execute more slowly than compiled languages. This problem can be mitigated by properly structuring the MATLAB program to maximize the performance of vectorized code, and by MATLAB's just-in-time compiler.

The second disadvantage is cost: a full copy of MATLAB is five to ten times more expensive than a conventional C++ or Fortran compiler. This relatively high cost is more than offset by the reduced time required for an engineer or scientist to create a working program, so MATLAB is cost-effective for businesses. However, it is too expensive for most individuals to consider purchasing. Fortunately, there is also an inexpensive Student Edition of MATLAB, which is a great tool for students wishing to learn the language. The Student Edition of MATLAB is essentially identical to the full edition, although there are fewer toolboxes available to it.

1.3 The MATLAB Environment

The fundamental unit of data in any MATLAB program is the **array**. An array is a collection of data values organized into rows and columns, and known by a single name. Individual data values within an array can be accessed by including the name of the array followed by subscripts in parentheses that identify the row and column of the particular value. Even scalars are treated as arrays by MATLAB—they are simply arrays with only one row and one column. We will learn how to create and manipulate MATLAB arrays in Section 1.4.

When MATLAB executes, it can display several types of windows that accept commands or display information. The three most important types of windows are Command Windows, where commands may be entered, Figure Windows, which display plots and graphs, and Edit Windows, which permit a user to create and modify MATLAB programs. We will see examples of all three types of windows in this section.

In addition, MATLAB can display other windows that provide help and that allow the user to examine the values of variables defined in memory. We will examine some of these additional windows here, and examine the others when we discuss how to debug MATLAB programs.

1.3.1 The MATLAB Desktop

When you start MATLAB Version 7, a special window called the MATLAB desktop appears. The desktop is a window that contains other windows showing MATLAB data, plus toolbars and a "Start" button similar to that used by Windows 2000 or XP. By default, most MATLAB tools are "docked" to the desktop, so that they appear inside the desktop window. However, the user can choose to "undock" any or all tools, making them appear in windows separate from the desktop.

The default configuration of the MATLAB desktop is shown in Figure 1.1. It integrates many tools for managing files, variables, and applications within the MATLAB environment.

The major tools within or accessible from the MATLAB desktop are:

- The Command Window
- The Command History Window
- The Start Button
- The Documents Window, including the Editor/Debugger and Array Editor
- The Figure Windows
- The Workspace Browser
- The Help Browser
- The Path Browser

The functions of these tools are discussed in later sections of this chapter.

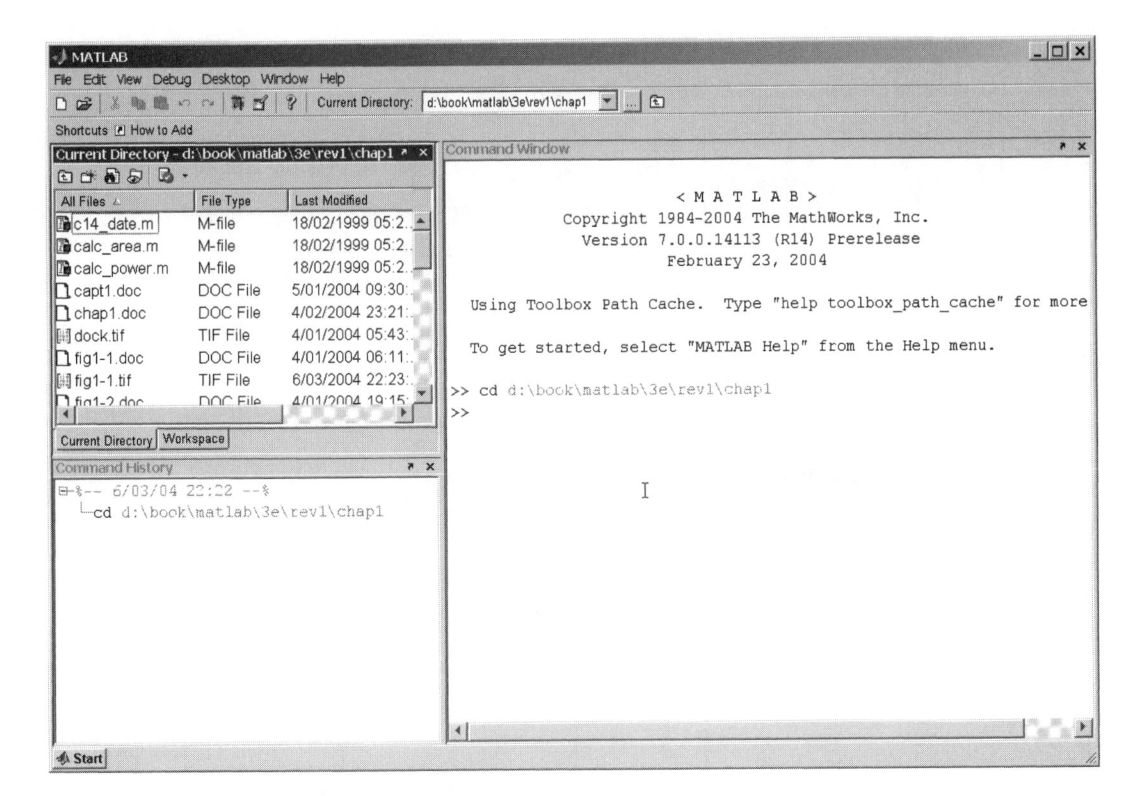

Figure 1.1 The default MATLAB desktop. The exact appearance of the desktop may differ slightly on different types of computers.

1.3.2 The Command Window

The right hand side of the default MATLAB desktop contains the **Command Window**. A user can enter interactive commands at the command prompt (») in the Command Window, and they will be executed on the spot.

As an example of a simple interactive calculation, suppose that you want to calculate the area of a circle with a radius of 2.5 m. This can be done in the MATLAB Command Window by typing:

```
» area = pi * 2.5^2
area =
   19.6350
```

MATLAB calculates the answer as soon as the Enter key is pressed, and stores the answer in a variable (really a 1×1 array) called `area`. The contents of the variable are displayed in the Command Window as shown in Figure 1.2, and the variable can be used in further calculations. (Note that π is predefined in MATLAB, so we can just use `pi` without first declaring it to be 3.141592 . . . .)

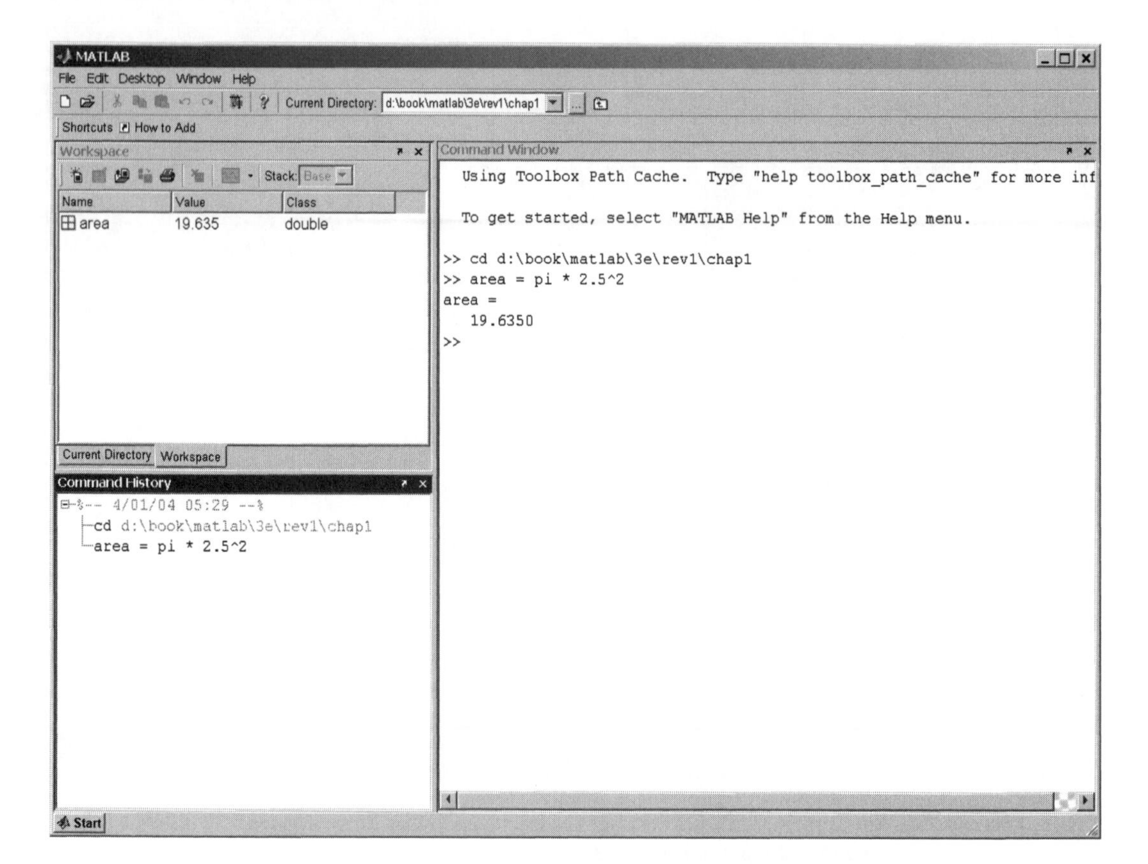

Figure 1.2 The Command Window appears on the right side of the desktop. Users enter commands and see responses here.

If a statement is too long to type on a single line, it may be continued on successive lines by typing an **ellipsis** (. . .) at the end of the first line, and then continuing on the next line. For example, the following two statements are identical.

```
x1 = 1 + 1/2 + 1/3 + 1/4 + 1/5 + 1/6;
```

and

```
x1 = 1 + 1/2 + 1/3 + 1/4 ...
       + 1/5 + 1/6;
```

Instead of typing commands directly in the Command Window, a series of commands can be placed into a file, and the entire file can be executed by typing its name in the Command Window. Such files are called **script files**. Script files (and functions, which we will see later) are also known as **M-files**, because they have a file extension of "`.m`".

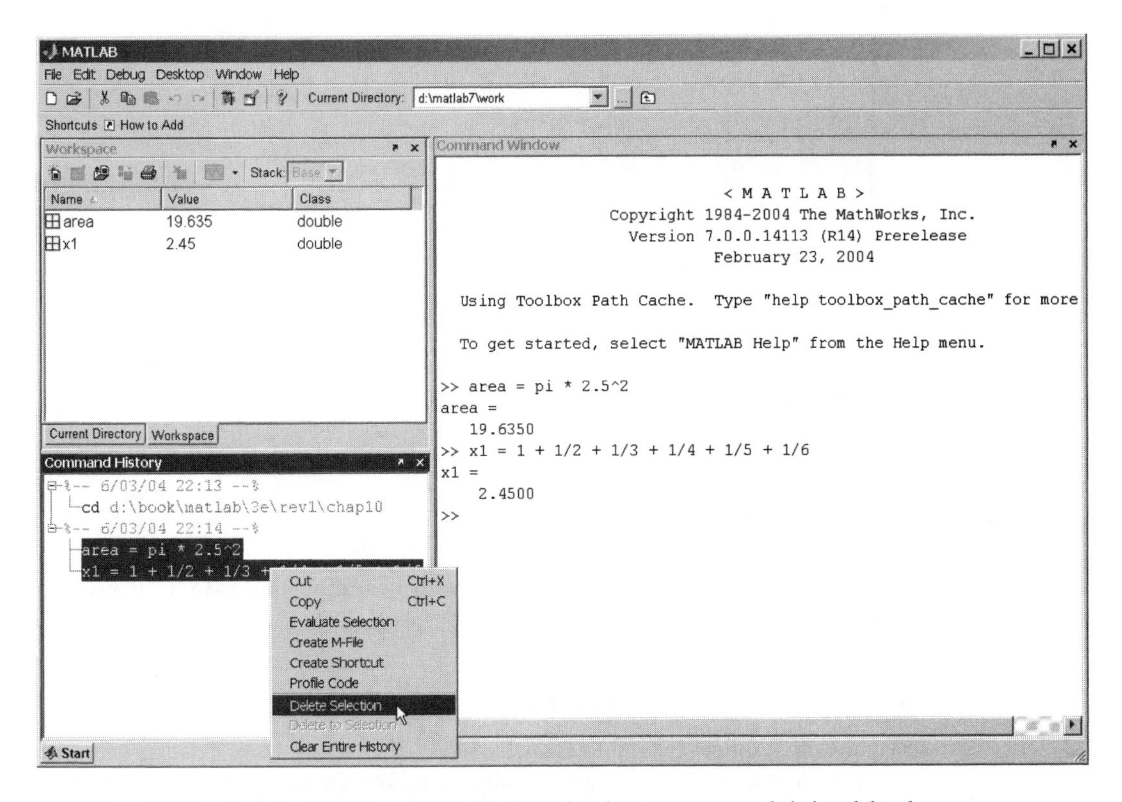

Figure 1.3 The Command History Window, showing two commands being deleted.

1.3.3 The Command History Window

The Command History window displays a list of the commands that a user has entered in the Command Window. The list of previous commands can extend back to previous executions of the program. Commands remain in the list until they are deleted. To reexecute any command, simply double-click it with the left mouse button. To delete one or more commands from the Command History window, select the commands and right-click them with the mouse. A popup menu will be displayed that allows the user to delete the items (see Figure 1.3).

1.3.4 The Start Button

The Start Button (see Figure 1.4) allows a user to access MATLAB tools, desktop tools, help files, etc. It works just like the Start button on a Windows desktop. To start a particular tool, just click on the Start Button and select the tool from the appropriate sub-menu.

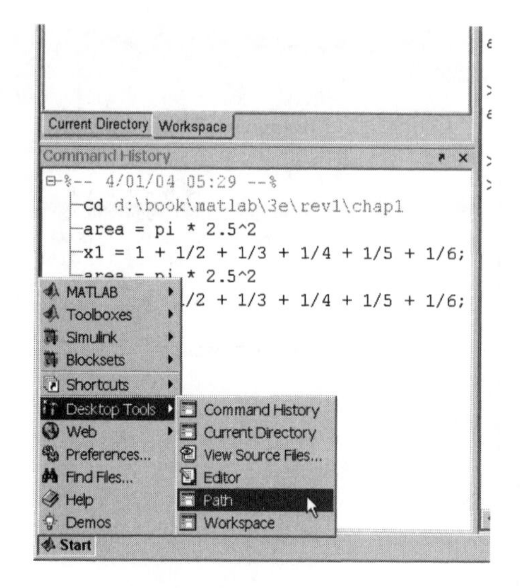

Figure 1.4 The Start Button, which allows a user to select from a wide variety of MATLAB and desktop tools.

1.3.5 The Edit/Debug Window

An **Edit Window** is used to create new M-files, or to modify existing ones. An Edit Window is created automatically when you create a new M-file or open an existing one. You can create a new M-file with the "File/New/M-file" selection from the desktop menu, or by clicking the toolbar icon. You can open an existing M-file file with the "File/Open" selection from the desktop menu, or by clicking the toolbar icon.

An Edit Window displaying a simple M-file called `calc_area.m` is shown in Figure 1.5. This file calculates the area of a circle given its radius, and displays the result. By default, the Edit Window is an independent window not docked to the desktop, as shown in Figure 1.5*(a)*. The Edit Window can also be docked to the MATLAB desktop. In that case, it appears within a container called the Documents Window, as shown in Figure 1.5*(b)*. We will learn how to dock and undock a window later in this chapter.

The Edit Window is essentially a programming text editor, with the MATLAB language features highlighted in different colors. Comments in an M-file file appear in green, variables and numbers appear in black, complete character strings appear in magenta, incomplete character strings appear in red, and language keywords appear in blue.

After an M-file is saved, it may be executed by typing its name in the Command Window. For the M-file in Figure 1.5, the results are:

```
» calc_area
The area of the circle is 19.635
```

The Edit Window also doubles as a debugger, as we shall see in Chapter 2.

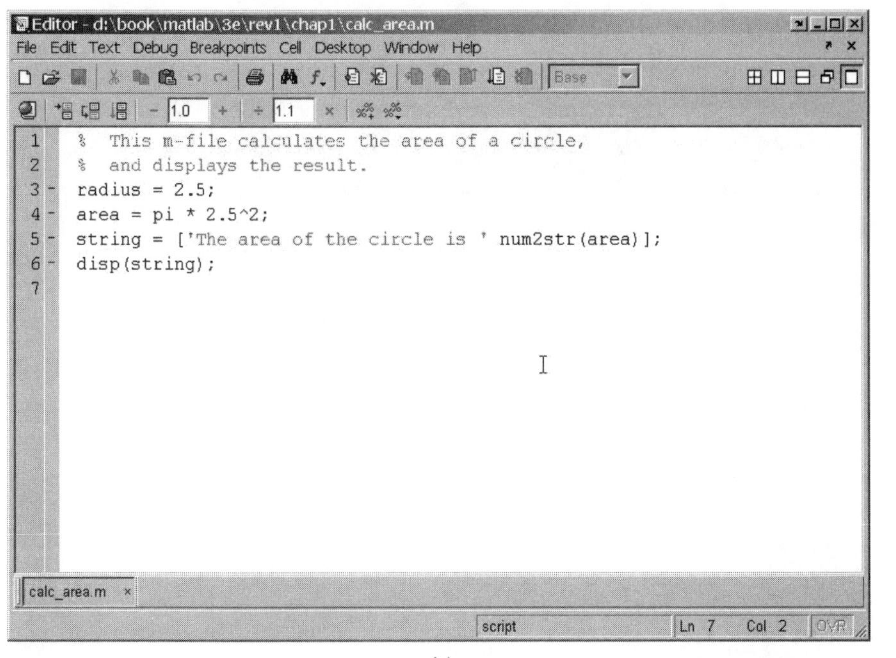

(a)

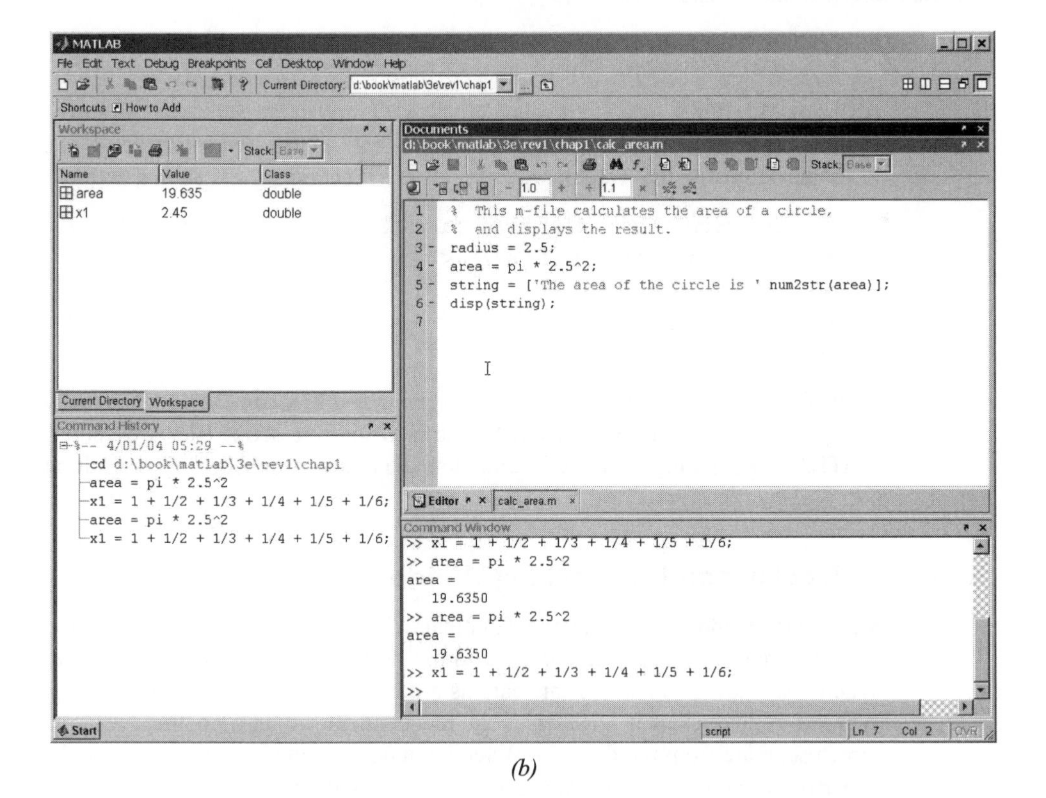

(b)

Figure 1.5 *(a)* The MATLAB Editor, displayed as an independent window. *(b)* The MATLAB Editor, docked to the MATLAB desktop.

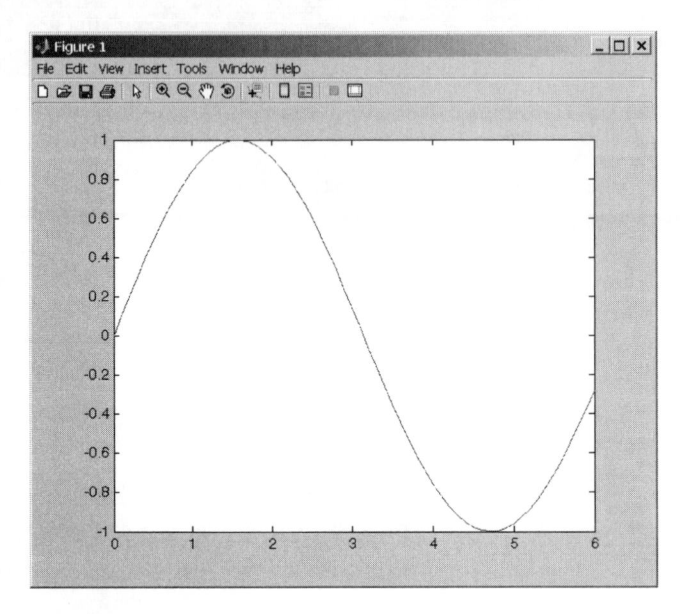

Figure 1.6 MATLAB plot of sin *x* versus *x*.

1.3.6 Figure Windows

A **Figure Window** is used to display MATLAB graphics. A figure can be a two- or three-dimensional plot of data, an image, or a graphical user interface (GUI). A simple script file that calculates and plots the function sin *x* is shown below:

```
% sin_x.m: This M-file calculates and plots the
% function sin(x) for 0 <= x <= 6.
x = 0:0.1:6;
y = sin(x);
plot(x,y);
```

If this file is saved under the name `sin_x.m`, then a user can execute the file by typing "`sin_x`" in the Command Window. When this script file is executed, MATLAB opens a Figure Window and plots the function sin *x* in it. The resulting plot is shown in Figure 1.6.

1.3.7 Docking and Undocking Windows

MATLAB windows such as the Command Window, the Edit Window, and Figure Windows can either be *docked* to the desktop, or they can be *undocked*. When a window is docked, it appears as a pane within the MATLAB desktop. When it is undocked it appears as an independent window on the computer screen separate from the desktop. When a window is docked to the desktop, the upper right-hand corner contains a small button with an arrow pointing up and to the right ( ). If this button is clicked, the window will become an independent window. When the window is an independent window, the upper right-hand corner contains a small button with an arrow pointing down and to the right ( ). If this button is clicked, the window will be re-docked with the desktop. Figure 1.5 shows the Edit Window in both its docked and undocked state. Note the undock and dock arrows in the upper right hand corner.

1.3.8 The MATLAB Workspace

A statement like

```
z = 10;
```

creates a variable named z, stores the value 10 in it, and saves it in a part of computer memory known as the **workspace**. A workspace is the collection of all the variables and arrays that can be used by MATLAB when a particular command, M-file, or function is executing. All commands executed in the Command Window (and all script files executed from the Command Window) share a common workspace, so they can all share variables. As we will see later, MATLAB functions differ from script files in that each function has its own separate workspace.

A list of the variables and arrays in the current workspace can be generated with the `whos` command. For example, after M-files `calc_area` and `sin_x` are executed, the workspace contains the following variables.

```
» whos
  Name          Size          Bytes   Class
  area          1x1               8   double array
  radius        1x1               8   double array
  string        1x32             64   char array
  x             1x61            488   double array
  y             1x61            488   double array

Grand total is 156 elements using 1056 bytes
```

Script file `calc_area` created variables `area`, `radius`, and `string`, while script file `sin_x` created variables `x` and `y`. Note that all of the variables are in the same workspace, so if two script files are executed in succession, the second script file can use variables created by the first script file.

The contents of any variable or array may be determined by typing the appropriate name in the Command Window. For example, the contents of `string` can be found as follows:

```
» string
string =
The area of the circle is 19.635
```

A variable can be deleted from the workspace with the `clear` command. The `clear` command takes the form

```
clear var1 var2 ...
```

where `var1` and `var2` are the names of the variables to be deleted. The command `clear variables` or simply `clear` deletes all variables from the current workspace.

1.3.9 The Workspace Browser

The contents of the current workspace can also be examined with a GUI-based Workspace Browser. The Workspace Browser appears by default in the upper left-hand corner of the desktop. It provides a graphic display of the same information as the `whos` command, and it also shows the actual contents of each array if the information is short enough to fit within the display area. The Workspace Browser is dynamically updated whenever the contents of the workspace change.

A typical Workspace Browser window is shown in Figure 1.7. As you can see, it displays the same information as the `whos` command. Double-clicking on any variable in the window will bring up the Array Editor, which allows the user to modify the information stored in the variable.

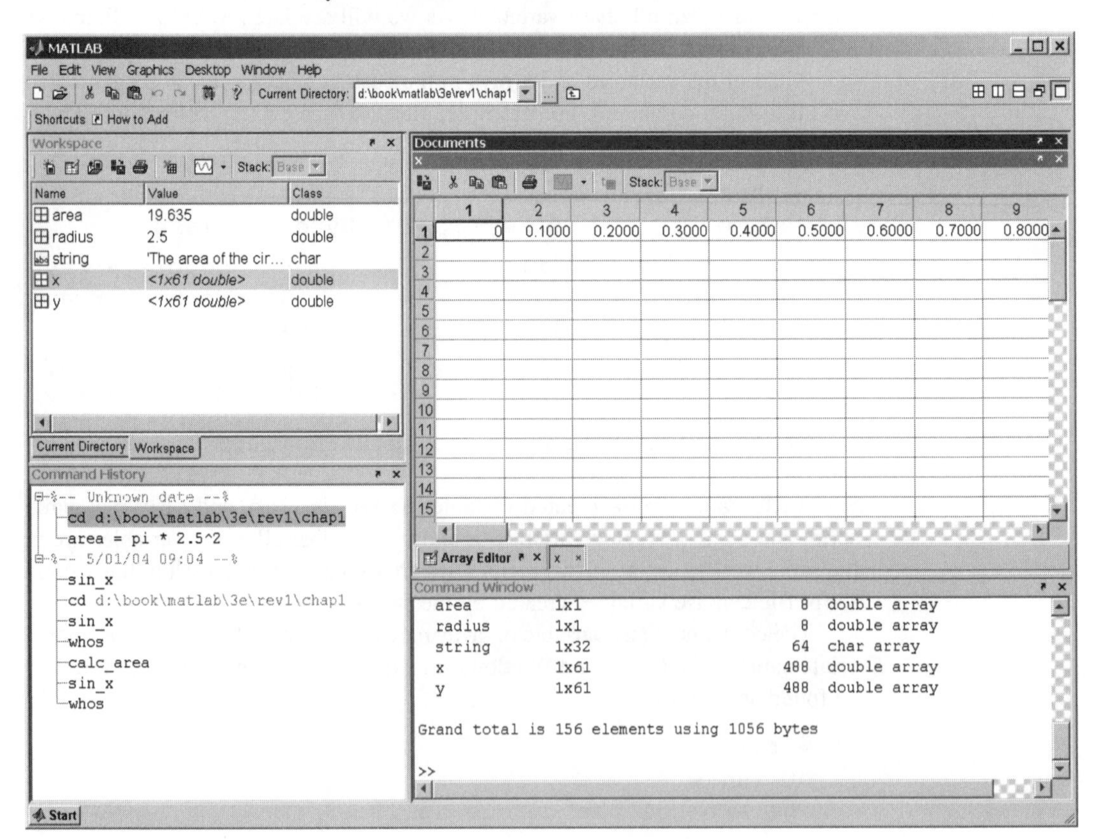

Figure 1.7 The Workspace Browser and the Array Editor. The Array Editor is invoked by double-clicking a variable in the Workspace Browser. It allows a user to change the values contained in a variable or array.

One or more variables may be deleted from the workspace by selecting them in the Workspace Browser with the mouse and pressing the delete key, or by right-clicking with the mouse and selecting the delete option.

1.3.10 Getting Help

There are three ways to get help in MATLAB. The preferred method is to use the Help Browser. The Help Browser can be started by selecting the ? icon from the desktop toolbar, or by typing `helpdesk` or `helpwin` in the Command Window. A user can get help by browsing the MATLAB documentation, or he or she can search for the details of a particular command. The Help Browser is shown in Figure 1.8.

There are also two command-line oriented ways to get help. The first way is to type `help` or `help` followed by a function name in the Command Window. If you just type `help`, MATLAB will display a list of possible help topics in the Command Window. If a specific function or a toolbox name is included, help will be provided for that particular function or toolbox.

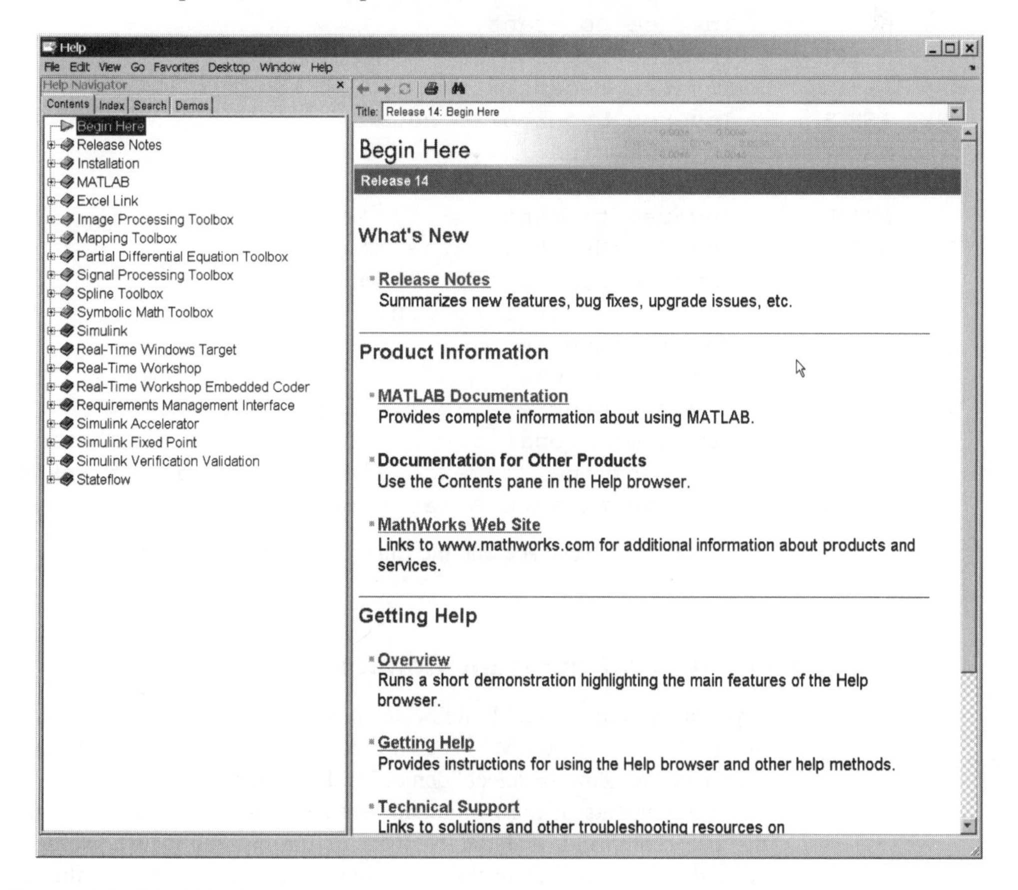

Figure 1.8 The Help Browser.

The second way to get help is the `lookfor` command. The `lookfor` command differs from the `help` command in that the `help` command searches for an exact function name match, while the `lookfor` command searches the quick summary information in each function for a match. This makes `lookfor` slower than `help`, but it improves the chances of getting back useful information. For example, suppose that you were looking for a function to take the inverse of a matrix. Since MATLAB does not have a function named `inverse`, the command "`help inverse`" will produce nothing. On the other hand, the command "`lookfor inverse`" will produce the following results:

```
» lookfor inverse
INVHILB    Inverse Hilbert matrix.
ACOS       Inverse cosine.
ACOSH      Inverse hyperbolic cosine.
ACOT       Inverse cotangent.
ACOTH      Inverse hyperbolic cotangent.
ACSC       Inverse cosecant.
ACSCH      Inverse hyperbolic cosecant.
ASEC       Inverse secant.
ASECH      Inverse hyperbolic secant.
ASIN       Inverse sine.
ASINH      Inverse hyperbolic sine.
ATAN       Inverse tangent.
ATAN2      Four quadrant inverse tangent.
ATANH      Inverse hyperbolic tangent.
ERFINV     Inverse error function.
INV        Matrix inverse.
PINV       Pseudoinverse.
IFFT       Inverse discrete Fourier transform.
IFFT2      Two-dimensional inverse discrete Fourier transform.
IFFTN      N-dimensional inverse discrete Fourier transform.
IPERMUTE   Inverse permute array dimensions.
```

From this list, we can see that the function of interest is named `inv`.

1.3.11 A Few Important Commands

If you are new to MATLAB, a few demonstrations may help to give you a feel for its capabilities. To run MATLAB's built-in demonstrations, type `demo` in the Command Window, or select "demos" from the Start button.

The contents of the Command Window can be cleared at any time using the `clc` command, and the contents of the current Figure Window can be cleared at any time using the `clf` command. The variables in the workspace can be cleared with the `clear` command. As we have seen, the contents of the workspace persist between the executions of separate commands and M-files, so it is possible for the results of one problem to have an effect on the next one that you may attempt to solve. To avoid this possibility, it is a good idea to issue the `clear` command at the start of each new independent calculation.

Another important command is the **abort** command. If an M-file appears to be running for too long, it may contain an infinite loop, and it will never terminate. In this case, the user can regain control by typing control-c (abbreviated `^c`) in the Command Window. This command is entered by holding down the control key while typing a "c". When MATLAB detects a `^c`, it interrupts the running program and returns a command prompt.

The exclamation point (`!`) is another important special character. Its special purpose is to send a command to the computer's operating system. Any characters after the exclamation point will be sent to the operating system and executed as though they had been typed at the operating system's command prompt. This feature lets you embed operating system commands directly into MATLAB programs.

Finally, it is possible to keep track of everything done during a MATLAB session with the **diary** command. The form of this command is

```
diary filename
```

After this command is typed, a copy of all input and most output typed in the Command Window is echoed in the diary file. This is a great tool for recreating events when something goes wrong during a MATLAB session. The command "`diary off`" suspends input into the diary file, and the command "`diary on`" resumes input again.

1.3.12 The MATLAB Search Path

MATLAB has a search path that it uses to find M-files. MATLAB's M-files are organized in directories on your file system. Many of these directories of M-files are provided along with MATLAB, and users may add others. If a user enters a name at the MATLAB prompt, the MATLAB interpreter attempts to find the name as follows:

1. It looks for the name as a variable. If it is a variable, MATLAB displays the current contents of the variable.
2. It checks to see if the name is an M-file in the current directory. If it is, MATLAB executes that function or command.
3. It checks to see if the name is an M-file in any directory in the search path. If it is, MATLAB executes that function or command.

Note that MATLAB checks for variable names first, so *if you define a variable with the same name as a MATLAB function or command, that function or command becomes inaccessible*. This is a common mistake made by novice users.

Programming Pitfalls

Never use a variable with the same name as a MATLAB function or command. If you do so, that function of command will become inaccessible.

Also, if there is more than one function or command with the same name, the *first* one found on the search path will be executed, and all of the others will be inaccessible. This is a common problem for novice users, since they sometimes create M-files files with the same names as standard MATLAB functions, making them inaccessible.

Programming Pitfalls

Never create an M-file with the same name as a MATLAB function or command.

MATLAB includes a special command (`which`) to help you find out just which version of a file is being executed and where it is located. This can be useful in finding filename conflicts. The format of this command is `which` *`functionname`*, where *`functionname`* is the name of the function that you are trying to locate. For example, the cross-product function `cross.m` can be located as follows:

```
» which cross
C:\MATLAB704\toolbox\matlab\specfun\cross.m
```

The MATLAB search path can be examined and modified at any time by selecting "Desktop Tools/Path" from the Start Button, or by typing `editpath` in the Command Window. The Path Tool is shown in Figure 1.9. It allows a user to add, delete, or change the order of directories in the path.

Other path-related functions include:

- `addpath` Add directory to MATLAB search path.
- `path` Display MATLAB search path.
- `path2rc` Adds current directory to MATLAB search path.
- `rmpath` Remove directory from MATLAB search path.

1.4 Using MATLAB as a Scratch Pad

In its simplest form, MATLAB can be used as a scratch pad to perform mathematical calculations. The calculations to be performed are typed directly into the Command Window, using the symbols +, –, *, /, and ^ for addition, subtraction, multiplication, division, and exponentiation respectively. After an expression is typed,

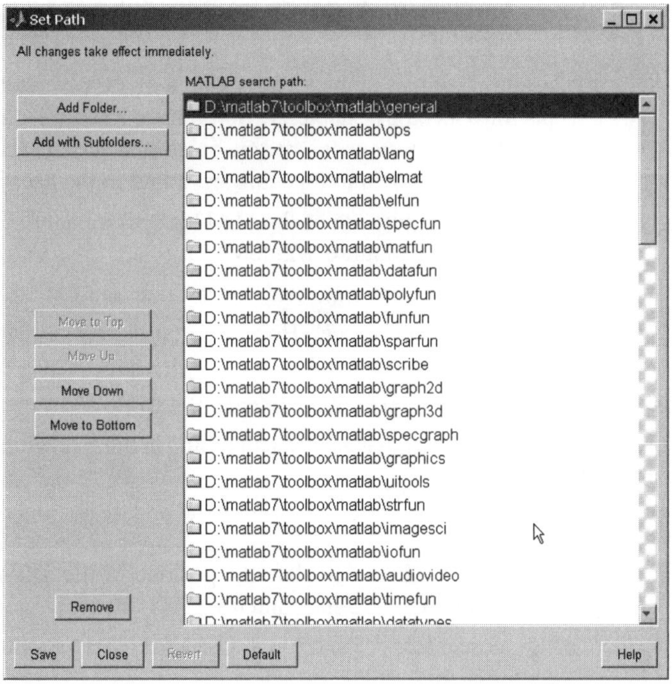

Figure 1.9 The Path Tool.

the results of the expression will be automatically calculated and displayed. For example, suppose we would like to calculate the volume of a cylinder of radius r and length l. The area of the circle at the base of the cylinder is given by the equation

$$A = \pi r^2 \tag{1-1}$$

and the total volume of the cylinder will be

$$V = Al \tag{1-2}$$

If the radius of the cylinder is 0.1 m and the length is 0.5 m, the volume of the cylinder can be found using the MATLAB statements (user inputs are shown in bold face):

```
» A = pi * 0.1^2
A =
    0.0314
» V = A * 0.5
V =
    0.0157
```

Note that `pi` is predefined to be the value 3.141592 . . . Also, note that the value stored in `A` was saved by MATLAB and re-used when we calculated `V`.

Quiz 1.1

This quiz provides a quick check to see if you have understood the concepts introduced in Chapter 1. If you have trouble with the quiz, reread the sections, ask your instructor, or discuss the material with a fellow student. The answers to this quiz are found in the back of the book.

1. What is the purpose of the MATLAB Command Window? The Edit Window? The Figure Window?
2. List the different ways that you get help in MATLAB.
3. What is a workspace? How can you determine what is stored in a MATLAB workspace?
4. How can you clear the contents of a workspace?
5. The distance traveled by a ball falling in the air is given by the equation

$$x = x_0 + v_0 t + \frac{1}{2} a t^2$$

   Use MATLAB to calculate the position of the ball at time $t = 5$ s if $x_0 = 10$ m, $v_0 = 15$ m/s, and $a = -9.81$ m/sec^2.

6. Suppose that $x = 3$ and $y = 4$. Use MATLAB to evaluate the following expression:

$$\frac{x^2 y^3}{(x - y)^2}$$

The following questions are intended to help you become familiar with MATLAB tools.

7. Execute the M-files `calc_area.m` and `sin_x.m` in the Command Window (these M-files are available from the book's Web site). Then use the Workspace Browser to determine what variables are defined in the current workspace.
8. Use the Array Editor to examine and modify the contents of variable `x` in the workspace. Then type the command `plot(x,y)` in the Command Window. What happens to the data displayed in the Figure Window?

1.5 Summary

In this chapter, we learned about the basic types of MATLAB windows, the workspace, and how to get on-line help. The MATLAB desktop appears when the program is started. It integrates many of the MATLAB tools in single location. These tools include the Command Window, the Command History Window, the

Start Button, the Workspace Browser, the Array Editor, and the Current Directory Viewer. The Command Window is the most important of the windows. It is the one in which all commands are typed and results are displayed.

The Edit/Debug window is used to create or modify M-files. It displays the contents of the M-file with the contents of the file color-coded according to function: comments, keywords, strings, and so forth. This window can be docked to the desktop, but by default it is independent.

The Figure Window is used to display graphics.

A MATLAB user can get help by using either the Help Browser or the command-line help functions `help` and `lookfor`. The Help Browser allows full access to the entire MATLAB documentation set. The command-line function `help` displays help about a specific function in the Command Window. Unfortunately, you must know the name of the function in order to get help about it. The function `lookfor` searches for a given string in the first comment line of every MATLAB function, and displays any matches.

When a user types a command in the Command Window, MATLAB searches for that command in the directories specified in the MATLAB path. It will execute the *first* M-file in the path that matches the command, and any further M-files with the same name will never be found. The Path Tool can be used to add, delete, or modify directories in the MATLAB path.

1.5.1 MATLAB Summary

The following summary lists all of the MATLAB special symbols described in this chapter, along with a brief description of each one.

Special Symbols	
+	Addition
-	Subtraction
*	Multiplication
/	Division
^	Exponentiation

1.6 Exercises

1.1 The following MATLAB statements plot the function $y(x) = 2e^{-0.2x}$ for the range $0 \leq x \leq 10$.

```
x = 0:0.1:10;
y = 2 * exp( -0.2 * x);
plot(x,y);
```

Use the MATLAB Edit Window to create a new empty M-file, type these statements into the file, and save the file with the name `test1.m`. Then, execute the program by typing the name `test1` in the Command Window. What result do you get?

1.2 Get help on the MATLAB function `exp` using: *(a)* The "`help exp`" command typed in the Command Window, and *(b)* the Help Browser.

1.3 Use the `lookfor` command to determine how to take the base-10 logarithm of a number in MATLAB.

1.4 Suppose that $u = 1$ and $v = 3$. Evaluate the following expressions using MATLAB.

(a) $\dfrac{4u}{3v}$

(b) $\dfrac{2v^{-2}}{(u + v)^2}$

(c) $\dfrac{v^3}{v^3 - u^3}$

(d) $\dfrac{4}{3}\pi v^2$

1.5 Use the MATLAB Help Browser to find the command required to show MATLAB's current directory. What is the current directory when MATLAB starts up?

1.6 Use the MATLAB Help Browser to find out how to create a new directory from within MATLAB. Then, create a new directory called `mynewdir` under the current directory. Add the new directory to the top of MATLAB's path.

1.7 Change the current directory to `mynewdir`. Then open an Edit Window and add the following lines:

```
% Create an input array from -2*pi to 2*pi
t = -2*pi:pi/10:2*pi;

% Calculate |sin(t)|
x = abs(sin(t));

% Plot result
plot(t,x);
```

Save the file with the name `test2.m`, and execute it by typing `test2` in the Command Window. What happens?

1.8 Close the Figure Window, and change back to the original directory that MATLAB started up in. Next type "`test2`" in the Command Window. What happens, and why?

CHAPTER 2

MATLAB Basics

In this chapter, we will introduce some basic elements of the MATLAB language. By the end of the chapter, you will be able to write simple but functional MATLAB programs and solve simple problem using MATLAB.

2.1 Variables and Arrays

The fundamental unit of data in any MATLAB program is the **array**. An array is a collection of data values organized into rows and columns and known by a single name (see Figure 2.1). Individual data values within an array are accessed by including the name of the array followed by subscripts in parentheses that identify the row and column of the particular value. Even scalars are treated as arrays by MATLAB—they are simply arrays with only one row and one column.

Arrays can be classified as either **vectors** or **matrices**. The term "vector" is usually used to describe an array with only one dimension, while the term "matrix" is usually used to describe an array with two or more dimensions. In this text, we will use the term "vector" when discussing one-dimensional arrays, and the term "matrix" when discussing arrays with two or more dimensions. If a particular discussion applies to both types of arrays, we will use the generic term "array."

The **size** of an array is specified by the number of rows and the number of columns in the array, with the number of rows mentioned first. The

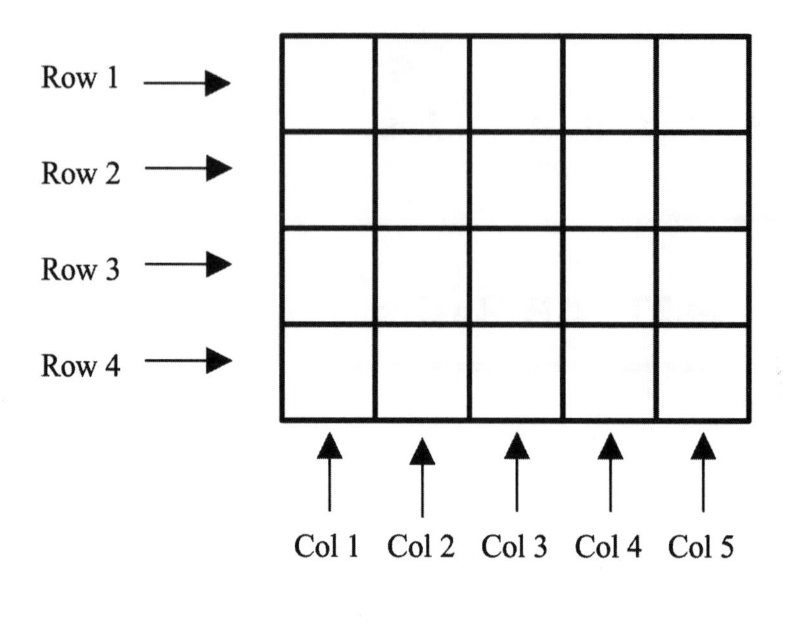

array `arr`

Figure 2.1 An array is a collection of data values organized into rows and columns.

total number of elements in the array will be the product of the number of rows and the number of columns. For example, the sizes of the following arrays are

Array	Size
$a = \begin{bmatrix} 1 & 2 \\ 3 & 4 \\ 5 & 6 \end{bmatrix}$	This is a 3 × 2 matrix, containing 6 elements.
$b = [1 \quad 2 \quad 3 \quad 4]$	This is a 1 × 4 array containing 4 elements, known as a **row vector**.
$c = \begin{bmatrix} 1 \\ 2 \\ 3 \end{bmatrix}$	This is a 3 × 1 array containing 3 elements, known as a **column vector**.

Individual elements in an array are addressed by the array name followed by the row and column of the particular element. If the array is a row or column vector, then only one subscript is required. For example, in the above arrays `a(2,1)` is 3 and `c(2)` = 2.

A MATLAB **variable** is a region of memory containing an array, which is known by a user-specified name. The contents of the array may be used or modified at any time by including its name in an appropriate MATLAB command.

MATLAB variable names must begin with a letter, followed by any combination of letters, numbers, and the underscore (_) character. Only the first 63 characters are significant; if more than 63 are used, the remaining characters will be ignored. If two variables are declared with names that differ only in the 64th character, MATLAB will treat them as the same variable. MATLAB will issue a warning if it has to truncate a long variable name to 63 characters.

Programming Pitfalls

Make sure that your variable names are unique in the first 63 characters. Otherwise, MATLAB will not be able to tell the difference between them.

When writing a MATLAB program, it is important to pick meaningful names for the variables. Meaningful names make a program *much* easier to read and to maintain. Names such as `day`, `month`, and `year` are quite clear even to a person seeing a program for the first time. Since spaces cannot be used in MATLAB variable names, underscore characters can be substituted to create meaningful names. For example, *exchange rate* might become `exchange_rate`.

Good Programming Practice

Always give your variables descriptive and easy-to-remember names. For example, a currency exchange rate could be given the name `exchange_rate`. This practice will make your programs clearer and easier to understand.

It is also important to include a **data dictionary** in the header of any MATLAB program that you write. A data dictionary lists the definition of each variable used in a program. The definition should include both a description of the contents of the item and the units in which it is measured. A data dictionary may seem unnecessary while the program is being written, but it is invaluable when you or another person have to go back and modify the program at a later time.

Good Programming Practice

Create a data dictionary for each program to make program maintenance easier.

The MATLAB language is case sensitive, which means that uppercase and lowercase letters are not the same. Thus the variables `name`, `NAME`, and `Name` are all different in MATLAB. You must be careful to use the same capitalization every time that variable name is used. While it is not required, it is customary to use all lower-case letters for ordinary variable names.

Good Programming Practice

Be sure to capitalize a variable exactly the same way each time that it is used. It is good practice to use only lowercase letters in variable names.

The most common types of MATLAB variables are `double` and `char`. Variables of type `double` consist of scalars or arrays of 64-bit double-precision floating-point numbers. They can hold real, imaginary, or complex values. The real and imaginary components of each variable can be positive or negative numbers in the range 10^{-308} to 10^{308}, with 15 to 16 significant decimal digits of accuracy. They are the principal numerical data type in MATLAB.

A variable of type `double` is automatically created whenever a numerical value is assigned to a variable name. The numerical values assigned to `double` variables can be real, imaginary, or complex. A real value is just a number. For example, the following statement assigns the real value 10.5 to the `double` variable `var`:

```
var = 10.5;
```

An imaginary value is defined by appending the letter `i` or `j` to a number. For example, `10i` and `-4j` are both imaginary values. The following statement assigns the imaginary value $4i$ to the `double` variable `var`:

```
var = 4i;
```

A complex value has both a real and an imaginary component. It is created by adding a real and an imaginary number together. For example, the following statement assigns the complex value $10 + 10i$ to variable `var`:

```
var = 10 + 10i;
```

Variables of type `char` consist of scalars or arrays of 16-bit values, each representing a single character. Arrays of this type are used to hold character strings. They are automatically created whenever a single character or a character string is assigned to a variable name. For example, the following statement creates a variable of type `char` whose name is `comment`, and stores the specified string in it. After the statement is executed, `comment` will be a 1×26 character array.

```
comment = 'This is a character string';
```

In a language such as C, the type of every variable must be explicitly declared in a program before it is used. These languages are said to be **strongly typed**. In contrast, MATLAB is a **weakly typed** language. Variables may be created at any time by simply assigning values to them, and the type of data assigned to the variable determines the type of variable that is created.

2.2 Initializing Variables in MATLAB

MATLAB variables are automatically created when they are initialized. There are three common ways to initialize a variable in MATLAB:

1. Assign data to the variable in an assignment statement.
2. Input data into the variable from the keyboard.
3. Read data from a file.

The first two ways are discussed here, and the third approach is discussed in Section 2.6.

2.2.1 Initializing Variables in Assignment Statements

The simplest way to initialize a variable is to assign it one or more values in an **assignment statement**. An assignment statement has the general form

```
var = expression
```

where `var` is the name of a variable, and *`expression`* is a scalar constant, an array, or combination of constants, other variables, and mathematical operations (+, −, etc.). The value of the expression is calculated using the normal rules of mathematics, and the resulting values are stored in named variable. Simple examples of initializing variables with assignment statements include

```
var = 40i;
var2 = var/5;
array = [1 2 3 4];
x = 1; y = 2;
```

The first example creates a scalar variable of type `double` and stores the imaginary number $40i$ in it. The second example creates a scalar variable and stores the result of the expression `var/5` in it. The third example creates a variable and stores a 4-element row vector in it. The last example shows that multiple assignment statements can be placed on a single line, provided that they are separated by semicolons or commas. Note that if any of the variables had already existed when the statements were executed, their old contents would have been lost.

As the third example shows, variables can also be initialized with arrays of data. Such arrays are constructed using brackets (`[]`) and semicolons. All of the elements of an array are listed in **row order**. In other words, the values in each row are listed from left to right, with the topmost row first and the bottommost row last. Individual values within a row are separated by blank spaces or commas, and the rows themselves are separated by semicolons or new lines. The following expressions are all legal arrays that can be used to initialize a variable:

Expression	Description
`[3.4]`	This expression creates a 1×1 array (a scalar) containing the value 3.4. The brackets are not required in this case.
`[1.0  2.0  3.0]`	This expression creates a 1×3 array containing the row vector [1 2 3].
`[1.0; 2.0; 3.0]`	This expression creates a 3×1 array containing the column vector $\begin{bmatrix} 1 \\ 2 \\ 3 \end{bmatrix}$.
`[1, 2, 3; 4, 5, 6]`	This expression creates a 2×3 array containing the matrix $\begin{bmatrix} 1 & 2 & 3 \\ 4 & 5 & 6 \end{bmatrix}$.
`[1, 2, 3` ` 4, 5, 6]`	This expression creates a 2×3 array containing the matrix $\begin{bmatrix} 1 & 2 & 3 \\ 4 & 5 & 6 \end{bmatrix}$. The end of the first line terminates the first row.
`[]`	This expression creates an **empty array**, which contains no rows and no columns. (Note that this is not the same as an array containing zeros.)

The number of elements in every row of an array must be the same, and the number of elements in every column must be the same. An expression such as

```
[1 2 3; 4 5];
```

is illegal because row 1 has three elements while row 2 has only two elements.

Programming Pitfalls

The number of elements in every row of an array must be the same, and the number of elements in every column must be the same. Attempts to define an array with different numbers of elements in its rows or different numbers of elements in its columns will produce an error when the statement is executed.

The expressions used to initialize arrays can include algebraic operations and all or portions of previously defined arrays. For example, the assignment statements

```
a = [0 1+7];
b = [a(2) 7 a];
```

will define an array a = [0 8] and an array b = [8 7 0 8].

Also, not all of the elements in an array must be defined when it is created. If a specific array element is defined and one or more of the elements before it are not, then the earlier elements will automatically be created and initialized to zero. For example, if c is not previously defined, the statement

```
c(2,3) = 5;
```

will produce the matrix c $= \begin{bmatrix} 0 & 0 & 0 \\ 0 & 0 & 5 \end{bmatrix}$. Similarly, an array can be extended by specifying a value for an element beyond the currently defined size. For example, suppose that array d = [1 2]. Then the statement

```
d(4) = 4;
```

will produce the array d = [1 2 0 4].

The semicolon at the end of each assignment statement shown above has a special purpose: it *suppresses the automatic echoing of values* that normally occurs whenever an expression is evaluated in an assignment statement. If an assignment statement is typed without the semicolon, the results of the statement are automatically displayed in the command window:

```
» e = [1, 2, 3; 4, 5, 6]
e =
   1     2     3
   4     5     6
```

If a semicolon is added at the end of the statement, the echoing disappears. Echoing is an excellent way to quickly check your work, but it seriously slows down the execution of MATLAB programs. For that reason, we normally suppress echoing at all times.

However, echoing the results of calculations makes a great quick-and-dirty debugging tool. If you are not certain what the results of a specific assignment statement are, just leave off the semicolon from that statement, and the results will be displayed in the Command Window as the statement is executed.

✱ Good Programming Practice

Use a semicolon at the end of all MATLAB assignment statements to suppress echoing of assigned values in the Command Window. This greatly speeds program execution.

✷ Good Programming Practice

If you need to examine the results of a statement during program debugging, you may remove the semicolon from that statement only so that its results are echoed in the Command Window.

2.2.2 Initializing with Shortcut Expressions

It is easy to create small arrays by explicitly listing each term in the array, but what happens when the array contains hundreds or even thousands of elements? It is just not practical to write out each element in the array separately!

MATLAB provides a special shortcut notation for these circumstances using the **colon operator**. The colon operator specifies a whole series of values by specifying the first value in the series, the stepping increment, and the last value in the series. The general form of a colon operator is

```
first:incr:last
```

where `first` is the first value in the series, `incr` is the stepping increment, and `last` is the last value in the series. If the increment is one, it may be omitted. For example, the expression 1:2:10 is a shortcut for a 1×5 row vector containing the values 1, 3, 5, 7, and 9.

```
» x = 1:2:10
x =
   1   3   5   7   9
```

With colon notation, an array can be initialized to have the hundred values $\frac{\pi}{100}$, $\frac{2\pi}{100}$, $\frac{3\pi}{100}$, . . . , π as follows:

```
angles = (0.01:0.01:1.00) * pi;
```

Shortcut expressions can be combined with the **transpose operator** (`'`) to initialize column vectors and more complex matrices. The transpose operator swaps the rows and columns of any array that it is applied to. Thus the expression

```
f = [1:4]';
```

generates a 4-element row vector [1 2 3 4], and then transposes it into the 4-element column vector `f` = $\begin{bmatrix} 1 \\ 2 \\ 3 \\ 4 \end{bmatrix}$. Similarly, the expressions

```
g = 1:4;
h = [g'  g'];
```

will produce the matrix h = $\begin{bmatrix} 1 & 1 \\ 2 & 2 \\ 3 & 3 \\ 4 & 4 \end{bmatrix}$.

2.2.3 Initializing with Built-In Functions

Arrays can also be initialized using built-in MATLAB functions. For example, the function `zeros` can be used to create an all-zero array of any desired size. There are several forms of the `zeros` function. If the function has a single scalar argument, it will produce a square array using the single argument as both the number of rows and the number of columns. If the function has two scalar arguments, the first argument will be the number of rows, and the second argument will be the number of columns. Since the `size` function returns two values containing the number of rows and columns in an array, it can be combined with the `zeros` function to generate an array of zeros that is the same size as another array. Some examples using the `zeros` function follow:

```
a = zeros(2);
b = zeros(2,3);
c = [1 2; 3 4];
d = zeros(size(c));
```

These statements generate the following arrays:

a = $\begin{bmatrix} 0 & 0 \\ 0 & 0 \end{bmatrix}$ b = $\begin{bmatrix} 0 & 0 & 0 \\ 0 & 0 & 0 \end{bmatrix}$

c = $\begin{bmatrix} 1 & 2 \\ 3 & 4 \end{bmatrix}$ d = $\begin{bmatrix} 0 & 0 \\ 0 & 0 \end{bmatrix}$

Similarly, the `ones` function can be used to generate arrays containing all ones, and the `eye` function can be used to generate arrays containing **identity matrices**, in which all on-diagonal elements are one, while all off-diagonal elements are zero. Table 2.1 (on the next page) contains list of common MATLAB functions useful for initializing variables.

2.2.4 Initializing Variables with Keyboard Input

It is also possible to prompt a user and initialize a variable with data that he or she types directly at the keyboard. This option allows a script file to prompt a user for input data values while it is executing. The `input` function displays a prompt string in the Command Window, and then waits for the user to type in a response. For example, consider the following statement:

```
my_val = input('Enter an input value:');
```

Table 2.1 MATLAB Functions Useful for Initializing Variables

Function	Purpose
`zeros(n)`	Generates an n × n matrix of zeros.
`zeros(n,m)`	Generates an n × m matrix of zeros.
`zeros(size(arr))`	Generates a matrix of zeros of the same size as `arr`.
`ones(n)`	Generates an n × n matrix of ones.
`ones(n,m)`	Generates an n × m matrix of ones.
`ones(size(arr))`	Generates a matrix of ones of the same size as `arr`.
`eye(n)`	Generates an n × n identity matrix.
`eye(n,m)`	Generates an n × m identity matrix.
`length(arr)`	Returns the length of a vector, or the longest dimension of a 2-D array.
`size(arr)`	Returns two values specifying the number of rows and columns in `arr`.

When this statement is executed, MATLAB prints out the string `'Enter an input value:'`, and then waits for the user to respond. If the user enters a single number, it may just be typed in. If the user enters an array, it must be enclosed in brackets. In either case, whatever is typed will be stored in variable `my_val` when the Enter key is pressed. If only the return key is entered, then an empty matrix will be created and stored in the variable.

If the `input` function includes the character `'s'` as a second argument, then the input data is returned to the user as a character string. Thus, the statement

```
» in1 = input('Enter data: ');
Enter data: 1.23
```

stores the value 1.23 into `in1`, while the statement

```
» in2 = input('Enter data: ','s');
Enter data: 1.23
```

stores the character string `'1.23'` into `in2`.

Quiz 2.1

This quiz provides a quick check to see if you have understood the concepts introduced in Sections 2.1 and 2.2. If you have trouble with the quiz, reread the sections, ask your instructor, or discuss the material with a fellow student. The answers to this quiz are found in the back of the book.

1. What is the difference between an array, a matrix, and a vector?

2. Answer the following questions for the array shown below.

$$\texttt{c} = \begin{bmatrix} 1.1 & -3.2 & 3.4 & 0.6 \\ 0.6 & 1.1 & -0.6 & 3.1 \\ 1.3 & 0.6 & 5.5 & 0.0 \end{bmatrix}$$

   (*a*) What is the size of `c`?
   (*b*) What is the value of `c(2,3)`?
   (*c*) List the subscripts of all elements containing the value 0.6.

3. Determine the size of the following arrays. Check your answers by entering the arrays into MATLAB and using the `whos` command or the Workspace Browser. Note that the later arrays may depend on the definitions of arrays defined earlier in this exercise.

   (*a*) `u = [10 20*i 10+20];`
   (*b*) `v = [-1; 20; 3];`
   (*c*) `w = [1 0 -9; 2 -2 0; 1 2 3];`
   (*d*) `x = [u' v];`
   (*e*) `y(3,3) = -7;`
   (*f*) `z = [zeros(4,1) ones(4,1) zeros(1,4)'];`
   (*g*) `v(4) = x(2,1);`

4. What is the value of `w(2,1)` above?
5. What is the value of `x(2,1)` above?
6. What is the value of `y(2,1)` above?
7. What is the value of `v(3)` after statement *(g)* is executed?

2.3 Multidimensional Arrays

As we have seen, MATLAB arrays can have one or more dimensions. One-dimensional arrays can be visualized as a series of values laid out in a row or column, with a single subscript used to select the individual array elements (Figure 2.2*a* on the next page). Such arrays are useful to describe data that is a function of one independent variable, such as a series of temperature measurements made at fixed intervals of time.

Some types of data are functions of more than one independent variable. For example, we might wish to measure the temperature at five different locations at four different times. In this case, our 20 measurements could logically be grouped into five different columns of four measurements each, with a separate column for each location (Figure 2.2*b*). In this case, we will use two subscripts to access a given element in the array: the first one to select the row and the second one to select the column. Such arrays are called **two-dimensional arrays**. The number of elements in a two-dimensional array will be the product of the number of rows and the number of columns in the array.

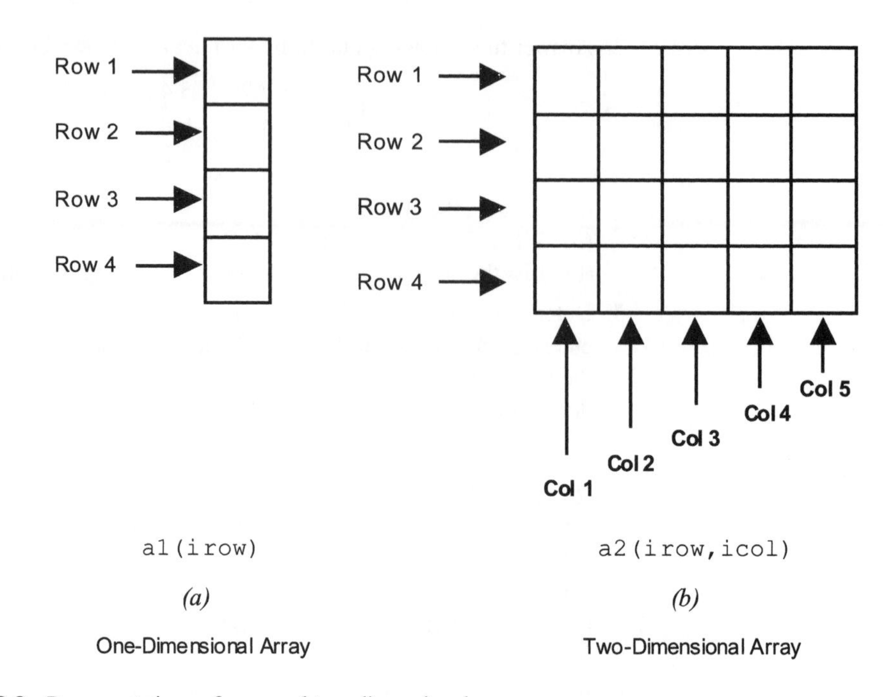

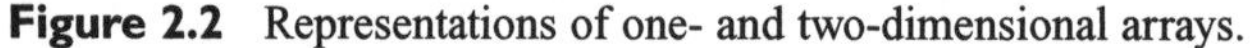

Figure 2.2 Representations of one- and two-dimensional arrays.

MATLAB allows us to create arrays with as many dimensions as necessary for any given problem. These arrays have one subscript for each dimension, and an individual element is selected by specifying a value for each subscript. The total number of elements in the array will be the product of the maximum value of each subscript. For example, the following two statements create a $2 \times 3 \times 2$ array `c`:

```
» c(:,:,1)=[1 2 3; 4 5 6];
» c(:,:,2)=[7 8 9; 10 11 12];
» whos c

  Name      Size     Bytes   Class

  c         2x3x2       96   double array
```

This array contains 12 elements ($2 \times 3 \times 2$). It contents can be displayed just like any other array.

```
» c
c(:,:,1) =
      1    2    3
      4    5    6
c(:,:,2) =
      7    8    9
     10   11   12
```

2.3.1 Storing Multidimensional Arrays in Memory

A two-dimensional array with m rows and n columns will contain m × n elements, and these elements will occupy m × n successive locations in the computer's memory. How are the elements of the array arranged in the computer's memory? MATLAB always allocates array elements in **column major order**. That is, MATLAB allocates the first column in memory, then the second, then the third, etc., until all of the columns have been allocated. Figure 2.3 (on the next page) illustrates this memory allocation scheme for a 4 × 3 array a. As we can see, element `a(1,2)` is really the fifth element allocated in memory. The order that elements are allocated in memory will become important when we discuss single-subscript addressing in the next section and low-level I/O functions in Chapter 8.

This same allocation scheme applies to arrays with more than two dimensions. The first array subscript is incremented most rapidly, the second subscript is incremented less rapidly, etc., and the last subscript in incremented most slowly. For example, in a 2 × 2 × 2 array, the elements would be allocated in the following order: (1,1,1), (2,1,1), (1,2,1), (2,2,1), (1,1,2), (2,1,2), (1,2,2), (2,2,2).

2.3.2 Accessing Multidimensional Arrays with One Dimension

One of MATLAB's peculiarities is that it will permit a user or programmer to treat a multidimensional array as though it were a one-dimensional array whose length is equal to the number of elements in the multidimensional array. If a multidimensional array is addressed with a single dimension, then the elements will be accessed in the order in which they were allocated in memory.

For example, suppose that we declare the 4 × 3 element array a as follows:

```
» a = [1 2 3; 4 5 6; 7 8 9; 10 11 12]
a =
     1     2     3
     4     5     6
     7     8     9
    10    11    12
```

Then the value of `a(5)` will be 2, which is the value of element `a(1,2)`, because `a(1,2)` was allocated fifth in memory.

Under normal circumstances, you should never use this feature of MATLAB. Addressing multidimensional arrays with a single subscript is a recipe for confusion.

✱ Good Programming Practice

Always use the proper number of dimensions when addressing a multidimensional array.

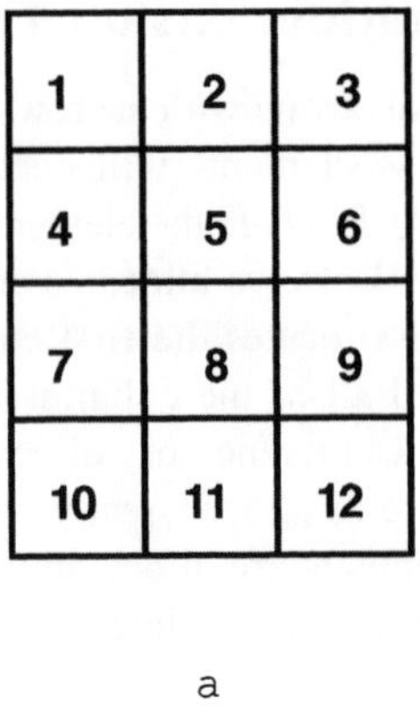

1	2	3
4	5	6
7	8	9
10	11	12

a

(a)

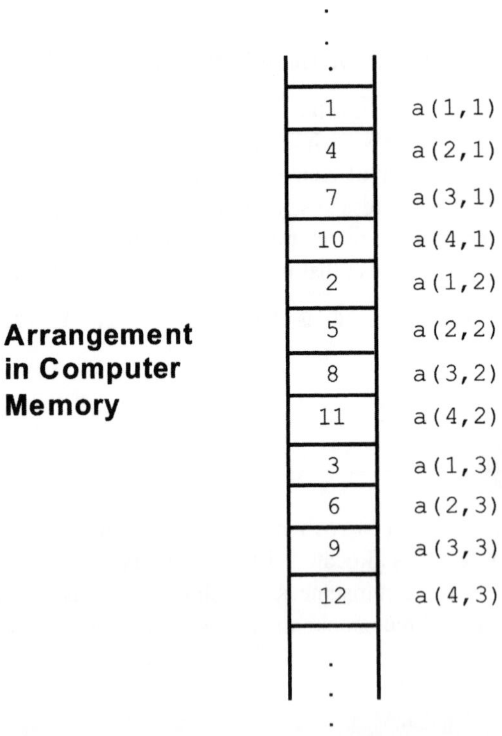

Arrangement in Computer Memory

1	a(1,1)
4	a(2,1)
7	a(3,1)
10	a(4,1)
2	a(1,2)
5	a(2,2)
8	a(3,2)
11	a(4,2)
3	a(1,3)
6	a(2,3)
9	a(3,3)
12	a(4,3)

(b)

Figure 2.3 *(a)* Data values for array a. *(b)* Layout of values in memory for array a.

2.4 Subarrays

It is possible to select and use subsets of MATLAB arrays as though they were separate arrays. To select a portion of an array, just include a list of all of the elements to be selected in the parentheses after the array name. For example, suppose array `arr1` is defined as follows:

```
arr1 = [1.1  -2.2  3.3  -4.4  5.5];
```

Then `arr1(3)` is just 3, `arr1([1 4])` is the array `[1.1 -4.4]`, and `arr1(1:2:5)` is the array `[1.1 3.3 5.5]`.

For a two-dimensional array, a colon can be used in a subscript to select all of the values of that subscript. For example, suppose

```
arr2 = [1  2  3;  -2  -3  -4;  3  4  5];
```

This statement would create an array `arr2` containing the values $\begin{bmatrix} 1 & 2 & 3 \\ -2 & -3 & -4 \\ 3 & 4 & 5 \end{bmatrix}$. With this definition, the subarray `arr2(1,:)` would be `[1 2 3]`, and the subarray `arr2(:,1:2:3)` would be $\begin{bmatrix} 1 & 3 \\ -2 & -4 \\ 3 & 5 \end{bmatrix}$.

2.4.1 The `end` Function

MATLAB includes a special function named `end` that is very useful for creating array subscripts. When used in an array subscript, `end` *returns the highest value taken on by that subscript.* For example, suppose that array `arr3` is defined as follows:

```
arr3 = [1  2  3  4  5  6  7  8];
```

Then `arr3(5:end)` would be the array `[5 6 7 8]`, and `array(end)` would be the value 8.

The value returned by `end` is always the highest value of a given subscript. If `end` appears in different subscripts, it can return *different* values within the same expression. For example, suppose that the 3×4 array `arr4` is defined as follows:

```
arr4 = [1  2  3  4;  5  6  7  8;  9  10  11  12];
```

Then the expression `arr4(2:end,2:end)` would return the array $\begin{bmatrix} 6 & 7 & 8 \\ 10 & 11 & 12 \end{bmatrix}$. Note that the first `end` returned the value 3, while the second `end` returned the value 4!

2.4.2 Using Subarrays on the Left-Hand Side of an Assignment Statement

It is also possible to use subarrays on the left-hand side of an assignment statement to update only some of the values in an array, as long as the **shape** (the number of rows and columns) of the values being assigned matches the shape of the subarray. If the shapes do not match, then an error will occur. For example, suppose that the 3 × 4 array `arr4` is defined as follows:

```
» arr4 = [1 2 3 4; 5 6 7 8; 9 10 11 12]
arr4 =
     1     2     3     4
     5     6     7     8
     9    10    11    12
```

Then the following assignment statement is legal, since the expressions on both sides of the equal sign have the same shape (2 × 2):

```
» arr4(1:2,[1 4]) = [20 21; 22 23]
arr4 =
    20     2     3    21
    22     6     7    23
     9    10    11    12
```

Note that the array elements (1,1), (1,4), (2,1), and (2,4) were updated. In contrast, the following expression is illegal because the two sides do not have the same shape.

```
» arr5(1:2,1:2) = [3 4]
??? In an assignment  A(matrix,matrix) = B, the
number of rows in B and the number of elements in
the A row index matrix must be the same.
```

Programming Pitfalls

For assignment statements involving subarrays, the *shapes of the subarrays on either side of the equal sign must match*. MATLAB will produce an error if they do not match.

There is a major difference in MATLAB between assigning values to a subarray and assigning values to an array. If values are assigned to a subarray, *only those values are updated, while all other values in the array remain unchanged*. On the other hand, if values are assigned to an array, *the entire contents of the array are deleted and replaced by the new values*. For example, suppose that the 3 × 4 array `arr4` is defined as follows:

```
» arr4 = [1 2 3 4; 5 6 7 8; 9 10 11 12]
arr4 =
      1    2    3    4
      5    6    7    8
      9   10   11   12
```

Then the following assignment statement replaces the *specified elements* of `arr4`:

```
» arr4(1:2,[1 4]) = [20 21; 22 23]
arr4 =
     20    2    3   21
     22    6    7   23
      9   10   11   12
```

In contrast, the following assignment statement replaces the *entire contents* of `arr4` with a 2 × 2 array:

```
» arr4 = [20 21; 22 23]
arr4 =
     20   21
     22   23
```

Good Programming Practice

Be sure to distinguish between assigning values to a subarray and assigning values to an array. MATLAB behaves differently in these two cases.

2.4.3 Assigning a Scalar to a Subarray

A scalar value on the right-hand side of an assignment statement always matches the shape specified on the left-hand side. The scalar value is copied into every element specified on the left-hand side of the statement. For example, assume that the 3 × 4 array `arr4` is defined as follows:

```
arr4 = [1 2 3 4; 5 6 7 8; 9 10 11 12];
```

Then the expression shown below assigns the value one to four elements of the array.

```
» arr4(1:2,1:2) = 1
arr4 =
      1    1    3    4
      1    1    7    8
      9   10   11   12
```

2.5 Special Values

MATLAB includes a number of pre-defined special values. These pre-defined values may be used at any time in MATLAB without initializing them first. A list of the most common pre-defined values is given in Table 2.2.

These predefined values are stored in ordinary variables, so they can be overwritten or modified by a user. If a new value is assigned to one of the predefined variables, then that new value will replace the default one in all later calculations. For example, consider the following statements that calculate the circumference of a 10-cm circle:

```
circ1 = 2 * pi * 10
pi = 3;
circ2 = 2 * pi * 10
```

In the first statement, `pi` has its default value of 3.14159 . . ., so `circ1` is 62.8319, which is the correct circumference. The second statement redefines `pi` to be 3, so in the third statement `circ2` is 60. Changing a predefined value in the program has created an incorrect answer and has also introduced a subtle and hard-to-find bug. Imagine trying to locate the source of such a hidden error in a 10,000-line program!

Table 2.2 Predefined Special Values

Function	Purpose
`pi`	Contains π to 15 significant digits.
`i, j`	Contain the value i ($\sqrt{-1}$).
`Inf`	This symbol represents machine infinity. It is usually generated as a result of a division by 0.
`NaN`	This symbol stands for Not-a-Number. It is the result of an undefined mathematical operation, such as the division of zero by zero.
`clock`	This special variable contains the current date and time in the form of a 6-element row vector containing the year, month, day, hour, minute, and second.
`date`	Contains the current date in a character string format, such as `24-Nov-2005`.
`eps`	This variable name is short for "epsilon." It is the smallest difference between two numbers that can be represented on the computer.
`ans`	A special variable used to store the result of an expression if that result is not explicitly assigned to some other variable.

Programming Pitfalls

Never redefine the meaning of a predefined variable in MATLAB. It is a recipe for disaster, producing subtle and hard-to-find bugs.

Quiz 2.2

This quiz provides a quick check to see if you have understood the concepts introduced in Sections 2.3 through 2.5. If you have trouble with the quiz, reread the sections, ask your instructor, or discuss the material with a fellow student. The answers to this quiz are found in the back of the book.

1. Assume that array c is defined as shown, and determine the contents of the following sub-arrays:

$$c = \begin{bmatrix} 1.1 & -3.2 & 3.4 & 0.6 \\ 0.6 & 1.1 & -0.6 & 3.1 \\ 1.3 & 0.6 & 5.5 & 0.0 \end{bmatrix}$$

   (*a*) `c(2,:)`
   (*b*) `c(:,end)`
   (*c*) `c(1:2,2:end)`
   (*d*) `c(6)`
   (*e*) `c(4:end)`
   (*f*) `c(1:2,2:4)`
   (*g*) `c([1 3],2)`
   (*h*) `c([2 2],[3 3])`

2. Determine the contents of array a after the following statements are executed.

   (*a*)
```
a = [1 2 3; 4 5 6; 7 8 9];
a([3 1],:) = a([1 3],:);
```
   (*b*)
```
a = [1 2 3; 4 5 6; 7 8 9];
a([1 3],:) = a([2 2],:);
```
   (*c*)
```
a = [1 2 3; 4 5 6; 7 8 9];
a = a([2 2],:);
```

3. Determine the contents of array a after the following statements are executed.

   (*a*)
```
a = eye(3,3);
b = [1 2 3];
a(2,:) = b;
```

(*b*)
```
a = eye(3,3);
b = [4 5 6];
a(:,3) = b';
```

(*c*)
```
a = eye(3,3);
b = [7 8 9];
a(3,:) = b([3 1 2]);
```

2.6 Displaying Output Data

There are several ways to display output data in MATLAB. This simplest way is one we have already seen—just leave the semicolon off of the end of a statement, and it will be echoed to the Command Window. We will now explore a few other ways to display data.

2.6.1 Changing the Default Format

When data is echoed in the Command Window, integer values are always displayed as integers, character values are displayed as strings, and other values are printed using a **default format**. The default format for MATLAB shows four digits after the decimal point, and it may be displayed in scientific notation with an exponent if the number is too large or too small. For example, the statements

```
x = 100.11
y = 1001.1
z = 0.00010011
```

produce the following output

```
x =
   100.1100
y =
   1.0011e+003
z =
   1.0011e-004
```

This default format can be changed in one of two ways: from the MATLAB window, or using the **format** command. If you are working in the Command Window, then you can change the format by selection the "File/Preferences/Command Window" menu option. This option will pop up a window that allows a user to change the way values are displayed in the Command Window (try it now!).

Alternatively, a user can use the `format` command to change the preferences. The format command changes the default format according to the values given in Table 2.3. The default format can be modified to display more significant digits of data, to force the display to be in scientific notation, to display data

Table 2.3 Output Display Formats

Format Command	Results	Example[1]
`format short`	4 digits after decimal (default format)	`12.3457`
`format long`	14 digits after decimal	`12.34567890123457`
`format short e`	5 digits plus exponent	`1.2346e+001`
`format short g`	5 total digits with or without exponent	`12.346`
`format long e`	15 digits plus exponent	`1.234567890123457e+001`
`format long g`	15 total digits with or without exponent	`12.3456789012346`
`format bank`	"dollars and cents" format	`12.35`
`format hex`	hexadecimal display of bits	`4028b0fcd32f707a`
`format rat`	approximate ratio of small integers	`1000/81`
`format compact`	suppress extra line feeds	
`format loose`	restore extra line feeds	
`format +`	only signs are printed	`+`

[1]The data value used for the example is `12.345678901234567` in all cases.

to two decimal digits, or to eliminate extra line feeds to get more data visible in the Command Window at a single time. Experiment with the commands in Table 2.3 for yourself.

Which of these ways to change the data format is better? If you are working directly at the computer, it is probably easier to use the menu item. On the other hand, if you are writing programs, it is probably better to use the `format` command, because it can be embedded directly into a program.

2.6.2 The `disp` Function

Another way to display data is with the `disp` function. The `disp` function accepts an array argument, and displays the value of the array in the Command Window. If the array is of type `char`, the character string contained in the array is printed out.

This function is often combined with the functions `num2str` (convert a number to a string) and `int2str` (convert an integer to a string) to create messages to be displayed in the Command Window. For example, the following MATLAB statements will display "The value of pi = 3.1416" in the Command Window. The first statement creates a string array containing the message, and the second statement displays the message.

```
str = ['The value of pi = ' num2str(pi)];
disp (str);
```

2.6.3 Formatted Output with the `fprintf` Function

An even more flexible way to display data is with the `fprintf` function. The `fprintf` function displays one or more values together with related text and lets the programmer control the way that the displayed value appears. The general form of this function when it is used to print to the Command Window is:

```
fprintf(format,data)
```

where `format` is a string describing the way the data is to be printed, and `data` is one or more scalars or arrays to be printed. The `format` is a character string containing text to be printed plus special characters describing the format of the data. For example, the function

```
fprintf('The value of pi is %f \n',pi)
```

will print out `'The value of pi is 3.141593'` followed by a line feed. The characters `%f` are called **conversion characters**; they indicate that the a value in the data list should be printed out in floating point format at that location in the format string. The characters `\n` are **escape characters**; they indicate that a line feed should be issued so that the following text starts on a new line. There are many types of conversion characters and escape characters that may be used in an `fprintf` function. A few of them are listed in Table 2.4, and a complete list can be found using the MATLAB Help Browser.

It is also possible to specify the width of the field in which a number will be displayed and the number of decimal places to display. This is done by specifying the width and precision after the `%` sign and before the `f`. For example, the function

```
fprintf('The value of pi is %6.2f \n',pi)
```

will print out `'The value of pi is   3.14'` followed by a line feed. The conversion characters `%6.2f` indicate that the first data item in the function should be printed out in floating point format in a field six characters wide, including two digits after the decimal point.

The `fprintf` function has one very significant limitation: *it displays only the real portion of a complex value*. This limitation can lead to misleading results

Table 2.4 Common Special Characters in `fprintf` Format Strings

Format String	Results
`%d`	Display value as an integer.
`%e`	Display value in exponential format.
`%f`	Display value in floating point format.
`%g`	Display value in either floating point or exponential format, whichever is shorter.
`\n`	Skip to a new line.

when calculations produce complex answers. In those cases, it is better to use the `disp` function to display answers.

For example, the following statements calculate a complex value `x` and display it using both `fprintf` and `disp`.

```
x = 2 * ( 1 - 2*i )^3;
str = ['disp:     x = ' num2str(x)];
disp(str);
fprintf('fprintf: x = %8.4f\n',x);
```

The results printed out by these statements are

```
disp:     x = -22+4i
fprintf: x = -22.0000
```

Note that the `fprintf` function ignored the imaginary part of the answer.

Programming Pitfalls

The `fprintf` function displays only the *real* part of a complex number, which can produce misleading answers when working with complex values.

2.7 Data Files

There are many ways to load and save data files in MATLAB. For the moment, we will consider only the **`load`** and **`save`** commands, which are the simplest ones to use.

The **`save`** command saves data from the current MATLAB workspace into a disk file. The most common form of this command is

```
save filename var1 var2 var3
```

where `filename` is the name of the file where the variables are saved and `var1`, `var2`, etc. are the variables to be saved in the file. By default, the file name will be given the extension "`mat`", and such data files are called MAT-files. If no variables are specified, the entire contents of the workspace are saved.

MATLAB saves MAT-files in a special compact format which preserves many details, including the name and type of each variable, the size of each array, and all data values. A MAT-file created on any platform (PC, Mac, Unix, or Linux) can be read on any other platform, so MAT-files are a good way to exchange data between computers if both computers run MATLAB. Unfortunately, the MAT-file is in a format that cannot be read by other programs. If data must be shared with other programs, then the `-ascii` option should be specified, and the data values will be written to the file as ASCII character strings separated by spaces. However, the special information such as variable names and types is lost when the data is saved in ASCII format, and the resulting data file will be much larger.

For example, suppose the array x is defined as

```
x=[1.23 3.14 6.28; -5.1 7.00 0];
```

The command "save x.dat x -ascii" will produce a file named x.dat containing the following data:

```
 1.2300000e+000  3.1400000e+000  6.2800000e+000
-5.1000000e+000  7.0000000e+000  0.0000000e+000
```

This data is in a format that can be read by spreadsheets or by programs written in other computer languages, so it makes it easy to share data between MATLAB programs and other applications.

✷ Good Programming Practice

If data must be exchanged between MATLAB and other programs, save the MATLAB data in ASCII format. If the data will be used only in MATLAB, save the data in MAT-file format.

MATLAB doesn't care what file extension is used for ASCII files. However, it is better for the user if a consistent naming convention is used, and an extension of "dat" is a common choice for ASCII files.

✷ Good Programming Practice

Save ASCII data files with a "dat" file extension to distinguish them from MAT-files, which have a "mat" file extension.

The **load** command is the opposite of the save command. It loads data from a disk file into the current MATLAB workspace. The most common form of this command is

```
load filename
```

where filename is the name of the file to be loaded. If the file is a MAT-file, then all of the variables in the file will be restored, with the names and types the same as before. If a list of variables is included in the command, then only those variables will be restored. If the given filename has no extent, or if the file extent is .mat, then the load command will treat the file as a MAT-file.

MATLAB can load data created by other programs in space-separated ASCII format. If the given filename has any file extent other than .mat, then the load command will treat the file as an ASCII file. The contents of an ASCII file will be converted into a MATLAB array having the same name as the file (without the file extent) that the data was loaded from. For example, suppose that an ASCII data file named `x.dat` contains the following data:

```
1.23     3.14     6.28
-5.1     7.00     0
```

Then the command "`load x.dat`" will create a 2 × 3 array named `x` in the current workspace, containing these data values.

The `load` statement can be forced to treat a file as a MAT-file by specifying the `-mat` option. For example, the statement

```
load -mat x.dat
```

would treat file `x.dat` as a MAT-file even though its file extent is not `.mat`. Similarly, the `load` statement can be forced to treat a file as an ASCII file by specifying the `-ascii` option. These options allow the user to load a file properly even if its file extent doesn't match the MATLAB conventions.

Quiz 2.3

This quiz provides a quick check to see if you have understood the concepts introduced in Sections 2.6 and 2.7. If you have trouble with the quiz, reread the sections, ask your instructor, or discuss the material with a fellow student. The answers to this quiz are found in the back of the book.

1. How would you tell MATLAB to display all real values in exponential format with 15 significant digits?
2. What do the following sets of statements do? What is the output from them?

   (*a*)
   ```
   radius = input('Enter circle radius:\n');
   area = pi * radius^2;
   str = ['The area is ' num2str(area)];
   disp(str);
   ```

   (*b*)
   ```
   value = int2str(pi);
   disp(['The value is ' value '!']);
   ```

3. What do the following sets of statements do? What is the output from them?

   ```
   value = 123.4567e2;
   fprintf('value = %e\n',value);
   fprintf('value = %f\n',value);
   fprintf('value = %g\n',value);
   fprintf('value = %12.4f\n',value);
   ```

2.8 Scalar and Array Operations

Calculations are specified in MATLAB with an assignment statement, whose general form is

```
variable_name = expression;
```

The assignment statement calculates the value of the expression to the right of the equal sign, and *assigns* that value to the variable named on the left of the equal sign. Note that the equal sign does not mean equality in the usual sense of the word. Instead, it means: *store the value of* `expression` *into location* `variable_name`. For this reason, the equal sign is called the **assignment operator**. A statement such as

```
ii = ii + 1;
```

is complete nonsense in ordinary algebra, but it makes perfect sense in MATLAB. It means: take the current value stored in variable `ii`, add one to it, and store the result back into variable `ii`.

2.8.1 Scalar Operations

The expression to the right of the assignment operator can be any valid combination of scalars, arrays, parentheses, and arithmetic operators. The standard arithmetic operations between two scalars are given in Table 2.5.

Parentheses may be used to group terms whenever desired. When parentheses are used, the expressions inside the parentheses are evaluated before the expressions outside the parentheses. For example, the expression `2 ^ ((8+2)/5)` is evaluated as shown below

```
2 ^ ((8+2)/5) = 2 ^ (10/5)
              = 2 ^ 2
              = 4
```

2.8.2 Array and Matrix Operations

MATLAB supports two types of operations between arrays, known as *array operations* and *matrix operations*. **Array operations** are operations performed

Table 2.5 Arithmetic Operations Between Two Scalars

Operation	Algebraic Form	MATLAB form
Addition	$a + b$	`a + b`
Subtraction	$a - b$	`a - b`
Multiplication	$a \times b$	`a * b`
Division	$\frac{a}{b}$	`a/b`
Exponentiation	a^b	`a ^ b`

between arrays on an **element-by-element basis**. That is, the operation is performed on corresponding elements in the two arrays. For example, if `a` = $\begin{bmatrix} 1 & 2 \\ 3 & 4 \end{bmatrix}$ and `b` = $\begin{bmatrix} -1 & 3 \\ -2 & 1 \end{bmatrix}$, then `a + b` = $\begin{bmatrix} 0 & 5 \\ 1 & 5 \end{bmatrix}$. Note that for these operations to work, *the number of rows and columns in both arrays must be the same.* If not, MATLAB will generate an error message.

Array operations may also occur between an array and a scalar. If the operation is performed between an array and a scalar, the value of the scalar is applied to every element of the array. For example, if `a` = $\begin{bmatrix} 1 & 2 \\ 3 & 4 \end{bmatrix}$, then `a + 4` = $\begin{bmatrix} 5 & 6 \\ 7 & 8 \end{bmatrix}$.

In contrast, **matrix operations** follow the normal rules of linear algebra, such as matrix multiplication. In linear algebra, the product `c = a × b` is defined by the equation

$$c(i,j) = \sum_{k=1}^{n} a(i,k)b(k,j)$$

For example, if `a` = $\begin{bmatrix} 1 & 2 \\ 3 & 4 \end{bmatrix}$ and `b` = $\begin{bmatrix} -1 & 3 \\ -2 & 1 \end{bmatrix}$, then `a × b`= $\begin{bmatrix} -5 & 5 \\ -11 & 13 \end{bmatrix}$. Note that for matrix multiplication to work, *the number of columns in matrix* `a` *must be equal to the number of rows in matrix* `b`.

MATLAB uses a special symbol to distinguish array operations from matrix operations. In the cases where array operations and matrix operations have a different definition, MATLAB uses a period before the symbol to indicate an array operation (for example, `.*`). A list of common array and matrix operations is given in Table 2.6 on the next page.

New users often confuse array operations and matrix operations. In some cases, substituting one for the other will produce an illegal operation, and MATLAB will report an error. In other cases, both operations are legal, and MATLAB will perform the wrong operation and come up with a wrong answer. The most common problem happens when working with square matrices. Both array multiplication and matrix multiplication are legal for two square matrices of the same size, but the resulting answers are totally different. Be careful to specify exactly what you want!

Programming Pitfalls

Be careful to distinguish between array operations and matrix operations in your MATLAB code. It is especially common to confuse array multiplication with matrix multiplication.

Table 2.6 Common Array and Matrix Operations

Operation	MATLAB form	Comments
Array Addition	`a + b`	Array addition and matrix addition are identical.
Array Subtraction	`a - b`	Array subtraction and matrix subtraction are identical.
Array Multiplication	`a .* b`	Element-by-element multiplication of `a` and `b`. Both arrays must be the same shape, or one of them must be a scalar.
Matrix Multiplication	`a * b`	Matrix multiplication of `a` and `b`. The number of columns in `a` must equal the number of rows in `b`.
Array Right Division	`a ./ b`	Element-by-element division of `a` and `b`: `a(i,j)/b(i,j)`. Both arrays must be the same shape, or one of them must be a scalar.
Array Left Division	`a .\ b`	Element-by-element division of `a` and `b`, but with `b` in the numerator: `b(i,j)/a(i,j)`. Both arrays must be the same shape, or one of them must be a scalar.
Matrix Right Division	`a / b`	Matrix division defined by `a * inv(b)`, where `inv(b)` is the inverse of matrix `b`.
Matrix Left Division	`a \ b`	Matrix division defined by `inv(a) * b`, where `inv(a)` is the inverse of matrix `a`.
Array Exponentiation	`a .^ b`	Element-by-element exponentiation of `a` and `b`: `a(i,j) ^ b(i,j)`. Both arrays must be the same shape, or one of them must be a scalar.

Example 2.1—Results Array Versus Matrix Operations

Assume that `a`, `b`, `c`, and `d` are defined as follows:

$$a = \begin{bmatrix} 1 & 0 \\ 2 & 1 \end{bmatrix} \qquad b = \begin{bmatrix} -1 & 2 \\ 0 & 1 \end{bmatrix}$$

$$c = \begin{bmatrix} 3 \\ 2 \end{bmatrix} \qquad d = 5$$

What is the result of each of the following expressions?

(a) `a + b`   (e) `a + c`
(b) `a .* b`   (f) `a + d`
(c) `a * b`   (g) `a .* d`
(d) `a * c`   (h) `a * d`

SOLUTION

(a) This is array or matrix addition: `a + b` $= \begin{bmatrix} 0 & 2 \\ 2 & 2 \end{bmatrix}$

(b) This is element-by-element array multiplication: `a. * b` $= \begin{bmatrix} -1 & 0 \\ 0 & 1 \end{bmatrix}$

(c) This is matrix multiplication: `a * b` $= \begin{bmatrix} -1 & 2 \\ -2 & 5 \end{bmatrix}$

(d) This is matrix multiplication: `a * c` $= \begin{bmatrix} 3 \\ 8 \end{bmatrix}$

(e) This operation is illegal, since `a` and `c` have different numbers of columns.

(f) This is addition of an array to a scalar: `a + d` $= \begin{bmatrix} 6 & 5 \\ 7 & 6 \end{bmatrix}$

(g) This is array multiplication: `a .* d` $= \begin{bmatrix} 5 & 0 \\ 10 & 5 \end{bmatrix}$

(h) This is matrix multiplication: `a * d` $= \begin{bmatrix} 5 & 0 \\ 10 & 5 \end{bmatrix}$ ◀

The matrix left division operation has a special significance that we must understand. A 3 × 3 set of simultaneous linear equations takes the form

$$\begin{aligned} a_{11}x_1 + a_{12}x_2 + a_{13}x_3 &= b_1 \\ a_{21}x_1 + a_{22}x_2 + a_{23}x_3 &= b_2 \\ a_{31}x_1 + a_{32}x_2 + a_{33}x_3 &= b_3 \end{aligned} \tag{2-1}$$

which can be expressed as

$$Ax = B \tag{2-2}$$

where $A = \begin{bmatrix} a_{11} & a_{12} & a_{13} \\ a_{21} & a_{22} & a_{23} \\ a_{31} & a_{32} & a_{33} \end{bmatrix}$, $B = \begin{bmatrix} b_1 \\ b_2 \\ b_3 \end{bmatrix}$, and $x = \begin{bmatrix} x_1 \\ x_2 \\ x_3 \end{bmatrix}$.

Equation (2-2) can be solved for x using linear algebra. The result is

$$x = A^{-1}B \tag{2-3}$$

Since the left division operator `A \ B` is defined to be `inv(A) x B`, the left division operator solves a system of simultaneous equations in a single statement!

✳ Good Programming Practice

Use the left division operator to solve systems of simultaneous equations.

2.9 Hierarchy of Operations

Often, many arithmetic operations are combined into a single expression. For example, consider the equation for the distance traveled by an object starting from rest and subjected to a constant acceleration:

```
distance = 0.5 * accel * time ^ 2
```

There are two multiplications and an exponentiation in this expression. In such an expression, it is important to know the order in which the operations are evaluated. If exponentiation is evaluated before multiplication, this expression is equivalent to

```
distance = 0.5 * accel * (time ^ 2)
```

But if multiplication is evaluated before exponentiation, this expression is equivalent to

```
distance = (0.5 * accel * time) ^ 2
```

These two equations have different results, and we must be able to unambiguously distinguish between them.

To make the evaluation of expressions unambiguous, MATLAB has established a series of rules governing the hierarchy or order in which operations are evaluated within an expression. The rules generally follow the normal rules of algebra. The order in which the arithmetic operations are evaluated is given in Table 2.7.

Table 2.7 Hierarchy of Arithmetic Operations

Precedence	Operation
1	The contents of all parentheses are evaluated, starting from the innermost parentheses and working outward.
2	All exponentials are evaluated, working from left to right.
3	All multiplications and divisions are evaluated, working from left to right.
4	All additions and subtractions are evaluated, working from left to right.

Example 2.2—Order of Evaluation of Operations

Variables a, b, c, and d have been initialized to the following values:

```
a = 3;     b = 2;     c = 5;     d = 3;
```

Evaluate the following MATLAB assignment statements:

```
(a) output = a*b+c*d;
(b) output = a*(b+c)*d;
(c) output = (a*b)+(c*d);
(d) output = a^b^d;
(e) output = a^(b^d);
```

SOLUTION

(a) Expression to evaluate:	`output = a*b+c*d;`
Fill in numbers:	`output = 3*2+5*3;`
First, evaluate multiplications and divisions from left to right:	`output = 6 +5*3;`
	`output = 6 + 15;`
Now evaluate additions:	`output = 21`
(b) Expression to evaluate:	`output = a*(b+c)*d;`
Fill in numbers:	`output = 3*(2+5)*3;`
First, evaluate parentheses:	`output = 3*7*3;`
Now, evaluate multiplications and divisions from left to right:	`output = 21*3;`
	`output = 63;`
(c) Expression to evaluate:	`output = (a*b)+(c*d);`
Fill in numbers:	`output = (3*2)+(5*3);`
First, evaluate parentheses:	`output = 6 + 15;`
Now evaluate additions:	`output = 21;`
(d) Expression to evaluate:	`output = a^b^d;`
Fill in numbers:	`output = 3^2^3;`
Evaluate exponentials from left to right:	`output = 9^3;`
	`output = 729;`
(e) Expression to evaluate:	`output = a^(b^d);`
Fill in numbers:	`output = 3^(2^3);`
First, evaluate parentheses:	`output = 3^8;`
Now, evaluate exponential:	`output = 6561;`

As we see above, the order in which operations are performed has a major effect on the final result of an algebraic expression.

It is important that every expression in a program be made as clear as possible. Any program of value must not only be written but also be maintained and modified when necessary. You should always ask yourself, "Will I easily understand this expression if I come back to it in six months? Can another programmer look at my code and easily understand what I am doing?" If there is any doubt in your mind, use extra parentheses in the expression to make it as clear as possible.

✱ Good Programming Practice

Use parentheses as necessary to make your equations clear and easy to understand.

If parentheses are used within an expression, the parentheses must be balanced. That is, there must be an equal number of open parentheses and close parentheses within the expression. It is an error to have more of one type than the other. Errors of this sort are usually typographical, and they are caught by the MATLAB interpreter when the command is executed. For example, the expression

```
(2 + 4)/2)
```

produces an error during when the expression is executed.

Quiz 2.4

This quiz provides a quick check to see if you have understood the concepts introduced in Sections 2.8 and 2.9. If you have trouble with the quiz, reread the sections, ask your instructor, or discuss the material with a fellow student. The answers to this quiz are found in the back of the book.

1. Assume that a, b, c, and d are defined as follows, and calculate the results of the following operations if they are legal. If an operation is illegal, explain why it is illegal.

$$\mathrm{a} = \begin{bmatrix} 2 & 1 \\ -1 & 2 \end{bmatrix} \qquad \mathrm{b} = \begin{bmatrix} 0 & -1 \\ 3 & 1 \end{bmatrix}$$

$$\mathrm{c} = \begin{bmatrix} 1 \\ 2 \end{bmatrix} \qquad \mathrm{d} = -3$$

(*a*) `result = a .* c;`
(*b*) `result = a * [c c];`
(*c*) `result = a .* [c c];`

(*d*) `result = a + b * c;`

(*e*) `result = a + b .* c;`

2. Solve for x in the equation $Ax = B$, where $A = \begin{bmatrix} 1 & 2 & 1 \\ 2 & 3 & 2 \\ -1 & 0 & 1 \end{bmatrix}$ and $B = \begin{bmatrix} 1 \\ 1 \\ 0 \end{bmatrix}$.

2.10 Built-In MATLAB Functions

In mathematics, a **function** is an expression that accepts one or more input values and calculates a single result from them. Scientific and technical calculations usually require functions that are more complex than the simple addition, subtraction, multiplication, division, and exponentiation operations that we have discussed so far. Some of these functions are very common and are used in many different technical disciplines. Others are rarer and specific to a single problem or a small number of problems. Examples of very common functions are the trigonometric functions, logarithms, and square roots. Examples of rarer functions include the hyperbolic functions, Bessel functions, and so forth. One of MATLAB's greatest strengths is that it comes with an incredible variety of built-in functions ready for use.

2.10.1 Optional Results

Unlike mathematical functions, MATLAB functions can return *more than one result* to the calling program. The function `max` is an example of such a function. This function normally returns the maximum value of an input vector, but it can also return a second argument containing the location in the input vector where the maximum value was found. For example, the statement

```
maxval = max ([1 -5 6 -3])
```

returns the result `maxval` = 6. However, if two variables are provided to store results in, the function returns *both* the maximum value *and* the location of the maximum value.

```
[maxval, index] = max ([1 -5 6 -3])
```

produces the results `maxval` = 6 and `index` = 3.

2.10.2 Using MATLAB Functions with Array Inputs

Many MATLAB functions are defined for one or more scalar inputs and produce a scalar output. For example, the statement `y = sin(x)` calculates the sine of `x` and stores the result in y. If these functions receive an array of input values, they will calculate an array of output values on an element-by-element basis. For example, if `x = [0 pi/2 pi 3*pi/2 2*pi]`, then the statement

```
y = sin(x)
```

will produce the result `y = [0 1 0 -1 0]`.

2.10.3 Common MATLAB Functions

A few of the most common and useful MATLAB functions are shown in Table 2.8. These functions will be used in many examples and homework problems. If you need to locate a specific function not on this list, you can search for the function alphabetically or by subject using the MATLAB Help Browser.

Note that unlike most computer languages, many MATLAB functions work correctly for both real and complex inputs. MATLAB functions automatically calculate the correct answer, even if the result is imaginary or complex. For example, the function `sqrt(-2)` will produce a runtime error in languages such as C or Fortran. In contrast, MATLAB correctly calculates the imaginary answer:

```
» sqrt(-2)
ans =
     0 + 1.4142i
```

2.11 Introduction to Plotting

MATLAB's extensive, device-independent plotting capabilities are one of its most powerful features. They make it very easy to plot any data at any time. To plot a data set, just create two vectors containing the x and y values to be plotted, and use the `plot` function.

For example, suppose that we wish to plot the function $y = x^2 - 10x + 15$ for values of x between 0 and 10. It takes only three statements to create this plot. The first statement creates a vector of x values between 0 and 10 using the colon operator. The second statement calculates the y values from the equation (note that we are using array operators here so that this equation is applied to each x value on an element-by-element basis). Finally, the third statement creates the plot.

```
x = 0:1:10;
y = x.^2 - 10.*x + 15;
plot(x,y);
```

When the `plot` function is executed, MATLAB opens a Figure Window and displays the plot in that window. The plot produced by these statements is shown in Figure 2.4 on page 56.

Table 2.8 Common MATLAB Functions.

Function	Description
	Mathematical functions
`abs(x)`	Calculates $\|x\|$.
`acos(x)`	Calculates $\cos^{-1} x$.
`angle(x)`	Returns the phase angle of the complex value *x*, in radians.
`asin(x)`	Calculates $\sin^{-1} x$.
`atan(x)`	Calculates $\tan^{-1} x$.
`atan2(y,x)`	Calculates $\tan^{-1} \frac{y}{x}$ over all four quadrants of the circle (results in *radians* in the range $-\pi \leq \tan^{-1}\frac{y}{x} \leq \pi$).
`cos(x)`	Calculates $\cos x$, with x in radians.
`exp(x)`	Calculates e^x.
`log(x)`	Calculates the natural logarithm $\log_e x$.
`[value,index] = max(x)`	Returns the maximum value in vector *x*, and optionally the location of that value.
`[value,index] = min(x)`	Returns the minimum value in vector *x*, and optionally the location of that value.
`mod(x,y)`	Remainder or modulo function.
`sin(x)`	Calculates $\sin x$, with x in radians.
`sqrt(x)`	Calculates the square root of *x*.
`tan(x)`	Calculates $\tan x$, with x in radians.
	Rounding functions
`ceil(x)`	Rounds *x* to the nearest integer towards positive infinity: `ceil(3.1) = 4` and `ceil(-3.1) = -3`.
`fix(x)`	Rounds *x* to the nearest integer towards zero: `fix(3.1) = 3` and `fix(-3.1) = -3`.
`floor(x)`	Rounds *x* to the nearest integer towards minus infinity: `floor(3.1) = 3` and `floor(-3.1) = -4`.
`round(x)`	Rounds *x* to the nearest integer.
	String conversion functions
`char(x)`	Converts a matrix of numbers into a character string. For ASCII characters the matrix should contain numbers ≤ 127.
`double(x)`	Converts a character string into a matrix of numbers.
`int2str(x)`	Converts *x* into an integer character string.
`num2str(x)`	Converts *x* into a character string.
`str2num(s)`	Converts character string *s* into a numeric array.

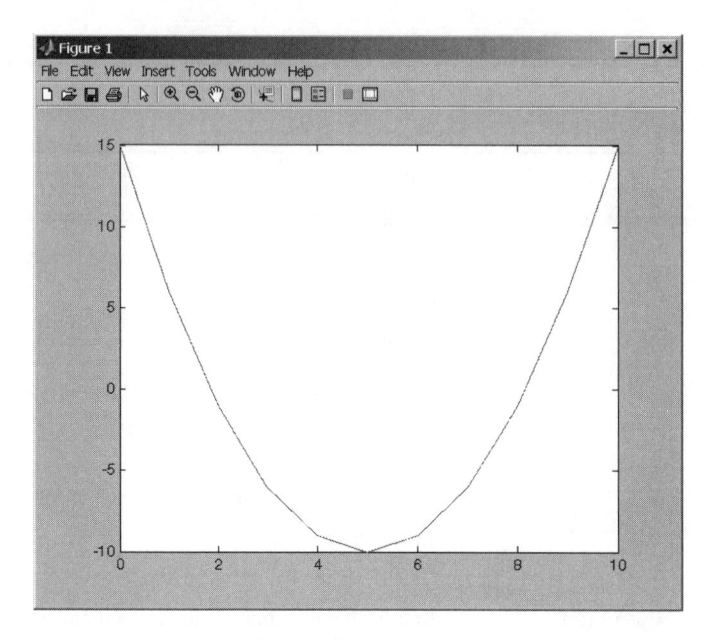

Figure 2.4 Plot of $y = x^2 - 10x + 15$ from 0 to 10.

2.11.1 Using Simple *xy* Plots

As we saw above, plotting is *very* easy in MATLAB. Any pair of vectors can be plotted versus each other as long as both vectors have the same length. However, the result is not a finished product, since there are no titles, axis labels, or grid lines on the plot.

Titles and axis labels can be added to a plot with the `title`, `xlabel`, and `ylabel` functions. Each function is called with a string containing the title or label to be applied to the plot. Grid lines can be added or removed from the plot with the `grid` command: `grid on` turns on grid lines, and `grid off` turns off grid lines. For example, the statements below generate a plot of the function $y = x^2 - 10x + 15$ with titles, labels, and gridlines. The resulting plot is shown in Figure 2.5.

```
x = 0:1:10;
y = x.^2 - 10.*x + 15;
plot(x,y);
title ('Plot of y = x.^2 - 10.*x + 15');
xlabel ('x');
ylabel ('y');
grid on;
```

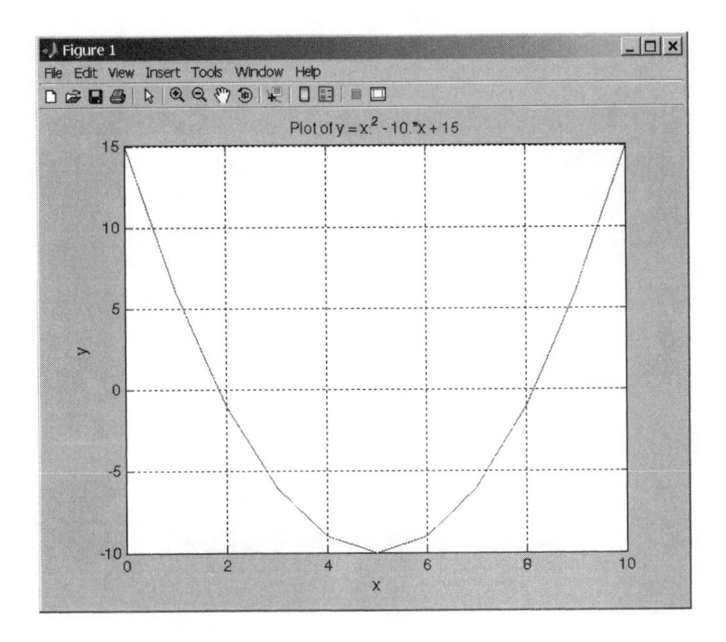

Figure 2.5 Plot of $y = x^2 - 10x + 15$ with a title, axis labels, and gridlines.

2.11.2 Printing a Plot

Once created, a plot can be printed on a printer with the `print` command, by clicking on the "print" icon in the Figure Window, or by selecting the "File/Print" menu option in the Figure Window.

The `print` command is especially useful because it can be included in a MATLAB program, allowing the program to automatically print graphical images. The form of the `print` command is:

```
print <options> <filename>
```

If no filename is included, this command prints a copy of the current figure on the system printer. If a filename is specified, the command prints a copy of the current figure to the specified file.

2.11.3 Exporting a Plot as a Graphical Image

The "File/Export" menu option on the Figure Window can be used to save a plot as a graphical image. In this case, the user selects the file name and the type of image from a standard dialog box (see Figure 2.6 on the next page). This is the easiest way to create an image from the Graphical User Interface.

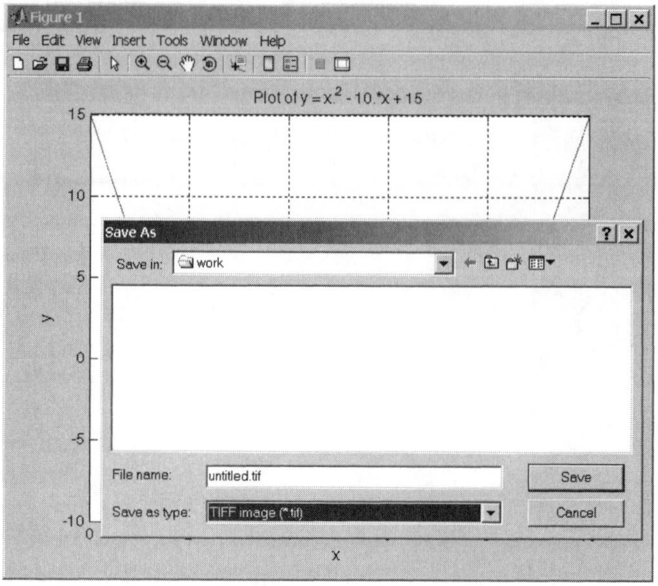

Figure 2.6 Exporting a plot as an image file using the File/Export menu item.

If we wish to export a plot as a graphic image from within a program, we can use the `print` command. The `print` command saves a plot as a graphical image by specifying appropriate options and a file name.

```
print <options> <filename>
```

There are many different options that specify the format of the output sent to a file. One very important option is `-dtiff`. This option specifies that the output will be to a file in Tagged Image File Format (TIFF). Since this format can be imported into all of the important word processors on PC, Mac, Unix, and Linux platforms, it is a great way to include MATLAB plots in a document. The following command will create a TIFF image of the current figure and store it in a file called `my_image.tif`:

```
print -dtiff my_image.tif
```

Other options allow image files to be created in other formats. Some of the most important image file formats are given in Table 2.9.

2.11.4 Multiple Plots

It is possible to plot multiple functions on the same graph by simply including more than one set of (x, y) values in the plot function. For example, suppose that we wanted to plot the function $f(x) = \sin 2x$ and its derivative on the same plot.

Table 2.9 `print` Options to Create Graphics Files

Option	Description
`-deps`	Creates a monochrome encapsulated postscript image.
`-depsc`	Creates a color encapsulated postscript image.
`-djpeg`	Creates a JPEG image.
`-dpng`	Creates a Portable Network Graphic color image.
`-dtiff`	Creates a compressed TIFF image.

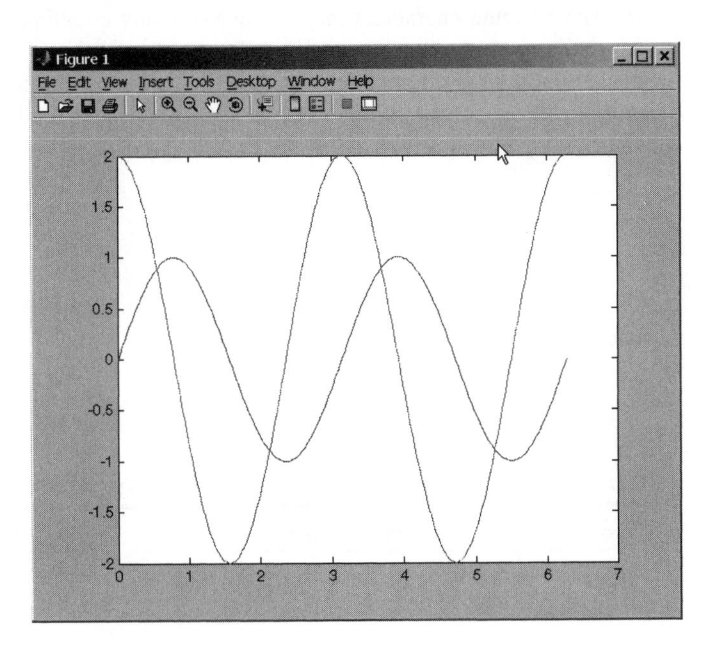

Figure 2.7 Plot of $f(x) = \sin 2x$ and $f(x) = 2 \cos 2x$ on the same axes.

The derivative of $f(x) = \sin 2x$ is:

$$\frac{d}{dt} \sin 2x = 2 \cos 2x \tag{2-4}$$

To plot both functions on the same axes, we must generate a set of x values and the corresponding y values for each function. Then to plot the functions, we would simply list both sets of (x, y) values in the plot function as shown below.

```
x = 0:pi/100:2*pi;
y1 = sin(2*x);
y2 = 2*cos(2*x);
plot(x,y1,x,y2);
```

The resulting plot is shown in Figure 2.7.

2.11.5 Line Color, Line Style, Marker Style, and Legends

MATLAB allows a programmer to select the color of a line to be plotted, the style of the line to be plotted, and the type of marker to be used for data points on the line. These traits may be selected using an attribute character string after the x and y vectors in the plot function.

The attribute character string can have up to three characters, with the first character specifying the color of the line, the second character specifying the style of the marker, and the last character specifying the style of the line. The characters for various colors, markers, and line styles are shown in Table 2.10.

The attribute characters may be mixed in any combination, and more than one attribute string may be specified if more than one pair of (x, y) vectors are included in a single `plot` function call. For example, the following statements will plot the function $y = x^2 - 10x + 15$ with a dashed red line, and include the actual data points as blue circles.

```
x = 0:1:10;
y = x.^2 - 10.*x + 15;
plot(x,y,'r--',x,y,'bo');
```

Legends may be created with the `legend` function. The basic form of this function is

```
legend('string1','string2',..., pos)
```

Table 2.10 Table of Plot Colors, Marker Styles, and Line Styles

Color		Marker Style		Line Style	
`y`	yellow	`.`	point	`-`	solid
`m`	magenta	`o`	circle	`:`	dotted
`c`	cyan	`x`	x-mark	`-.`	dash-dot
`r`	red	`+`	plus	`--`	dashed
`g`	green	`*`	star	<none>	no line
`b`	blue	`s`	square		
`w`	white	`d`	diamond		
`k`	black	`v`	triangle (down)		
		`^`	triangle (up)		
		<	triangle (left)		
		>	triangle (right)		
		`p`	pentagram		
		`h`	hexagram		
		<none>	no marker		

Table 2.11 Values of `pos` in the `legend` command

Value	Legend Location
'NW'	Above and to the left
'NL'	Above top left corner
'NC'	Above center of top edge
'NR'	Above top right corner
'NE'	Above and to right
'TW'	At top and to left
'TL'	Top left corner
'TC'	At top center
'TR'	Top right corner
'TE'	At top and to right
'MW'	At middle and to left
'ML'	Middle left edge
'MC'	Middle and center
'MR'	Middle right edge
'ME'	At middle and to right
'BW'	At bottom and to left
'BL'	Bottom left corner
'BC'	At bottom center
'BR'	Bottom right corner
'BE'	At bottom and to right
'SW'	Below and to left
'SL'	Below bottom left corner
'SC'	Below center of bottom edge
'SR'	Below bottom right corner
'SE'	Below and to right

where `string1`, `string2`, etc. are the labels associated with the lines plotted, and `pos` is an string specifying where to place the legend. The possible values for `pos` are given in Table 2.11, and are shown graphically in Figure 2.8 (on the next page).

The command `legend off` will remove an existing legend[1].

An example of a complete plot is shown in Figure 2.9 (on the next page), and the statements to produce that plot are shown below. They plot the function $f(x) = \sin 2x$ and its derivative on the same axes, with a solid black line for $f(x)$

[1]Before MATLAB 7.0, the `pos` parameter took a number in the range 0–4 to specify the location of a legend. This usage is now obsolete, but it is still supported for backwards compatibility.

NW	NL	NC	NR	NE
TW	TL	TC	TR	TE
MW	ML	MC	MR	ME
BW	BL	BC	BR	BE
SW	SL	SC	SR	SE

Limits of Plot Axes

Figure 2.8 Possible locations for a plot legend.

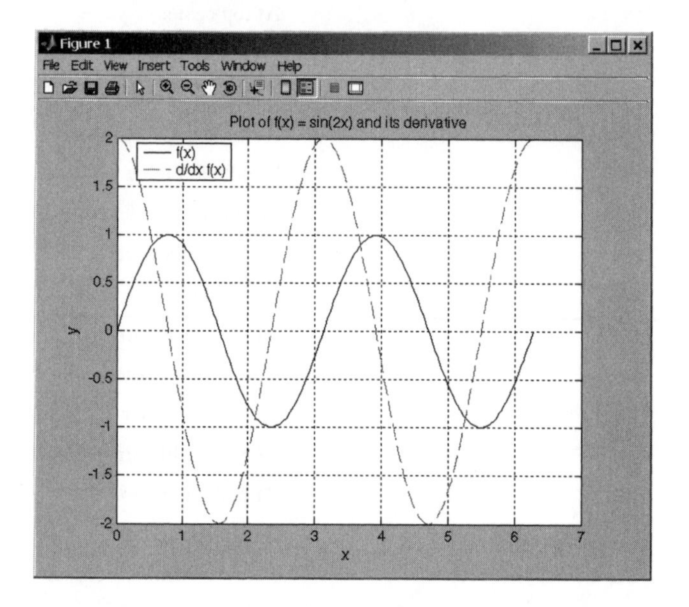

Figure 2.9 A complete plot with title, axis labels, legend, grid, and multiple line styles.

and a dashed red line for its derivative. The plot includes a title, axis labels, a legend in the top left corner of the plot, and grid lines.

```
x = 0:pi/100:2*pi;
y1 = sin(2*x);
y2 = 2*cos(2*x);
plot(x,y1,'k-',x,y2,'b–');
title ('Plot of f(x) = sin(2x) and its derivative');
xlabel ('x');
ylabel ('y');
legend ('f(x)','d/dx f(x)')
grid on;
```

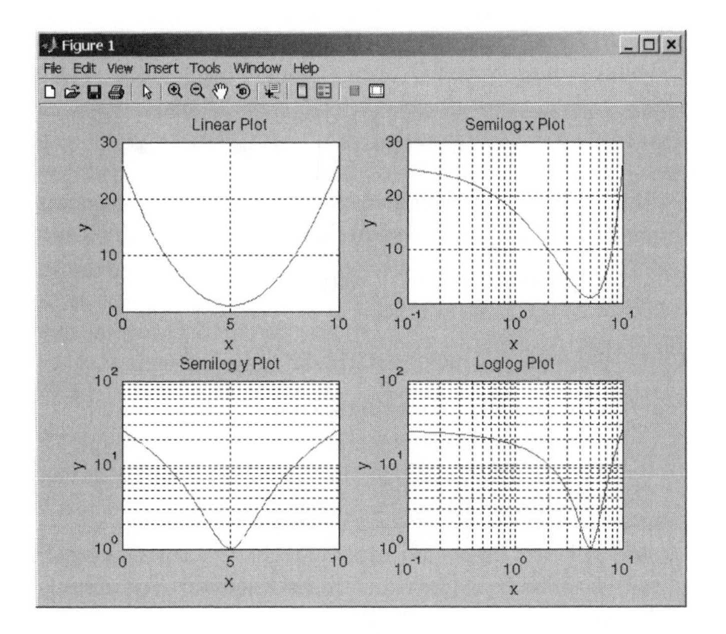

Figure 2.10 Comparison of linear, semilog x, semilog y, and log-log plots.

2.11.6 Logarithmic Scales

It is possible to plot data on logarithmic scales as well as linear scales. There are four possible combinations of linear and logarithmic scales on the x and y axes, and each combination is produced by a separate function.

1. The `plot` function plots both x and y data on linear axes.
2. The `semilogx` function plots x data on logarithmic axes and y data on linear axes.
3. The `semilogy` function plots x data on linear axes and y data on logarithmic axes.
4. The `loglog` function plots both x and y data on logarithmic axes.

All of these functions have identical calling sequences—the only difference is the type of axis used to plot the data. Examples of each plot are shown in Figure 2.10.

2.12 Examples

The following examples illustrate problem-solving with MATLAB.

Example 2.3—Temperature Conversion

Design a MATLAB program that reads an input temperature in degrees Fahrenheit, converts it to an absolute temperature in kelvin, and writes out the result.

SOLUTION The relationship between temperature in degrees Fahrenheit (°F) and temperature in kelvin (K) can be found in any physics textbook. It is

$$T\text{ (in kelvin)} = \left[\frac{5}{9}T\text{ (in °F)} - 32.0\right] + 273.15 \qquad (2\text{-}5)$$

The physics books also give us sample values on both temperature scales, which we can use to check the operation of our program. Two such values are:

- The boiling point of water `212° F` `373.15 K`
- The sublimation point of dry ice `-110° F` `194.26 K`

Our program must perform the following steps:

1. Prompt the user to enter an input temperature in °F.
2. Read the input temperature.
3. Calculate the temperature in kelvin from Equation (2-5).
4. Write out the result, and stop.

We will use function `input` to get the temperature in degrees Fahrenheit and function `fprintf` to print the answer. The resulting program is shown below.

```
%   Script file: temp_conversion
%
%   Purpose:
%     To convert an input temperature from degrees Fahrenheit to
%     an output temperature in kelvin.
%
%   Record of revisions:
%        Date          Programmer              Description of change
%        ====          ==========              =====================
%      01/03/05      S. J. Chapman             Original code
%
% Define variables:
%    temp_f     -- Temperature in degrees Fahrenheit
%    temp_k     -- Temperature in kelvin

% Prompt the user for the input temperature.
temp_f = input('Enter the temperature in degrees Fahrenheit: ');


% Convert to kelvin.
temp_k = (5/9) * (temp_f - 32) + 273.15;

% Write out the result.
fprintf('%6.2f degrees Fahrenheit = %6.2f kelvin.\n', ...
        temp_f,temp_k);
```

To test the completed program, we will run it with the known input values given above. Note that user inputs appear in bold face below.

```
» temp_conversion
Enter the temperature in degrees Fahrenheit: 212
212.00 degrees Fahrenheit = 373.15 kelvin.
» temp_conversion
Enter the temperature in degrees Fahrenheit: -110
-110.00 degrees Fahrenheit = 194.26 kelvin.
```

The results of the program match the values from the physics book.

In the foregoing program, we echoed the input values and printed the output values together with their units. The results of this program make sense only if the units (degrees Fahrenheit and Kelvin) are included together with their values. As a general rule, the units associated with any input value should always be printed along with the prompt that requests the value, and the units associated with any output value should always be printed along with that value.

Good Programming Practice

Always include the appropriate units with any values that you read or write in a program.

The preceding program exhibits many of the good programming practices that we have described in this chapter. It includes a data dictionary defining the meanings of all of the variables in the program. It also uses descriptive variable names, and appropriate units are attached to all printed values.

Example 2.4—Electrical Engineering: Maximum Power Transfer to a Load

Figure 2.11 (on the next page) shows a voltage source $V = 120$ V with an internal resistance R_S of 50 Ω supplying a load of resistance R_L. Find the value of load resistance R_L that will result in the maximum possible power being supplied by the source to the load. How much power be supplied in this case? Also, plot the power supplied to the load as a function of the load resistance R_L.

SOLUTION In this program, we need to vary the load resistance R_L and compute the power supplied to the load at each value of R_L. The power supplied to the load resistance is given by the equation

$$P_L = I^2 R_L \tag{2-6}$$

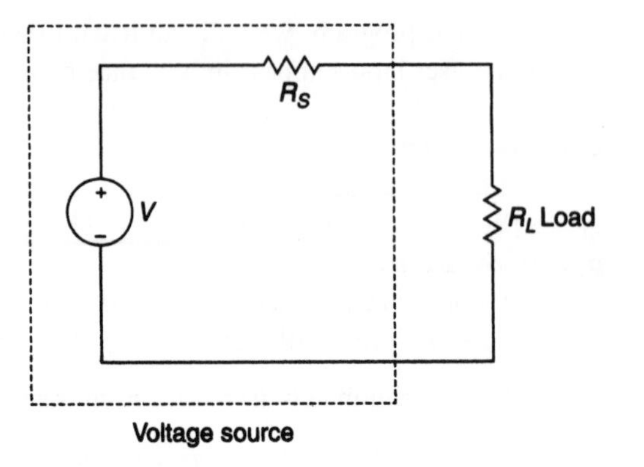

Figure 2.11 A voltage source with a voltage V and an internal resistance R_S supplying a load of resistance R_L.

where I is the current supplied to the load. The current supplied to the load can be calculated by Ohm's Law:

$$I = \frac{V}{R_{\text{TOT}}} = \frac{V}{R_S + R_L} \tag{2-7}$$

The program must perform the following steps:

1. Create an array of possible values for the load resistance R_L. The array will vary R_L from 1 Ω to 100 Ω in 1 Ω steps.
2. Calculate the current for each value of R_L.
3. Calculate the power supplied to the load for each value of R_L.
4. Plot the power supplied to the load for each value of R_L, and determine the value of load resistance resulting in the maximum power.

The final MATLAB program is shown below.

```
%  Script file: calc_power.m
%
%  Purpose:
%    To calculate and plot the power supplied to a load as
%    as a function of the load resistance.
%
%  Record of revisions:
%      Date         Programmer           Description of change
%      ====         ==========           =====================
%    01/03/05     S. J. Chapman          Original code
%
% Define variables:
%   amps        -- Current flow to load (amps)
%   pl          -- Power supplied to load (watts)
%   rl          -- Resistance of the load (ohms)
%   rs          -- Internal resistance of the power source (ohms)
%   volts       -- Voltage of the power source (volts)
```

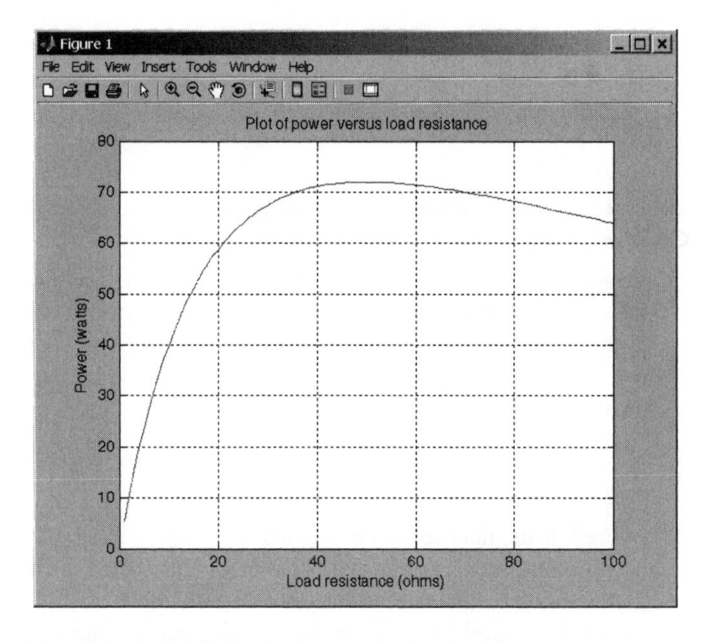

Figure 2.12 Plot of power supplied to load versus load resistance.

```
% Set the values of source voltage and internal resistance
volts = 120;
rs = 50;

% Create an array of load resistances
rl = 1:1:100;

% Calculate the current flow for each resistance
amps = volts ./ ( rs + rl );

% Calculate the power supplied to the load
pl = (amps .^ 2) .* rl;

% Plot the power versus load resistance
plot(rl,pl);
title('Plot of power versus load resistance');
xlabel('Load resistance (ohms)');
ylabel('Power (watts)');
grid on;
```

When this program is executed, the resulting plot is shown in Figure 2.12. From this plot, we can see that the maximum power is supplied to the load when the load's resistance is 50 Ω. The power supplied to the load at this resistance is 72 watts.

Note the use of the array operators `.*`, `.^`, and `./` in the above program. These operators cause the arrays `amps` and `pl` to be calculated on an element-by-element basis.

Example 2.5—Carbon 14 Dating

A radioactive isotope of an element is a form of the element which is not stable. Instead, it spontaneously decays into another element over a period of time. Radioactive decay is an exponential process. If Q_0 is the initial quantity of a radioactive substance at time $t = 0$, then the amount of that substance which will be present at any time t in the future is given by

$$Q(t) = Q_0 e^{-\lambda t} \tag{2-8}$$

where λ is the radioactive decay constant.

Because radioactive decay occurs at a known rate, it can be used as a clock to measure the time since the decay started. If we know the initial amount of the radioactive material Q_0 present in a sample, and the amount of the material Q left at the current time, we can solve for t in Equation (2-8) to determine how long the decay has been going on. The resulting equation is

$$t_{\text{decay}} = -\frac{1}{\lambda} \log_e \frac{Q}{Q_0} \tag{2-9}$$

Equation (2-9) has practical applications in many areas of science. For example, archaeologists use a radioactive clock based on carbon 14 to determine the time that has passed since a once-living thing died. Carbon 14 is continually taken into the body while a plant or animal is living, so the amount of it present in the body at the time of death is assumed to be known. The decay constant λ of carbon 14 is well known to be 0.00012097/year, so if the amount of carbon 14 remaining now can be accurately measured, then Equation (2-9) can be used to determine how long ago the living thing died. The amount of carbon 14 remaining as a function of time is shown in Figure 2.13.

Write a program that reads the percentage of carbon 14 remaining in a sample, calculates the age of the sample from it, and prints out the result with proper units.

SOLUTION Our program must perform the following steps:

1. Prompt the user to enter the percentage of carbon 14 remaining in the sample.
2. Read in the percentage.
3. Convert the percentage into the fraction $\frac{Q}{Q_0}$.
4. Calculate the age of the sample in years using Equation (2-9).
5. Write out the result, and stop.

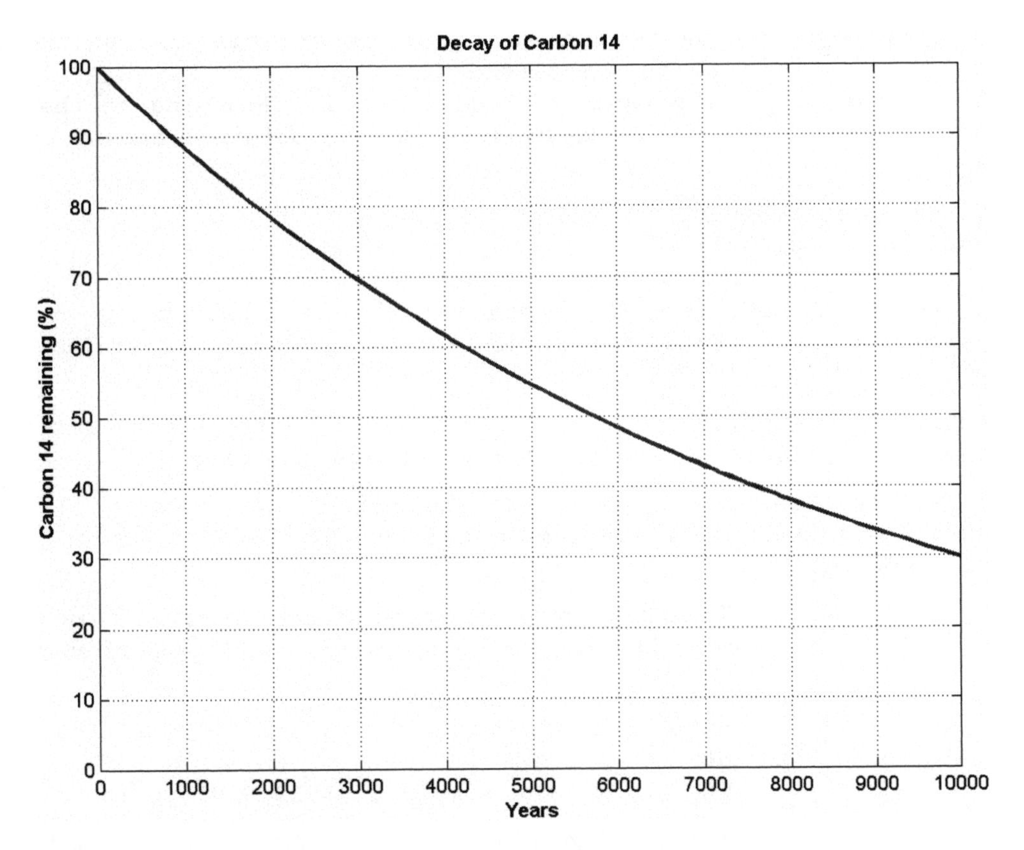

Figure 2.13 The radioactive decay of carbon 14 as a function of time. Notice that 50% of the original carbon 14 is left after about 5,730 years have elapsed.

The resulting code is shown below.

```
%  Script file: c14_date.m
%
%  Purpose:
%    To calculate the age of an organic sample from the percentage
%    of the original carbon 14 remaining in the sample.
%
%  Record of revisions:
%      Date          Programmer              Description of change
%      ====          ==========              =====================
%    01/03/05     S. J. Chapman            Original code
%
% Define variables:
%   age          -- The age of the sample in years
%   lamda        -- The radioactive decay constant for carbon-14,
%                   in units of 1/years.
```

```
%    percent    -- The percentage of carbon 14 remaining at the time
%                  of the measurement
%    ratio      -- The ratio of the carbon 14 remaining at the time
%                  of the measurement to the original amount of
%                  carbon 14.

% Set decay constant for carbon-14
lamda = 0.00012097;

% Prompt the user for the percentage of C-14 remaining.
percent = input('Enter the percentage of carbon 14 remaining:\n');

% Perform calculations
ratio = percent/100;                % Convert to fractional ratio
age = (-1.0/lamda) * log(ratio); % Get age in years

% Tell the user about the age of the sample.
string = ['The age of the sample is ' num2str(age) ' years.'];
disp(string);
```

To test the completed program, we will calculate the time it takes for half of the carbon 14 to disappear. This time is known as the *half-life* of carbon 14.

```
» c14_date
Enter the percentage of carbon 14 remaining:
50
The age of the sample is 5729.9097 years.
```

The *CRC Handbook of Chemistry and Physics* states that the half-life of carbon 14 is 5730 years, so output of the program agrees with the reference book.

2.13 Debugging MATLAB Programs

There is an old saying that the only sure things in life are death and taxes. We can add one more certainty to that list: if you write a program of any significant size, it won't work the first time you try it! Errors in programs are known as **bugs**, and the process of locating and eliminating them is known as **debugging**. Given that we have written a program and it is not working, how do we debug it?

Three types of errors are found in MATLAB programs. The first type of error is a **syntax error**. Syntax errors are errors in the MATLAB statement itself, such as spelling errors or punctuation errors. These errors are detected by the MATLAB compiler the first time that an M-file is executed. For example, the statement

```
x = (y + 3)/2);
```

contains a syntax error because it has unbalanced parentheses. If this statement appears in an M-file named `test.m`, the following message appears when `test` is executed:

```
» test
??? x = (y + 3)/2)
                 |
Missing operator, comma, or semi-colon.

Error in ==> d:\book\matlab\chap1\test.m
On line 2  ==>
```

The second type of error is the **run-time error**. A run-time error occurs when an illegal mathematical operation is attempted during program execution (for example, attempting to divide by 0). These errors cause the program to return `Inf` or `NaN`, which is then used in further calculations. The results of a program that contains calculations using `Inf` or `NaN` are usually invalid.

The third type of error is a **logical error**. Logical errors occur when the program compiles and runs successfully but produces the wrong answer.

The most common mistakes made during programming are *typographical errors*. Some typographical errors create invalid MATLAB statements. These errors produce syntax errors that are caught by the compiler. Other typographical errors occur in variable names. For example, the letters in some variable names might have been transposed, or an incorrect letter might be typed. The result will be a new variable, and MATLAB simply creates the new variable the first time that it is referenced. MATLAB cannot detect this type of error. Typographical errors can also produce logical errors. For example, if variables `vel1` and `vel2` are both used for velocities in the program, then one of them might be inadvertently used instead of the other one at some point. You must check for that sort of error by manually inspecting the code.

Sometimes a program will start to execute, but run-time errors or logical errors occur during execution. In this case, there is either something wrong with the input data or something wrong with the logical structure of the program. The first step in locating this sort of bug should be to *check the input data to the program*. Either remove semicolons from input statements or add extra output statements to verify that the input values are what you expect them to be.

If the variable names seem to be correct and the input data is correct, then you are probably dealing with a logical error. You should check each of your assignment statements.

1. If an assignment statement is very long, break it into several smaller assignment statements. Smaller statements are easier to verify.
2. Check the placement of parentheses in your assignment statements. It is a very common error to have the operations in an assignment statement evaluated in the wrong order. If you have any doubts as to the order in which the variables are being evaluated, add extra sets of parentheses to make your intentions clear.

3. Make sure that you have initialized all of your variables properly.
4. Be sure that any functions you use are in the correct units. For example, the input to trigonometric functions must be in units of radians, not degrees.

If you are still getting the wrong answer, add output statements at various points in your program to see the results of intermediate calculations. If you can locate the point where the calculations go bad, then you know just where to look for the problem, which is 95 percent of the battle.

If you still cannot find the problem after all of the preceding steps, explain what you are doing to another student or to your instructor, and let him or her look at the code. It is very common for a people to see just what they expect to see when they look at their own code. Another person can often quickly spot an error that you have overlooked time after time.

Good Programming Practice

To reduce your debugging effort, make sure that during your program design you

1. Initialize all variables.
2. Use parentheses to make the functions of assignment statements clear.

MATLAB includes a special debugging tool called a *symbolic debugger*. A symbolic debugger is a tool that allows you to walk through the execution of your program one statement at a time and to examine the values of any variables at each step along the way. Symbolic debuggers allow you to see all of the intermediate results without having to insert a lot of output statements into your code. We will learn how to use MATLAB's symbolic debugger in Chapter 3.

2.14 Summary

In this chapter, we have presented many of the fundamental concepts required to write functional MATLAB programs. We learned about the basic types of MATLAB windows, the workspace, and how to get on-line help.

We introduced two data types: `double` and `char`. We also introduced assignment statements, arithmetic calculations, intrinsic functions, input/output statements, and data files.

The order in which MATLAB expressions are evaluated follows a fixed hierarchy, with operations at a higher level evaluated before operations at lower levels. The hierarchy of operations is summarized in Table 2.12.

The MATLAB language includes an extremely large number of built-in functions to help us solve problems. This list of functions is *much* richer than the list

Table 2.12 Hierarchy of Operations

Precedence	Operation
1	The contents of all parentheses are evaluated, starting from the innermost parentheses and working outward.
2	All exponentials are evaluated, working from left to right.
3	All multiplications and divisions are evaluated, working from left to right.
4	All additions and subtractions are evaluated, working from left to right.

of functions found in other languages (e.g., Fortran or C++), and it includes device-independent plotting capabilities. A few of the common intrinsic functions are summarized in Table 2.8, and many others will be introduced throughout the remainder of the book. A complete list of all MATLAB functions is available through the on-line Help Browser.

2.14.1 Summary of Good Programming Practice

Every MATLAB program should be designed so that another person who is familiar with MATLAB can easily understand it. This is very important, since a good program may be used for a long period of time. Over that time, conditions will change, and the program will need to be modified to reflect the changes. The program modifications may be done by someone other than the original programmer. The programmer making the modifications must understand the original program well before attempting to change it.

It is much harder to design clear, understandable, and maintainable programs than it is to simply write programs. To do so, a programmer must develop the discipline to properly document his or her work. In addition, the programmer must be careful to avoid known pitfalls along the path to good programs. The following guidelines will help you to develop good programs:

1. Use meaningful variable names whenever possible. Use names that can be understood at a glance, like `day`, `month`, and `year`.
2. Create a data dictionary for each program to make program maintenance easier.
3. Use only lower-case letters in variable names, so that there won't be errors due to capitalization differences in different occurrences of a variable name.
4. Use a semicolon at the end of all MATLAB assignment statements to suppress echoing of assigned values in the Command Window. If you

need to examine the results of a statement during program debugging, you may remove the semicolon from that statement only.

5. If data must be exchanged between MATLAB and other programs, save the MATLAB data in ASCII format. If the data will be used only in MATLAB, save the data in MAT-file format.
6. Save ASCII data files with a "`dat`" file extent to distinguish them from MAT-files, which have a "`mat`" file extent.
7. Use parentheses as necessary to make your equations clear and easy to understand.
8. Always include the appropriate units with any values that you read or write in a program.

2.14.2 MATLAB Summary

The following summary lists all of the MATLAB special symbols, commands, and functions described in this chapter, along with a brief description of each one.

Special Symbols	
`[ ]`	Array constructor
`( )`	Forms subscripts
`' '`	Marks the limits of a character string
`,`	1. Separates subscripts or matrix elements 2. Separates assignment statements on a line
`;`	1. Suppresses echoing in Command Window 2. Separates matrix rows 3. Separates assignment statements on a line
`%`	Marks the beginning of a comment
`:`	Colon operator, used to create shorthand lists
`+`	Array and matrix addition
`-`	Array and matrix subtraction
`.*`	Array multiplication
`*`	Matrix multiplication
`./`	Array right division
`.\`	Array left division
`/`	Matrix right division
`\`	Matrix left division
`.^`	Array exponentiation
`'`	Transpose operator

Commands and Functions

`...`	Continues a MATLAB statement on the following line.
`abs(x)`	Calculates the absolute value of x.
`ans`	Default variable used to store the result of expressions not assigned to another variable.
`acos(x)`	Calculates the inverse cosine of x. The resulting angle is in radians between 0 and π.
`asin(x)`	Calculates the inverse sine of x. The resulting angle is in radians between $-\pi/2$ and $\pi/2$.
`atan(x)`	Calculates the inverse tangent of x. The resulting angle is in radians between $-\pi/2$ and $\pi/2$.
`atan2(y,x)`	Calculates the inverse tangent of y/x, valid over the entire circle. The resulting angle is in radians between $-\pi$ and π.
`ceil(x)`	Rounds x to the nearest integer towards positive infinity: `floor(3.1) = 4` and `floor(-3.1) = -3`.
`char`	Converts a matrix of numbers into a character string. For ASCII characters the matrix should contain numbers ≤ 127.
`clock`	Current time.
`cos(x)`	Calculates cosine of x, where x is in radians.
`date`	Current date.
`disp`	Displays data in command window.
`doc`	Opens HTML Help Desk directly at a particular function description.
`double`	Converts a character string into a matrix of numbers.
`eps`	Represents machine precision.
`exp(x)`	Calculates e^x.
`eye(n,m)`	Generates an identity matrix.
`fix(x)`	Rounds x to the nearest integer towards zero: `fix(3.1) = 3` and `fix(-3.1) = -3`.
`floor(x)`	Rounds x to the nearest integer towards minus infinity: `floor(3.1) = 3` and `floor(-3.1) = -4`.
`format +`	Print + and − signs only.
`format bank`	Print in "dollars and cents" format.
`format compact`	Suppress extra linefeeds in output.
`format hex`	Print hexadecimal display of bits.
`format long`	Print with 14 digits after the decimal.
`format long e`	Print with 15 digits plus exponent.
`format long g`	Print with 15 digits with or without exponent.
`format loose`	Print with extra linefeeds in output.
`format rat`	Print as an approximate ratio of small integers.
`format short`	Print with 4 digits after the decimal.

(continued)

Commands and Functions (Continued)

`format short e`	Print with 5 digits plus exponent.
`format short g`	Print with 5 digits with or without exponent.
`fprintf`	Print formatted information.
`grid`	Add/remove a grid from a plot.
`i`	$\sqrt{-1}$
`Inf`	Represents machine infinity (∞).
`input`	Writes a prompt and reads a value from the keyboard.
`int2str`	Converts x into an integer character string.
`j`	$\sqrt{-1}$
`legend`	Adds a legend to a plot.
`length(arr)`	Returns the length of a vector, or the longest dimension of a 2-D array.
`load`	Load data from a file.
`log(x)`	Calculates the natural logarithm of x.
`loglog`	Generates a log-log plot.
`lookfor`	Look for a matching term in the one-line MATLAB function descriptions.
`max(x)`	Returns the maximum value in vector x, and optionally the location of that value.
`min(x)`	Returns the minimum value in vector x, and optionally the location of that value.
`mod(n,m)`	Remainder or modulo function.
`NaN`	Represents not-a-number.
`num2str(x)`	Converts x into a character string.
`ones(n,m)`	Generates an array of ones.
`pi`	Represents the number π.
`plot`	Generates a linear xy plot.
`print`	Prints a Figure Window.
`round(x)`	Rounds x to the nearest integer.
`save`	Saves data from workspace into a file.
`semilogx`	Generates a log-linear plot.
`semilogy`	Generates a linear-log plot.
`sin(x)`	Calculates sine of x, where x is in radians.
`size`	Get number of rows and columns in an array.
`sqrt`	Calculates the square root of a number.
`str2num`	Converts a character string into a number.
`tan(x)`	Calculates tangent of x, where x is in radians.
`title`	Adds a title to a plot.
`zeros`	Generate an array of zeros.

2.15 Exercises

2.1 Answer the following questions for the array shown below.

$$\text{array1} = \begin{bmatrix} 1.1 & 0.0 & 2.1 & -3.5 & 6.0 \\ 0.0 & 1.1 & -6.6 & 2.8 & 3.4 \\ 2.1 & 0.1 & 0.3 & -0.4 & 1.3 \\ -1.4 & 5.1 & 0.0 & 1.1 & 0.0 \end{bmatrix}.$$

(a) What is the size of `array1`?
(b) What is the value of `array1(4,1)`?
(c) What is the size and value of `array1(:,1:2)`?
(d) What is the size and value of `array1([1 3],end)`?

2.2 Are the following MATLAB variable names legal or illegal? Why?

(a) `dog1`
(b) `1dog`
(c) `Do_you_know_the_way_to_san_jose`
(d) `_help`
(e) `What's_up?`

2.3 Determine the size and contents of the following arrays. Note that the later arrays may depend on the definitions of arrays defined earlier in this exercise.

(a) `a = 1:2:5;`
(b) `b = [a' a' a'];`
(c) `c = b(1:2:3,1:2:3);`
(d) `d = a + b(2,:);`
(e) `w = [zeros(1,3) ones(3,1)' 3:5'];`
(f) `b([1 3],2) = b([3 1],2);`

2.4 Assume that array `array1` is defined as shown, and determine the contents of the following sub-arrays:

$$\text{array1} = \begin{bmatrix} 1.1 & 0.0 & 2.1 & -3.5 & 6.0 \\ 0.0 & 1.1 & -6.6 & 2.8 & 3.4 \\ 2.1 & 0.1 & 0.3 & -0.4 & 1.3 \\ -1.4 & 5.1 & 0.0 & 1.1 & 0.0 \end{bmatrix}$$

(a) `array1(3,:)`
(b) `array1(:,3)`
(c) `array1(1:2:3,[3 3 4])`
(d) `array1([1 1],:)`

2.5 Assume that `value` has been initialized to 10π, and determine what is printed out by each of the following statements.

```
disp (['value = ' num2str(value)]);
disp (['value = ' int2str(value)]);
fprintf('value = %e\n',value);
fprintf('value = %f\n',value);
fprintf('value = %g\n',value);
fprintf('value = %12.4f\n',value);
```

2.6 Assume that `a`, `b`, `c`, and `d` are defined as follows, and calculate the results of the following operations if they are legal. If an operation is, explain why it is illegal.

$$\texttt{a} = \begin{bmatrix} 2 & -2 \\ -1 & 2 \end{bmatrix} \qquad \texttt{b} = \begin{bmatrix} 1 & -1 \\ 0 & 2 \end{bmatrix}$$

$$\texttt{c} = \begin{bmatrix} 1 \\ -2 \end{bmatrix} \qquad \texttt{d} = \text{eye(2)}$$

(a) `result = a + b;`
(b) `result = a * d;`
(c) `result = a .* d;`
(d) `result = a * c;`
(e) `result = a .* c;`
(f) `result = a \ b;`
(g) `result = a .\ b;`
(h) `result = a .^ b;`

2.7 Evaluate each of the following expressions.

(a) `11/5 + 6`
(b) `(11/5) + 6`
(c) `11/(5 + 6)`
(d) `3 ^ 2 ^ 3`
(e) `3 ^ (2 ^ 3)`
(f) `(3 ^ 2) ^ 3`
(g) `round(-11/5) + 6`
(h) `ceil(-11/5) + 6`
(i) `floor(-11/5) + 6`

2.8 Use MATLAB to evaluate each of the following expressions.

(a) $(3 - 5i)(-4 + 6i)$
(b) $\cos^{-1}(1.2)$

2.9 Solve the following system of simultaneous equations for x:

$$
\begin{aligned}
-2.0\,x_1 + 5.0\,x_2 + 1.0\,x_3 + 3.0\,x_4 + 4.0\,x_5 - 1.0\,x_6 &= 0.0 \\
2.0\,x_1 - 1.0\,x_2 - 5.0\,x_3 - 2.0\,x_4 + 6.0\,x_5 + 4.0\,x_6 &= 1.0 \\
-1.0\,x_1 + 6.0\,x_2 - 4.0\,x_3 - 5.0\,x_4 + 3.0\,x_5 - 1.0\,x_6 &= -6.0 \\
4.0\,x_1 + 3.0\,x_2 - 6.0\,x_3 - 5.0\,x_4 - 2.0\,x_5 - 2.0\,x_6 &= 10.0 \\
-3.0\,x_1 + 6.0\,x_2 + 4.0\,x_3 + 2.0\,x_4 - 6.0\,x_5 + 4.0\,x_6 &= -6.0 \\
2.0\,x_1 + 4.0\,x_2 + 4.0\,x_3 + 4.0\,x_4 + 5.0\,x_5 - 4.0\,x_6 &= -2.0
\end{aligned}
$$

2.10 **Position and Velocity of a Ball** If a stationary ball is released at a height h_0 above the surface of the Earth with a vertical velocity v_0, the position and velocity of the ball as a function of time will be given by the equations

$$h(t) = \frac{1}{2}gt^2 + v_0t + h_0 \tag{2-10}$$

$$v(t) = gt + v_0 \tag{2-11}$$

where g is the acceleration due to gravity (-9.81 m/s^2), h is the height above the surface of the Earth (assuming no air friction), and v is the vertical component of velocity. Write a MATLAB program that prompts a user for the initial height of the ball in meters and velocity of the ball in meters per second, and plots the height and velocity as a function of time. Be sure to include proper labels in your plots.

2.11 The distance between two points (x_1, y_1) and (x_2, y_2) on a Cartesian coordinate plane (Figure 2.14) is given by the equation

$$d = \sqrt{(x_1 - x_2)^2 + (y_1 - y_2)^2} \tag{2-12}$$

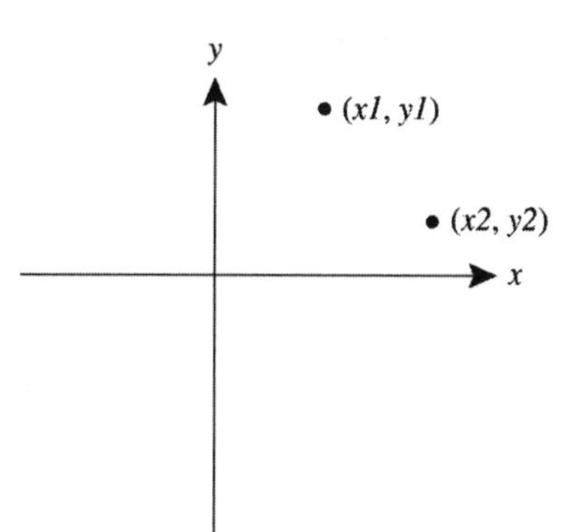

Figure 2.14 Distance between two points on a Cartesian plane.

Write a program to calculate the distance between any two points (x_1, y_1) and (x_2, y_2) specified by the user. Use good programming practices in your program. Use the program to calculate the distance between the points (2, 3) and (8, −5).

2.12 **Decibels** Engineers often measure the ratio of two power measurements in *decibels,* or dB. The equation for the ratio of two power measurements in decibels is

$$\text{dB} = 10 \log_{10} \frac{P_2}{P_1} \tag{2-13}$$

where P_2 is the power level being measured, and P_1 is some reference power level.

(a) Assume that the reference power level P_1 is 1 milliwatt, and write a program that accepts an input power P_2 and converts it into dB with respect to the 1 mW reference level. (Engineers have a special unit for dB power levels with respect to a 1 mW reference: dBm.) Use good programming practices in your program.

(b) Write a program that creates a plot of power in watts versus power in dBm with respect to a 1 mW reference level. Create both a linear *xy* plot and a log-linear *xy* plot.

2.13 **Hyperbolic cosine** The hyperbolic cosine function is defined by the equation

$$\cosh x = \frac{e^x + e^{-x}}{2} \tag{2-14}$$

Write a program to calculate the hyperbolic cosine of a user-supplied value `x`. Use the program to calculate the hyperbolic cosine of 3.0. Compare the answer that your program produces to the answer produced by the MATLAB intrinsic function `cosh(x)`. Also, use MATLAB to plot the function `cosh(x)`. What is the smallest value that this function can have? At what value of `x` does it occur?

2.14 **Energy Stored in a Spring** The force required to compress a linear spring is given by the equation

$$F = kx \tag{2-15}$$

where F is the force in newtons and k is the spring constant in newtons per meter. The potential energy stored in the compressed spring is given by the equation

$$E = \frac{1}{2} kx^2 \tag{2-16}$$

where E is the energy in joules. The following information is available for four springs:

	Spring 1	Spring 2	Spring 3	Spring 4
Force (N)	20	24	22	20
Spring constant k (N/m)	500	600	700	800

Determine the compression of each spring, and the potential energy stored in each spring. Which spring has the most energy stored in it?

2.15 **Radio Receiver** A simplified version of the front end of an AM radio receiver is shown in Figure 2.15. This receiver consists of an *RLC* tuned circuit containing a resistor, capacitor, and an inductor connected in series. The *RLC* circuit is connected to an external antenna and ground as shown in the picture.

The tuned circuit allows the radio to select a specific station out of all the stations transmitting on the AM band. At the resonant frequency of the circuit, essentially all of the signal V_0 appearing at the antenna appears across the resistor, which represents the rest of the radio. In other words, the radio receives its strongest signal at the resonant frequency. The resonant frequency of the LC circuit is given by the equation

$$f_0 = \frac{1}{2\pi\sqrt{LC}} \tag{2-17}$$

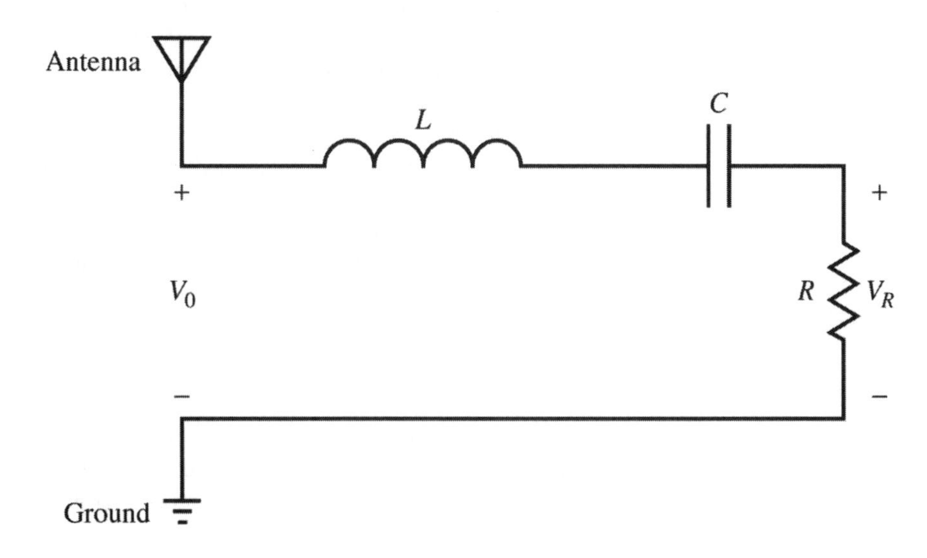

Figure 2.15 A simplified version of the front end of an AM radio receiver.

where L is inductance in henrys (H) and C is capacitance in farads (F). Write a program that calculates the resonant frequency of this radio set given specific values of L and C. Test your program by calculating the frequency of the radio when $L = 0.1$ mH and $C = 0.25$ nF.

2.16 **Radio Receiver** The average (rms) voltage across the resistive load in Figure 2.15 varies as a function of frequency according to Equation (2-18).

$$V_R = \frac{R}{\sqrt{R^2 + \left(\omega L - \frac{1}{\omega C}\right)^2}} V_0 \tag{2-18}$$

where $\omega = 2\pi f$ and f is the frequency in hertz. Assume that $L = 0.1$ mH, $C = 0.25$ nF, $R = 50\ \Omega$, and $V_0 = 10$ mV.

(a) Plot the rms voltage on the resistive load as a function of frequency. At what frequency does the voltage on the resitive load peak? What is the voltage on the load at this frequency? This frequency is called the resonant frequency f_0 of the circuit.
(b) If the frequency is changed to 10% greater than the resonant frequency, what is the voltage on the load? How selective is this radio receiver?
(c) At what frequencies will the voltage on the load drop to half of the voltage at the resonant frequency?

2.17 Suppose two signals were received at the antenna of the radio receiver described in the previous problem. One signal has a strength of 1 V at a frequency of 1,000 kHz, and the other signal has a strength of 1 V at 950 kHz. Calculate the voltage V_R that will be received for each of these signals. How much power will the first signal supply to the resistive load R? How much power will the second signal supply to the resistive load R? Express the ratio of the power supplied by signal 1 to the power supplied by signal 2 in decibels (see Problem 2.12 for the definition of a decibel). How much is the second signal enhanced or suppressed compared to the first signal? (*Note:* The power supplied to the resistive load can be calculated from the equation $P = V_R^2/R$).

2.18 **Aircraft Turning Radius** An object moving in a circular path at a constant tangential velocity v is shown in Figure 2.16. The radial acceleration required for the object to move in the circular path is given by Equation (2-19)

$$a = \frac{v^2}{r} \tag{2-19}$$

where a is the centripetal acceleration of the object in m/s^2, v is the tangential velocity of the object in m/s, and r is the turning radius in meters. Suppose that the object is an aircraft, and answer the following questions about it:

(a) Suppose that the aircraft is moving at Mach 0.85, or 85% of the speed of sound. If the centripetal acceleration is 2 g, what is the

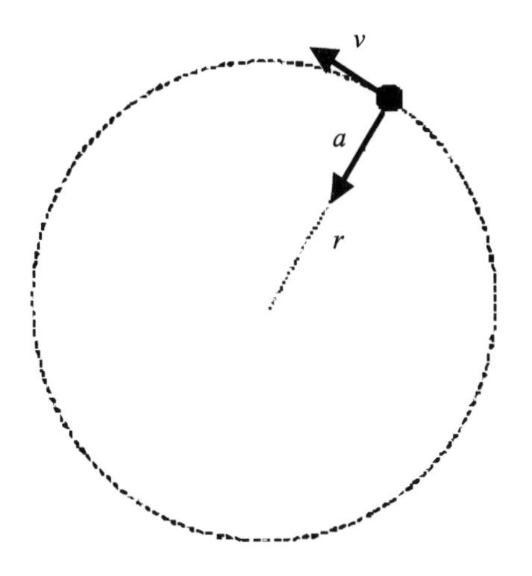

Figure 2.16 An object moving in uniform circular motion due to the centripetal acceleration *a*.

turning radius of the aircraft? (Note: For this problem, you may assume that Mach 1 is equal to 340 m/s, and that 1 g = 9.81 m/s^2).

(b) Suppose that the speed of the aircraft increases to Mach 1.5. What is the turning radius of the aircraft now?

(c) Plot the turning radius as a function of aircraft speed for speeds between Mach 0.5 and Mach 2.0, assuming that the acceleration remains 2 g.

(d) Suppose that the maximum acceleration that the pilot can stand is 7 g. What is the minimum possible turning radius of the aircraft at Mach 1.5?

(e) Plot the turning radius as a function of centripetal acceleration for accelerations between 2 g and 8 g, assuming a constant speed of Mach 0.85.

CHAPTER 3

Branching Statements and Program Design

In the previous chapter, we developed several complete working MATLAB programs. However, all of the programs were very simple, consisting of a series of MATLAB statements that were executed one after another in a fixed order. Such programs are called *sequential* programs. They read input data, process it to produce a desired answer, print out the answer, and quit. There is no way to repeat sections of the program more than once during program execution, and there is no way to selectively execute only certain portions of the program depending on values of the input data.

In the next two chapters, we will introduce a number of MATLAB statements that allow us to control the order in which statements are executed in a program. There are two broad categories of control statements: **branches**, which select specific sections of the code to execute, and **loops**, which cause specific sections of the code to be repeated. Branches are discussed in this chapter, and loops are discussed in Chapter 4.

With the introduction of branches and loops, our programs are going to become more complex, and it will get easier to make mistakes. To help avoid programming errors, we will introduce a formal program design procedure based upon the technique known as top-down design. We will also introduce a common algorithm development tool known as pseudocode.

3.1 Introduction to Top-Down Design Techniques

Suppose that you are an engineer working in industry, and that you need to write a program to solve some problem. How do you begin?

When given a new problem, there is a natural tendency to sit down at a keyboard and start programming without "wasting" a lot of time thinking about the problem first. It is often possible to get away with this "on-the-fly" approach to programming for very small problems, such as many of the examples in this book. In the real world, however, problems are larger, and a programmer attempting this approach will become hopelessly bogged down. For larger problems, it pays to completely think out the problem and the approach you are going to take to it before writing a single line of code.

We will introduce a formal program design process in this section and then apply that process to every major application developed in the remainder of the book. For some of the simple examples that we will be doing, the design process will seem like overkill. However, as the problems that we solve get larger and larger, the process becomes more and more essential to successful programming.

When I was an undergraduate, one of my professors was fond of saying, "Programming is easy. It's knowing what to program that's hard." His point was forcefully driven home to me after I left university and began working in industry on larger-scale software projects. I found that the most difficult part of my job was to *understand the problem* I was trying to solve. Once I really understood the problem, it became easy to break the problem apart into smaller, more easily manageable pieces with well-defined functions, and then to tackle those pieces one at a time.

Top-down design is the process of starting with a large task and breaking it down into smaller, more easily understandable pieces (subtasks) which perform a portion of the desired task. Each subtask may in turn be subdivided into smaller subtasks if necessary. Once the program is divided into small pieces, each piece can be coded and tested independently. We do not attempt to combine the subtasks into a complete task until each of the subtasks has been verified to work properly by itself.

The concept of top-down design is the basis of our formal program design process. We will now introduce the details of the process, which is illustrated in Figure 3.1. The steps involved are:

1. *Clearly state the problem that you are trying to solve.*
   Programs are usually written to fill some perceived need, but that need may not be articulated clearly by the person requesting the program. For example, a user may ask for a program to solve a system of simultaneous linear equations. This request is not clear enough to allow a programmer to design a program to meet the need; he or she must first know much more about the problem to be solved. Is the system of equations to be solved real or complex? What is the maximum number of equations and unknowns that the program must handle? Are there any symmetries in the equations which might be exploited to make the task easier? The program designer will have to talk with the user requesting the program, and the two of them will have to come up with a clear statement of exactly what they are trying to accomplish. A clear statement of the problem will prevent misunderstandings, and it will also help the program designer to

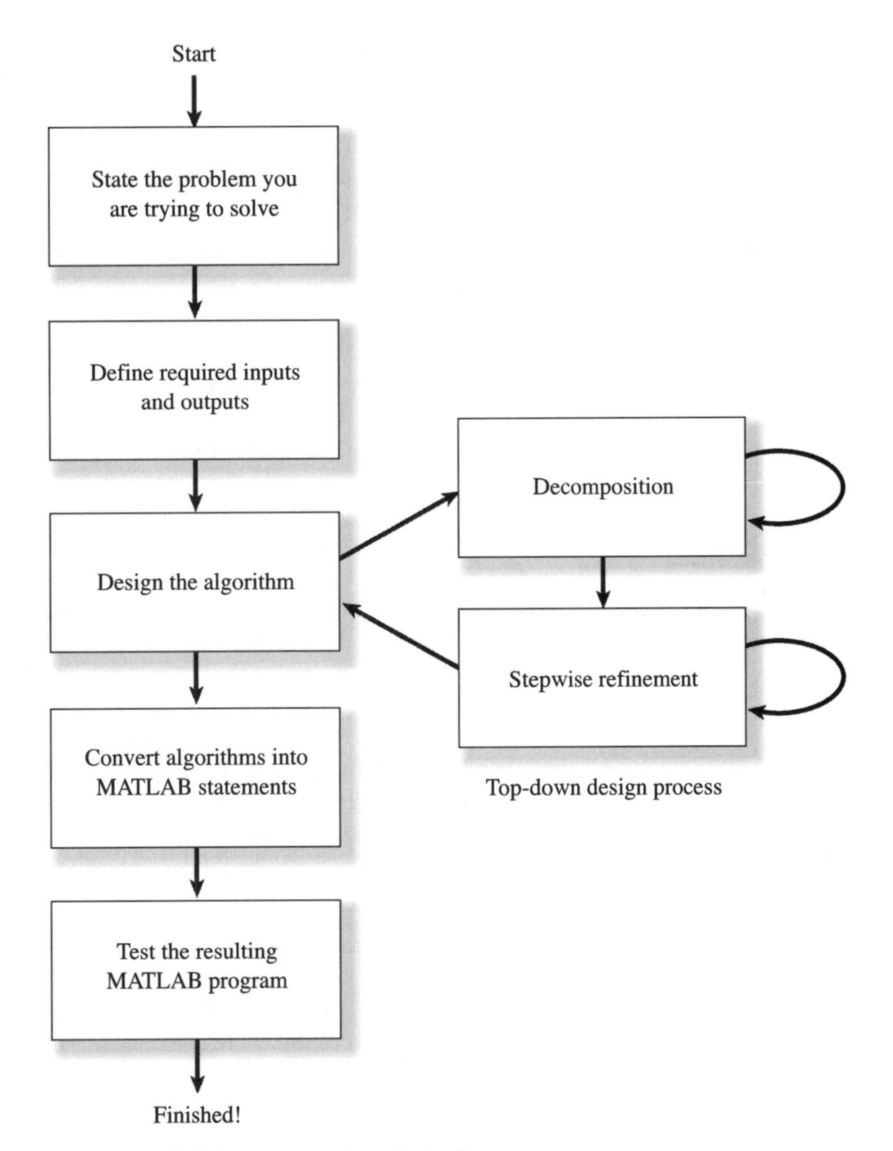

Figure 3.1 The program design process used in this book.

properly organize his or her thoughts. In the example we were describing, a proper statement of the problem might have been:

Design and write a program to solve a system of simultaneous linear equations having real coefficients and with up to 20 equations in 20 unknowns.

2. *Define the inputs required by the program and the outputs to be produced by the program.*
   The inputs to the program and the outputs produced by the program must be specified so that the new program will properly fit into the overall

processing scheme. In the above example, the coefficients of the equations to be solved are probably in some pre-existing order, and our new program needs to be able to read them in that order. Similarly, it needs to produce the answers required by the programs which may follow it in the overall processing scheme, and to write out those answers in the format needed by the programs following it.

3. *Design the algorithm that you intend to implement in the program.*
   An **algorithm** is a step-by-step procedure for finding the solution to a problem. It is at this stage in the process that top-down design techniques come into play. The designer looks for logical divisions within the problem, and divides it up into subtasks along those lines. This process is called *decomposition*. If the subtasks are themselves large, the designer can break them up into even smaller sub-subtasks. This process continues until the problem has been divided into many small pieces, each of which does a simple, clearly understandable job.

   After the problem has been decomposed into small pieces, each piece is further refined through a process called *stepwise refinement*. In stepwise refinement, a designer starts with a general description of what the piece of code should do and then defines the functions of the piece in greater and greater detail until they are specific enough to be turned into MATLAB statements. Stepwise refinement is usually done with **pseudocode**, which is described in the following section.

   It is often helpful to solve a simple example of the problem by hand during the algorithm development process. If the designer understands the steps that he or she went through in solving the problem by hand, then he or she will be in better able to apply decomposition and stepwise refinement to the problem.

4. *Turn the algorithm into MATLAB statements.*
   If the decomposition and refinement process was carried out properly, this step will be very simple. All the programmer will have to do is to replace pseudocode with the corresponding MATLAB statements on a one-for-one basis.

5. *Test the resulting MATLAB program.*
   This step is the real killer. The components of the program must first be tested individually, if possible, and then the program as a whole must be tested. When testing a program, we must verify that it works correctly for *all legal input data sets*. It is very common for a program to be written, tested with some standard data set, and released for use, only to find that it produces the wrong answers (or crashes) with a different input data set. If the algorithm implemented in a program includes different branches, we must test all of the possible branches to confirm that the program operates correctly under every possible circumstance.

Large programs typically go through a series of tests before they are released for general use (see Figure 3.2). The first stage of testing is sometimes called **unit testing**. During unit testing, the individual subtasks of the program are tested separately to confirm that they work correctly. After the unit testing is completed, the program goes through a series of *builds* during which the individual subtasks are combined to produce the final program. The first build of the program typically

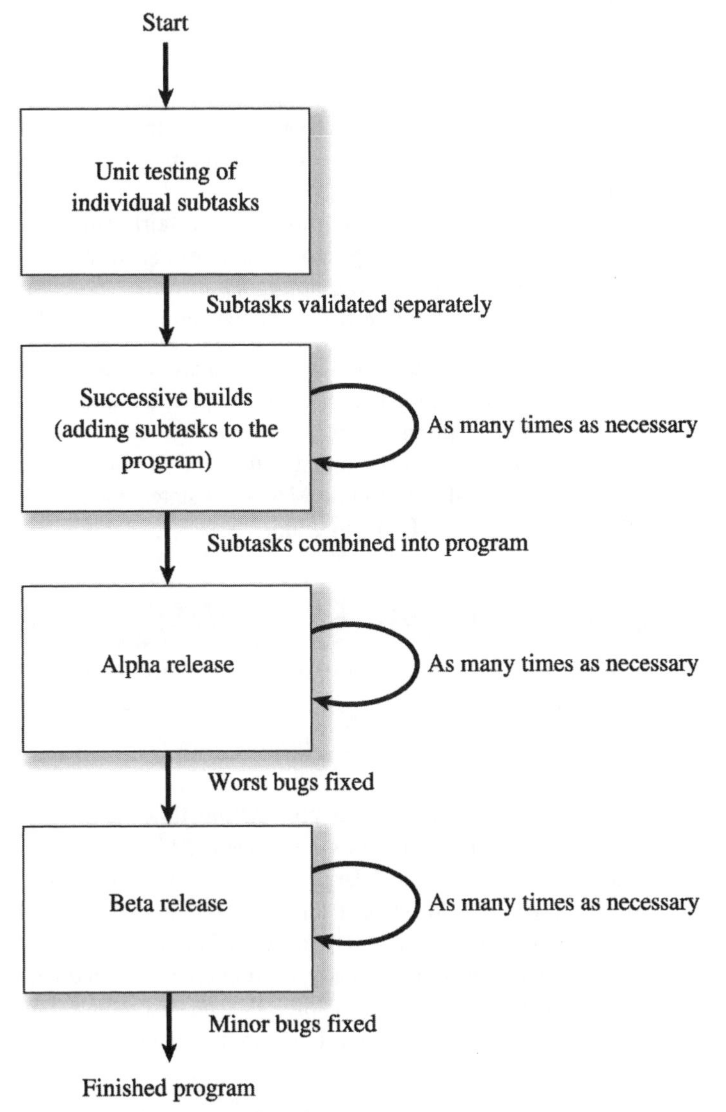

Figure 3.2 A typical testing process for a large program.

includes only a few of the subtasks. It is used to check the interactions among those subtasks and the functions performed by the combinations of the subtasks. In successive builds, more and more subtasks are added, until the entire program is complete. Testing is performed on each build, and any errors (bugs) that are detected are corrected before moving on to the next build.

Testing continues even after the program is complete. The first complete version of the program is usually called the **alpha release**. It is exercised by the programmers and others very close to them in as many different ways as possible, and the bugs discovered during the testing are corrected. When the most serious bugs have been removed from the program, a new version called the **beta release** is prepared. The beta release is normally given to "friendly" outside users who have a need for the program in their normal day-to-day jobs. These users put the program through its paces under many different conditions and with many different input data sets, and they report any bugs that they find to the programmers. When those bugs have been corrected, the program is ready to be released for general use.

Because the programs in this book are fairly small, we will not go through the sort of extensive testing described above. However, we will follow the basic principles in testing all of our programs.

The program design process may be summarized as follows:

1. Clearly state the problem that you are trying to solve.
2. Define the inputs required by the program and the outputs to be produced by the program.
3. Design the algorithm that you intend to implement in the program.
4. Turn the algorithm into MATLAB statements.
5. Test the MATLAB program.

Good Programming Practice

Follow the steps of the program design process to produce reliable, understandable MATLAB programs.

In a large programming project, the time actually spent programming is surprisingly small. In his book *The Mythical Man-Month*[1], Frederick P. Brooks, Jr. suggests that in a typical large software project, one-third of the time is spent planning what to do (steps 1 through 3), one-sixth of the time is spent actually writing the program (step 4), and fully half of the time is spent in testing and debugging the program! Clearly, anything that we can do to reduce the testing and debugging time will be very helpful. We can best reduce the testing and debugging time by doing a very careful job in the planning phase and by using good

[1]Frederick P. Brooks Jr., *The Mythical Man-Month, Anniversary Edition,* Addison-Wesley, 1995.

programming practices. Good programming practices will reduce the number of bugs in the program and will make the ones that do creep in easier to find.

3.2 Use of Pseudocode

As a part of the design process, it is necessary to describe the algorithm that you intend to implement. The description of the algorithm should be in a standard form that is easy for both you and other people to understand, and the description should aid you in turning your concept into MATLAB code. The standard forms that we use to describe algorithms are called **constructs** (or sometimes structures), and an algorithm described using these constructs is called a structured algorithm. When the algorithm is implemented in a MATLAB program, the resulting program is called a **structured program**.

The constructs used to build algorithms can be described in a special way called pseudocode. **Pseudocode** is a hybrid mixture of MATLAB and English. It is structured like MATLAB, with a separate line for each distinct idea or segment of code, but the descriptions on each line are in English. Each line of the pseudocode should describe its idea in plain, easily understandable English. Pseudocode is very useful for developing algorithms, since it is flexible and easy to modify. It is especially useful since pseudocode can be written and modified with the same editor or word processor used to write the MATLAB program—no special graphical capabilities are required.

For example, the pseudocode for the algorithm in Example 1.3 is:

```
Prompt user to enter temperature in degrees Fahrenheit
Read temperature in degrees Fahrenheit (temp_f)
temp_k in kelvins ← (5/9) * (temp_f - 32) + 273.15
Write temperature in kelvins
```

Notice that a left arrow (←) is used instead of an equal sign (=) to indicate that a value is stored in a variable, since this avoids any confusion between assignment and equality. Pseudocode is intended to aid you in organizing your thoughts before converting them into MATLAB code.

3.3 The Logical Data Type

The `logical` data type is a special type of data that can have one of only two possible values: `true` or `false`. These values are produced by the two special functions `true` and `false`. They are also produced by two types of MATLAB operators: relational operators and logic operators.

Logical values are stored in a single byte of memory, so they take up much less space than numbers, which usually occupy 8 bytes.

The operation of many MATLAB branching constructs is controlled by logical variables or expressions. If the result of a variable or expression is true, then one section of code is executed. If not, then a different section of code is executed.

To create a `logical` variable, just assign a logical value it to in an assignment statement. For example, the statement

```
a1 = true;
```

creates a logical variable `a1` containing the logical value `true`. If this variable is examined with the `whos` command, we can see that it has the logical data type:

```
» whos a1
  Name        Size                    Bytes  Class
  a1          1x1                         1  logical array
```

Unlike programming languages such as Java, C++, and Fortran, it is legal in MATLAB to mix numerical and logical data in expressions. If a logical value is used in a place where a numerical value is expected, `true` values are converted to 1 and `false` values are converted to 0, and then used as numbers. If a numerical value is used in a place where a logical value is expected, nonzero values are converted to `true` and 0 values are converted to `false`, and then used as logical values.

It is also possible to explicitly convert numerical values to logical values, and vice versa. The `logical` function converts numerical data to logical data, and the `real` function converts logical data to numerical data.

3.3.1 Relational Operators

Relational operators are operators with two numerical or string operands that yield a `logical` result, depending on the relationship between the two operands. The general form of a relational operator is

$$a_1 \text{ op } a_2$$

where a_1 and a_2 are arithmetic expressions, variables, or strings, and op is one of the relational operators given in Table 3.1.

If the relationship between a_1 and a_2 expressed by the operator is true, the operation returns a true value; otherwise, the operation returns false.

Table 3.1 Relational Operators

Operator	Operation
==	Equal to
~=	Not equal to
>	Greater than
>=	Greater than or equal to
<	Less than
<=	Less than or equal to

Some relational operations and their results are given below:

Operation	**Result**
`3 < 4`	`true (1)`
`3 <= 4`	`true (1)`
`3 == 4`	`false (0)`
`3 > 4`	`false (0)`
`4 <= 4`	`true (1)`
`'A' < 'B'`	`true (1)`

The last relational operation is true because characters are evaluated in alphabetical order.

Note that both `true` and `1` are shown as the result of true operations, and both `false` and `0` are shown as the result of false operations. MATLAB is a bit schizophrenic about how the results of logical operations are displayed. When a relational operator is evaluated in the Command Window, the result of the operation will be displayed as a 0 or 1. When it is displayed in the Workspace Browser, the same value will be show as `false` or `true` (see Figure 3.3 on the next page).

Relational operators may be used to compare a scalar value with an array. For example, if `a` $= \begin{bmatrix} 1 & 0 \\ -2 & 1 \end{bmatrix}$ and `b = 0`, then the expression `a > b` will yield the logical array $\begin{bmatrix} \texttt{true} & \texttt{false} \\ \texttt{false} & \texttt{true} \end{bmatrix}$ (shown as $\begin{bmatrix} 1 & 0 \\ 0 & 1 \end{bmatrix}$ in the Command Window). Relational operators may also be used to compare two arrays, as long as both arrays have the same size. For example, if `a` $= \begin{bmatrix} 1 & 0 \\ -2 & 1 \end{bmatrix}$ and `b` $= \begin{bmatrix} 0 & 2 \\ -2 & -1 \end{bmatrix}$, then the expression `a >= b` will yield the logical array $\begin{bmatrix} \texttt{true} & \texttt{false} \\ \texttt{true} & \texttt{true} \end{bmatrix}$ (shown as $\begin{bmatrix} 1 & 0 \\ 1 & 1 \end{bmatrix}$ in the Command Window). If the arrays have different sizes, a runtime error will result.

Note that since strings are really arrays of characters, *relational operators can only compare two strings if they are of equal lengths*. If they are of unequal lengths, the comparison operation will produce an error. We will learn of a more general way to compare strings in Chapter 6.

The equivalence relational operator is written with two equal signs, while the assignment operator is written with a single equal sign. These are very different operators that beginning programmers often confuse. The `==` symbol is a

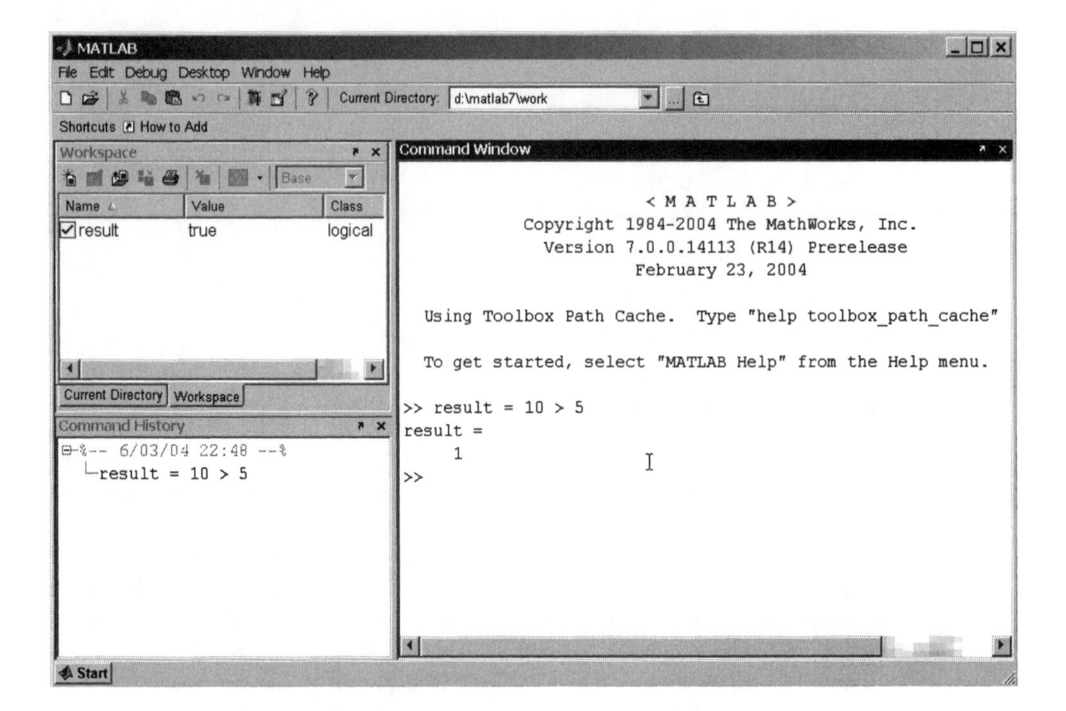

Figure 3.3 The result of a relational operator is a true or false value that can be stored in a `logical` variable. In the example shown here, the result of the operator 10 > 5 is displayed as a 1 on the Command Window, and as a `true` in the Workspace Browser.

comparison operation that returns a logical result, while the = symbol *assigns* the value of the expression to the right of the equal sign to the variable on the left of the equal sign. It is a very common mistake for beginning programmers to use a single equal sign when trying to do a comparison.

Programming Pitfalls

Be careful not to confuse the equivalence relational operator (==) with the assignment operator (=).

In the hierarchy of operations, relational operators are evaluated after all arithmetic operators have been evaluated. Therefore, the following two expressions are equivalent (both are true).

```
7 + 3  <  2 + 11
(7 + 3) < (2 + 11)
```

3.3.2 A Caution About the == and ~= Operators

The equivalence operator (==) returns a `true` value (1) when the two values being compared are equal, and a `false` (0) when the two values being compared are different. Similarly, non-equivalence operator (~=) returns a `false` (0) when the two values being compared are equal, and a `true` (1) when the two values being compared are different. These operators are generally safe to use for comparing strings, but they can sometimes produce surprising results when two numeric values are compared. Due to **roundoff errors** during computer calculations, two theoretically equal numbers can differ slightly, causing an equality or inequality test to fail.

For example, consider the following two numbers, both of which should be equal to 0.0.

```
a = 0;
b = sin(pi);
```

Since these numbers are theoretically the same, the relational operation `a == b` *should* produce a 1. In fact, the results of this MATLAB calculation are

```
» a = 0;
» b = sin(pi);
» a == b
ans =
          0
```

MATLAB reports that `a` and `b` are different because a slight roundoff error in the calculation of `sin(pi)` makes the result 1.2246×10^{-16} instead of exactly zero. The two theoretically equal values differ slightly due to roundoff error!

Instead of comparing two numbers for *exact* equality, you should set up your tests to determine if the two numbers *nearly* equal to each other within some accuracy that takes into account the roundoff error expected for the numbers being compared. The test

```
» abs(a - b) < 1.0E-14
ans =
          1
```

produces the correct answer, despite the roundoff errors in calculating `a` and `b`.

Good Programming Practice

Be cautious about testing for equality with numeric values, since roundoff errors may cause two variables that should be equal to fail a test for equality. Instead, test to see if the variables are *nearly* equal within the roundoff error to be expected on the computer you are working with.

3.3.3 Logic Operators

Logic operators are operators with one or two logical operands that yield a logical result. There are five binary logic operators: AND (& and &&), inclusive OR (| and ||), and exclusive OR (xor), and one unary operator: NOT (~). The general form of a binary logic operation is

$$l_1 \text{ op } l_2$$

and the general form of a unary logic operation is

$$\text{op } l_1$$

where l_1 and l_2 are expressions or variables, and op is one of the following logic operators shown in Table 3.2.
If the relationship between l_1 and l_2 expressed by the operator is true, then the operation returns a value of `true` (displayed as 1 in the Command Window); otherwise, the operation returns a value of `false` (0 in the Command Window).

The results of the operators are summarized in **truth tables**, which show the result of each operation for all possible combinations of l_1 and l_2. Table 3.3 shows the truth tables for all logic operators.

Logical ANDs

The result of an AND operator is `true` if an only if both input operands are `true`. If either or both operands are `false`, the result is `false`, as shown in Table 3.3.

Table 3.2 Logic Operators

Operator	Operation
&	Logical AND
&&	Logical AND with shortcut evaluation
|	Logical Inclusive OR
||	Logical Inclusive OR with shortcut evaluation
xor	Logical Exclusive OR
~	Logical NOT

Table 3.3 Truth Tables for Logic Operators

Inputs		and		or		xor	not
l_1	l_2	l_1 & l_2	l_1 && l_2	l_1 | l_2	l_1 || l_2	xor(l_1, l_2)	~l_1
false	false	false	false	false	false	false	true
false	true	false	false	true	true	true	true
true	false	false	false	true	true	true	false
true	true	true	true	true	true	false	false

Note that there are two logical AND operators: `&&` and `&`. Why are there two AND operators, and what is the difference between them? The basic difference between `&&` and `&` is that `&&` supports *short-circuit evaluations* (or *partial evaluations*), while `&` doesn't. That is, `&&` will evaluate expression l_1 and immediately return a `false` value if l_1 is `false`. If l_1 is `false`, the operator never evaluates l_2, because the result of the operator will be `false` regardless of the value of l_2. In contrast, the `&` operator always evaluates both l_1 and l_2 before returning an answer.

A second difference between `&&` and `&` is that `&&` only works between scalar values, while `&` works with either scalar or array values, as long as the sizes of the arrays are compatible.

When should you use `&&` and when should you use `&` in a program? Most of the time, it doesn't matter which AND operation is used. If you are comparing scalars, and it is not necessary to always evaluate l_2, then use the `&&` operator. The partial evaluation will make the operation faster in the cases where the first operand is `false`.

Sometimes it is important to use shortcut expressions. For example, suppose that we wanted to test for the situation where the ratio of two variables `a` and `b` is greater than 10. The code to perform this test is:

```
x = a/b > 10.0
```

This code normally works fine, but what about the case where `b` is zero? In that case, we would be dividing by zero, which produces an `Inf` instead of a number. The test could be modified to avoid this problem as follows:

```
x = (b ~= 0) && (a/b > 10.0)
```

This expression uses partial evaluation, so if `b` = 0, the expression `a/b > 10.0` will never be evaluated, and no `Inf` will occur.

Good Programming Practice

Use the `&` AND operator if it is necessary to ensure that both operands are evaluated in an expression, or if the comparison is between arrays. Otherwise, use the `&&` AND operator, since the partial evaluation will make the operation faster in the cases where the first operand is `false`. The `&` operator is preferred in most practical cases.

Logical Inclusive ORs

The result of an inclusive OR operator is `true` if either of the input operands are `true`. If both operands are `false`, the result is `false`, as shown in Table 3.3.

Note that there are two inclusive OR operators: `||` and `|`. Why are there two inclusive OR operators, and what is the difference between them? The basic difference between || and | is that || supports partial evaluations, while | doesn't. That is, || will evaluate expression l_1 and immediately return a `true` value if l_1 is `true`. If l_1 is `true`, the operator never evaluates l_2, because the result of the operator will be `true` regardless of the value of l_2. In contrast, the | operator always evaluates both l_1 and l_2 before returning an answer.

A second difference between || and | is that || only works between scalar values, while | works with either scalar or array values, as long as the sizes of the arrays are compatible.

When should you use || and when should you use | in a program? Most of the time, it doesn't matter which OR operation is used. If you are comparing scalars, and it is not necessary to always evaluate l_2, use the || operator. The partial evaluation will make the operation faster in the cases where the first operand is `true`.

✱ Good Programming Practice

Use the | inclusive OR operator if it is necessary to ensure that both operands are evaluated in an expression or if the comparison is between arrays. Otherwise, use the || operator, since the partial evaluation will make the operation faster in the cases where the first operand is `true`. The | operator is preferred in most practical cases.

Logical Exclusive OR

The result of an exclusive OR operator is `true` if and only if one operand is `true` and the other one is `false`. If both operands are `true` or both operands are `false`, then the result is `false`, as shown in Table 3.3. Note that both operands must always be evaluated in order to calculate the result of an exclusive OR.

The logical exclusive OR operation is implemented as a function. For example,

```
a = 10;
b = 0;
x = xor(a, b);
```

This result is `true`. The value of `a` is nonzero, so it will be converted to `true`. The value of `b` is zero, so it will be converted to `false`. Therefore, the result of the `xor` operation will be true.

Logical NOT

The NOT operator (~) is a unary operator, having only one operand. The result of a NOT operator is `true` if its operand is `false`, and `false` if its operand is `true`, as shown in Table 3.3.

Using Numeric Data with Logic Operators

Real numeric data can also be use with logic operators. Since logic operators expect logical input values, MATLAB converts nonzero values to `true` and zero values to `false` before performing the operation. Thus, the result of `~5` is `false` (0 in the Command Window) and the result of `~0` is `true` (1 in the Command Window).

Logic operators may be used to compare a scalar value with an array. For example, if $a = \begin{bmatrix} \texttt{true} & \texttt{false} \\ \texttt{false} & \texttt{true} \end{bmatrix}$ and `b = false`, then the expression `a & b` will yield the result $\begin{bmatrix} \texttt{false} & \texttt{false} \\ \texttt{false} & \texttt{false} \end{bmatrix}$ (displayed as $\begin{bmatrix} 0 & 0 \\ 0 & 0 \end{bmatrix}$ in the Command Window). Logic operators may also be used to compare two arrays, as long as both arrays have the same size. For example, if $a = \begin{bmatrix} \texttt{true} & \texttt{false} \\ \texttt{false} & \texttt{true} \end{bmatrix}$ and $b = \begin{bmatrix} \texttt{true} & \texttt{true} \\ \texttt{false} & \texttt{false} \end{bmatrix}$, then the expression `a | b` will yield the result $\begin{bmatrix} \texttt{true} & \texttt{true} \\ \texttt{false} & \texttt{true} \end{bmatrix}$ (displayed as $\begin{bmatrix} 1 & 1 \\ 0 & 1 \end{bmatrix}$ in the Command Window). If the arrays have different sizes, a runtime error will result.

Logic operators may *not* be used with complex or imaginary numeric data. For example, an expression such as "`2i & 2i`" will produce an error when it is evaluated.

Hierarchy of Operations

In the hierarchy of operations, logic operators are evaluated *after all arithmetic operations and all relational operators have been evaluated.* The order in which the operators in an expression are evaluated is:

1. All arithmetic operators are evaluated first in the order previously described.
2. All relational operators (==, ~=, >, >=, <, <=) are evaluated, working from left to right.
3. All ~ operators are evaluated.
4. All & and && operators are evaluated, working from left to right.
5. All |, ||, and `xor` operators are evaluated, working from left to right.

As with arithmetic operations, parentheses can be used to change the default order of evaluation. Examples of some logic operators and their results are given below.

Example 3.1

Assume that the following variables are initialized with the values shown, and calculate the result of the specified expressions:

```
value1 = true
value2 = false
value3 = 1
```

```
value4 = -10
value5 = 0
value6 = [1 2; 0 1]
```

	Expression	Result	Comment
(*a*)	`~value1`	`false`	
(*b*)	`~value3`	`false`	The number 1 is converted to `true` before operation is applied.
(*c*)	`value1 | value2`	`true`	
(*d*)	`value1 & value2`	`false`	
(*e*)	`value4 & value5`	`false`	−10 is converted to `true` and 0 is converted to `false` before the operation is applied.
(*f*)	`~(value4 & value5)`	`true`	−10 is converted to `true` and 0 is converted to `false` before the operation is applied.
(*g*)	`value1 + value4`	−9	`value1` is converted to the number 1 before the addition is performed.
(*h*)	`value1 + (~value4)`	1	The logical `value1` is converted to the number 1 before the addition is performed. The number `value4` is converted to `true` before the NOT is performed. Then `~value4` is evaluated to be `false`. This `false` value is converted to 0 before the addition, so the final result is $1 + 0 = 1$.
(*i*)	`value3 && value6`	Illegal	The `&&` operator must be used with scalar operands.
(*j*)	`value3 & value6`	[`true` `true`; `false` `true`]	AND between a scalar and an array operand.

◀

The ~ operator is evaluated before other logic operators. Therefore, the parentheses in part (*f*) of the above example were required. If they had been absent, the expression in part (*f*) would have been evaluated in the order `(~value4) & value5`.

3.3.4 Logical Functions

MATLAB includes a number of logical functions that return `true` whenever the condition they test for is true, and `false` whenever the condition they test for is false. These functions can be used with relational and logic operator to control the operation of branches and loops.

A few of the more important logical functions are given in Table 3.4.

Table 3.4 Selected MATLAB Logical Functions.

Function	Purpose
`ischar(a)`	Returns `true` if `a` is a character array and `false` otherwise.
`isempty(a)`	Returns `true` if `a` is an empty array and `false` otherwise.
`isinf(a)`	Returns `true` if the value of `a` is infinite (`Inf`) and `false` otherwise.
`isnan(a)`	Returns `true` if the value of `a` is `NaN` (not a number) and `false` otherwise.
`isnumeric(a)`	Returns `true` if `a` is a numeric array and `false` otherwise.
`logical`	Converts numerical values to logical values: if a value is nonzero, it is converted to `true`. If it is zero, it is converted to `false`.

Quiz 3.1

This quiz provides a quick check to see if you have understood the concepts introduced in Section 3.3. If you have trouble with the quiz, reread the sections, ask your instructor, or discuss the material with a fellow student. The answers to this quiz are found in the back of the book.

Assume that `a`, `b`, `c`, and `d` are as defined and evaluate the following expressions.

```
a = 20;        b = -2;
c = 0;         d = 1;
```

1. `a > b`
2. `b > d`
3. `a > b && c > d`
4. `a == b`
5. `a && b > c`
6. `~~b`

Assume that `a`, `b`, `c`, and `d` are as defined and evaluate the following expressions.

$$a = 2; \qquad b = \begin{bmatrix} 1 & -2 \\ 0 & 10 \end{bmatrix};$$

$$c = \begin{bmatrix} 0 & 1 \\ 2 & 0 \end{bmatrix}; \qquad d = \begin{bmatrix} -2 & 1 & 2 \\ 0 & 1 & 0 \end{bmatrix};$$

7. `~(a > b)`
8. `a > c && b > c`
9. `c <= d`

10. `logical(d)`

11. `a * b > c`

12. `a * (b > c)`

Assume that `a`, `b`, `c`, and `d` are as defined. Explain the order in which each of the following expressions are evaluated and specify the results in each case:

```
a = 2;          b = 3;
c = 10;         d = 0;
```

13. `a*b^2 > a*c`

14. `d || b > a`

15. `(d | b) > a`

Assume that `a`, `b`, `c`, and `d` are as defined and evaluate the following expressions.

```
a = 20;         b = -2;
c = 0;          d = 'Test';
```

16. `isinf(a/b)`

17. `isinf(a/c)`

18. `a > b && ischar(d)`

19. `isempty(c)`

20. `(~a) & b`

21. `(~a) + b`

3.4 Branches

Branches are MATLAB statements that permit us to select and execute specific sections of code (called *blocks*) while skipping other sections of code. They are variations of the `if` construct, the `switch` construct, and the `try/catch` construct.

3.4.1 The `if` Construct

The `if` construct has the form

```
if control_expr_1
    Statement 1   }
    Statement 2   } Block 1
    ...           }
elseif control_expr_2
    Statement 1   }
    Statement 2   } Block 2
    ...           }
```

```
else
    Statement 1
    Statement 2     } Block 3
    ...
end
```

where the control expressions are logical expressions that control the operation of the `if` construct. If *control_expr_1* is true (nonzero), then the program executes the statements in Block 1, and skips to the first executable statement following the `end`. Otherwise, the program checks for the status of *control_expr_2*. If *control_expr_2* is true (nonzero), then the program executes the statements in Block 2, and skips to the first executable statement following the `end`. If all control expressions are zero, then the program executes the statements in the block associated with the `else` clause.

There can be any number of `elseif` clauses (0 or more) in an `if` construct, but there can be at most one `else` clause. The control expression in each clause will be tested only if the control expressions in every clause above are false (0). Once one of the expressions proves to be true and the corresponding code block is executed, the program skips to the first executable statement following the `end`. If all control expressions are false, then the program executes the statements in the block associated with the `else` clause. If there is no `else` clause, then execution continues after the `end` statement without executing any part of the `if` construct.

Note that the MATLAB keyword `end` in this construct is *completely different* from the MATLAB function `end` that we used in Chapter 2 to return the highest value of a given subscript. MATLAB tells the difference between these two uses of `end` from the context in which the word appears within an M-file.

In most circumstances, *the control expressions will be some combination of relational and logic operators*. As we learned earlier in this chapter, relational and logic operators produce a true (1) when the corresponding condition is true and a false (0) when the corresponding condition is false. When an operator is true, its result is nonzero, and the corresponding block of code will be executed.

As an example of an `if` construct, consider the solution of a quadratic equation of the form

$$ax^2 + bx + c = 0 \tag{3-1}$$

The solution to this equation is

$$x = \frac{-b \pm \sqrt{b^2 - 4ac}}{2a} \tag{3-2}$$

The term $b^2 - 4ac$ is known as the *discriminant* of the equation. If $b^2 - 4ac > 0$, then there are two distinct real roots to the quadratic equation. If $b^2 - 4ac = 0$, then there is a single repeated root to the equation, and if $b^2 - 4ac < 0$, there are two complex roots to the quadratic equation.

Suppose that we wanted to examine the discriminant of a quadratic equation and to tell a user whether the equation has two complex roots, two identical real roots, or two distinct real roots. In pseudocode, this construct would take the form

```
if (b^2 - 4*a*c) < 0
   Write msg that equation has two complex roots.
elseif (b**2 - 4.*a*c) == 0
   Write msg that equation has two identical real roots.
else
   Write msg that equation has two distinct real roots.
end
```

The MATLAB statements to do this are

```
if (b^2 - 4*a*c) < 0
   disp('This equation has two complex roots.');
elseif (b^2 - 4*a*c) == 0
   disp('This equation has two identical real roots.');
else
   disp('This equation has two distinct real roots.');
end
```

For readability, the blocks of code within an `if` construct are usually indented by two or three spaces, but this is not actually required.

Good Programming Practice

Always indent the body of an `if` construct by two or more spaces to improve the readability of the code. Note that indentation is automatic if you use the MATLAB editor to write your programs.

It is possible to write a complete `if` construct on a single line by separating the parts of the construct by commas or semicolons. Thus the following two constructs are identical:

```
if x < 0
   y = abs(x);
end
```

and

```
if x < 0; y = abs(x); end
```

However, this should only be done for very simple constructs.

3.4.2 Examples Using `if` Constructs

We will now look at two examples that illustrate the use of `if` constructs.

Example 3.2—The Quadratic Equation

Write a program to solve for the roots of a quadratic equation, regardless of type.

SOLUTION We will follow the design steps outlined earlier in the chapter.

1. **State the problem**
   The problem statement for this example is very simple. We want to write a program that will solve for the roots of a quadratic equation, whether they are distinct real roots, repeated real roots, or complex roots.

2. **Define the inputs and outputs**
   The inputs required by this program are the coefficients a, b, and c of the quadratic equation

$$ax^2 + bx + c = 0 \tag{3-1}$$

   The output from the program will be the roots of the quadratic equation, whether they are distinct real roots, repeated real roots, or complex roots.

3. **Design the algorithm**
   This task can be broken down into three major sections, whose functions are input, processing, and output:

```
Read the input data
Calculate the roots
Write out the roots
```

We will now break each of the above major sections into smaller, more detailed pieces. There are three possible ways to calculate the roots, depending on the value of the discriminant, so it is logical to implement this algorithm with a three-branched `if` construct. The resulting pseudocode is:

```
Prompt the user for the coefficients a, b, and c.
Read a, b, and c
discriminant <- b^2 - 4 * a * c
if discriminant > 0
   x1 <- ( -b + sqrt(discriminant) )/( 2 * a )
   x2 <- ( -b - sqrt(discriminant) )/( 2 * a )
   Write msg that equation has two distinct real roots.
   Write out the two roots.

elseif discriminant == 0
   x1 <- -b/( 2 * a )
   Write msg that equation has two identical real roots.
   Write out the repeated root.

else
   real_part <- -b/( 2 * a )
```

```
        imag_part <- sqrt ( abs ( discriminant ) )/( 2 * a )
        Write msg that equation has two complex roots.
        Write out the two roots.
    end
```

4. **Turn the algorithm into MATLAB statements.**
   The final MATLAB code is shown in below:

```
%  Script file: calc_roots.m
%
%  Purpose:
%    This program solves for the roots of a quadratic equation
%    of the form a*x**2 + b*x + c = 0. It calculates the answers
%    regardless of the type of roots that the equation possesses.
%
%  Record of revisions:
%      Date         Programmer            Description of change
%      ====         ==========            =====================
%    01/02/05     S. J. Chapman           Original code
%
%  Define variables:
%    a              -- Coefficient of x^2 term of equation
%    b              -- Coefficient of x term of equation
%    c              -- Constant term of equation
%    discriminant   -- Discriminant of the equation
%    imag_part      -- Imag part of equation (for complex roots)
%    real_part      -- Real part of equation (for complex roots)
%    x1             -- First solution of equation (for real roots)
%    x2             -- Second solution of equation (for real roots)

% Prompt the user for the coefficients of the equation
disp ('This program solves for the roots of a quadratic ');
disp ('equation of the form A*X^2 + B*X + C = 0. ');
a = input ('Enter the coefficient A: ');
b = input ('Enter the coefficient B: ');
c = input ('Enter the coefficient C: ');

% Calculate discriminant
discriminant = b^2 - 4 * a * c;

% Solve for the roots, depending on the value of the discriminant
if discriminant > 0 % there are two real roots, so...

   x1 = ( -b + sqrt(discriminant) )/( 2 * a );
   x2 = ( -b - sqrt(discriminant) )/( 2 * a );
   disp ('This equation has two real roots:');
   fprintf ('x1 = %f\n', x1);
   fprintf ('x2 = %f\n', x2);
```

```
elseif discriminant == 0   % there is one repeated root, so...

   x1 = ( -b )/( 2 * a );
   disp ('This equation has two identical real roots:');
   fprintf ('x1 = x2 = %f\n', x1);

else % there are complex roots, so ...

   real_part = ( -b )/( 2 * a );
   imag_part = sqrt ( abs ( discriminant ) )/( 2 * a );
   disp ('This equation has complex roots:');
   fprintf('x1 = %f +i %f\n', real_part, imag_part );
   fprintf('x1 = %f -i %f\n', real_part, imag_part );

end
```

5. **Test the program.**
   Next, we must test the program using real input data. Since there are three possible paths through the program, we must test all three paths before we can be certain that the program is working properly. From Equation (3-2), it is possible to verify the solutions to the equations given below:

$$x^2 + 5x + 6 = 0 \qquad x = -2, \text{ and } x = -3$$
$$x^2 + 4x + 4 = 0 \qquad x = -2$$
$$x^2 + 2x + 5 = 0 \qquad x = -1 \pm i2$$

   If this program is executed three times with the above coefficients, the results are as shown below (user inputs are shown in bold face):

```
» calc_roots
This program solves for the roots of a quadratic
equation of the form A*X^2 + B*X + C = 0.
Enter the coefficient A: 1
Enter the coefficient B: 5
Enter the coefficient C: 6
This equation has two real roots:
x1 = -2.000000
x2 = -3.000000
» calc_roots
This program solves for the roots of a quadratic
equation of the form A*X^2 + B*X + C = 0.
Enter the coefficient A: 1
Enter the coefficient B: 4
Enter the coefficient C: 4
This equation has two identical real roots:
x1 = x2 = -2.000000
» calc_roots
This program solves for the roots of a quadratic
equation of the form A*X^2 + B*X + C = 0.
```

```
Enter the coefficient A: 1
Enter the coefficient B: 2
Enter the coefficient C: 5
This equation has complex roots:
x1 = -1.000000 +i 2.000000
x1 = -1.000000 -i 2.000000
```

The program gives the correct answers for our test data in all three possible cases.

Example 3.3—Evaluating a Function of Two Variables

Write a MATLAB program to evaluate a function $f(x, y)$ for any two user-specified values x and y. The function $f(x, y)$ is defined as follows.

$$f(x, y) = \begin{cases} x + y & x \geq 0 \text{ and } y \geq 0 \\ x + y^2 & x \geq 0 \text{ and } y < 0 \\ x^2 + y & x < 0 \text{ and } y \geq 0 \\ x^2 + y^2 & x < 0 \text{ and } y < 0 \end{cases}$$

SOLUTION The function $f(x, y)$ is evaluated differently depending on the signs of the two independent variables x and y. To determine the proper equation to apply, it will be necessary to check for the signs of the x and y values supplied by the user.

1. **State the problem**
   This problem statement is very simple: Evaluate the function $f(x, y)$ for any user-supplied values of x and y.

2. **Define the inputs and outputs**
   The inputs required by this program are the values of the independent variables x and y. The output from the program will be the value of the function $f(x, y)$.

3. **Design the algorithm**
   This task can be broken down into three major sections, whose functions are input, processing, and output:

   ```
   Read the input values x and y
   Calculate f(x,y)
   Write out f(x,y)
   ```

   We will now break each of the above major sections into smaller, more detailed pieces. There are four possible ways to calculate the function $f(x, y)$, depending upon the values of x and y, so it is logical to implement this algorithm with a four-branched `if` statement. The resulting pseudocode is:

```
Prompt the user for the values x and y.
Read x and y
if x ≥ 0 and y ≥ 0
   fun ← x + y
elseif x ≥ 0 and y < 0
   fun ← x + y^2
elseif x < 0 and y ≥ 0
   fun ← x^2 + y
else
   fun ← x^2 + y^2
end
Write out f(x,y)
```

4. **Turn the algorithm into MATLAB statements.**
   The final MATLAB code is shown below.

```
%  Script file: funxy.m
%
%  Purpose:
%    This program solves the function f(x,y) for a
%    user-specified x and y, where f(x,y) is defined as:
%
%
%                 ┌x + y               x >= 0 and y >= 0
%                 │x + y^2             x >= 0 and y < 0
%       f(x,y) =  │x^2 + y             x < 0  and y >= 0
%                 └x^2 + y^2           x < 0  and y < 0
%
%
%  Record of revisions:
%      Date         Programmer          Description of change
%      ====         ==========          =====================
%    01/03/05     S. J. Chapman         Original code
%
% Define variables:
%   x      -- First independent variable
%   y      -- Second independent variable
%   fun    -- Resulting function

% Prompt the user for the values x and y
x = input ('Enter the x coefficient: ');
y = input ('Enter the y coefficient: ');

% Calculate the function f(x,y) based upon
% the signs of x and y.
```

```
if x >= 0 && y >= 0
   fun = x + y;
elseif x >= 0 && y < 0
   fun = x + y^2;
elseif x < 0 && y >= 0
   fun = x^2 + y;
else
   fun = x^2 + y^2;
end
% Write the value of the function.
disp (['The value of the function is ' num2str(fun)]);
```

5. **Test the program.**
   Next, we must test the program using real input data. Since there are four possible paths through the program, we must test all four paths before we can be certain that the program is working properly. To test all four possible paths, we will execute the program with the four sets of input values $(x, y) = (2, 3)$, $(2, -3)$, $(-2, 3)$, and $(-2, -3)$. Calculating by hand, we see that

$$
\begin{aligned}
f(2, 3) &= 2 + 3 = 5 \\
f(2, -3) &= 2 + (-3)^2 = 11 \\
f(-2, 3) &= (-2)^2 + 3 = 7 \\
f(-2, -3) &= (-2)^2 + (-3)^2 = 13
\end{aligned}
$$

   If this program is compiled, and then run four times with the above values, the results are:

```
» funxy
Enter the x coefficient: 2
Enter the y coefficient: 3
The value of the function is 5
» funxy
Enter the x coefficient: 2
Enter the y coefficient: -3
The value of the function is 11
» funxy
Enter the x coefficient: -2
Enter the y coefficient: 3
The value of the function is 7
» funxy
Enter the x coefficient: -2
Enter the y coefficient: -3
The value of the function is 13
```

The program gives the correct answers for our test values in all four possible cases.

3.4.3 Notes Concerning the Use of the `if` Constructs

The `if` construct is very flexible. It must have one `if` statement and one `end` statement. In between, it can have any number of `elseif` clauses, and may also have one `else` clause. With this combination of features, it is possible to implement any desired branching construct.

In addition, `if` constructs may be **nested**. Two `if` constructs are said to be nested if one of them lies entirely within a single code block of the other one. The following two `if` constructs are properly nested.

```
if x > 0
    ...
    if y < 0
        ...
    end
    ...
end
```

The MATLAB interpreter always associates a given `end` statement with the most recent `if` statement, so the first `end` above closes the `if y < 0` statement, while the second `end` closes the `if x > 0` statement. This works well for a properly written program, but can cause the interpreter to produce confusing error messages in cases where the programmer makes a coding error. For example, suppose that we have a large program containing a construct like the one shown below.

```
...
if (test1)
    ...
    if (test2)
        ...
        if (test3)
            ...
        end
        ...
    end
    ...
end
```

This program contains three nested `if` constructs that may span hundreds of lines of code. Now suppose that the first `end` statement is accidentally deleted during an editing session. When that happens, the MATLAB interpreter will automatically associate the second `end` with the innermost `if (test3)` construct, and the third `end` with the middle `if (test2)`. When the interpreter reaches the end of the file, it will notice that the first `if (test1)` construct was never ended, and it will generate an error message saying that there is a missing `end`. Unfortunately, it can't tell *where* the problem occurred, so we will have to go back and manually search the entire program to locate the problem.

It is sometimes possible to implement an algorithm using either multiple `elseif` clauses or nested `if` statements. In that case, a programmer may choose whichever style he or she prefers.

Example 3.4—Assigning Letter Grades

Suppose that we are writing a program which reads in a numerical grade and assigns a letter grade to it according to the following table:

```
95 < grade          A
86 < grade ≤ 95     B
76 < grade ≤ 86     C
66 < grade ≤ 76     D
 0 < grade ≤ 66     F
```

Write an `if` construct that will assign the grades as described above using (*a*) multiple `elseif` clauses and (*b*) nested `if` constructs.

SOLUTION

(a) One possible structure using `elseif` clauses is

```
if grade > 95.0
   disp('The grade is A.');
elseif grade > 86.0
   disp('The grade is B.');
elseif grade > 76.0
   disp('The grade is C.');
elseif grade > 66.0
   disp('The grade is D.');
else
   disp('The grade is F.');
end
```

(b) One possible structure using nested `if` constructs is

```
if grade > 95.0
   disp('The grade is A.');
else
   if grade > 86.0
      disp('The grade is B.');
   else
      if grade > 76.0
         disp('The grade is C.');
      else
         if grade > 66.0
            disp('The grade is D.');
         else
            disp('The grade is F.');
```

```
            end
        end
    end
end
```

It should be clear from the foregoing example that if there are a lot of mutually exclusive options, a single `if` construct with multiple `elseif` clauses will be simpler than a nested `if` construct.

✷ Good Programming Practice

For branches in which there are many mutually exclusive options, use a single `if` construct with multiple `elseif` clauses in preference to nested `if` constructs.

3.4.4 The switch Construct

The `switch` construct is another form of branching construct. It permits a programmer to select a particular code block to execute based on the value of a single integer, character, or logical expression. The general form of a `switch` construct is:

```
switch (switch_expr)
case case_expr_1,
   Statement 1      }
   Statement 2      }  Block 1
   ...              }
case case_expr_2,
   Statement 1      }
   Statement 2      }  Block 2
   ...              }
...
otherwise,
   Statement 1      }
   Statement 2      }  Block n
   ...              }
end
```

If the value of *switch_expr* is equal to *case_expr_1*, then the first code block will be executed, and the program will jump to the first statement following the end of the `switch` construct. Similarly, if the value of *switch_expr* is equal to *case_expr_2*, then the second code block will be executed, and the program will jump to the first statement following the end of the `switch` construct. The same idea applies for any other cases in the construct. The `otherwise` code block is optional. If it is present, it will be executed whenever the value of *switch_expr* is outside the range of all of the case selectors. If it is not present and the value of *switch_expr* is outside the range of all of the case selectors, then none of the code blocks will be executed. The pseudocode for the case construct looks just like its MATLAB implementation.

If many values of the *switch_expr* should cause the same code to execute, all of those values may be included in a single block by enclosing them in brackets, as shown below. If the switch expression matches any of the case expressions in the list, then the block will be executed.

```
switch (switch_expr)
case {case_expr_1, case_expr_2, case_expr_3},
   Statement 1          }
   Statement 2          }  Block 1
   ...                  }
otherwise,
   Statement 1          }
   Statement 2          }  Block n
   ...                  }
end
```

The *switch_expr* and each *case_expr* may be either numerical or string values.

Note that at most one code block can be executed. After a code block is executed, execution skips to the first executable statement after the `end` statement. Thus if the switch expression matches more than one case expression, *only the first one of them will be executed.*

Let's look at a simple example of a `switch` construct. The following statements determine whether an integer between 1 and 10 is even or odd, and print out an appropriate message. It illustrates the use of a list of values as case selectors, and also the use of the `otherwise` block.

```
switch (value)
case {1,3,5,7,9},
   disp('The value is odd.');
case {2,4,6,8,10},
   disp('The value is even.');
otherwise,
   disp('The value is out of range.');
end
```

3.4.5 The `try/catch` Construct

The `try/catch` construct is a special form of branching construct designed to trap errors. Ordinarily, when a MATLAB program encounters an error while running, the program aborts. The `try/catch` construct modifies this default behavior. If an error occurs in a statement in the `try` block of this construct, then instead of aborting, the code in the `catch` block is executed and the program keeps running. This allows a programmer to handle errors within the program without causing the program to stop.

The general form of a `try/catch` construct is:

```
try
   Statement 1
   Statement 2        } Try Block
   ...
catch
   Statement 1
   Statement 2        } Catch Block
   ...
end
```

When a `try/catch` construct is reached, the statements in the `try` block of a will be executed. If no error occurs, the statements in the `catch` block will be skipped, and execution will continue at the first statement following the end of the construct. On the other hand, if an error *does* occur in the `try` block, the program will stop executing the statements in the `try` block, and immediately execute the statements in the catch block.

An example program containing a `try/catch` construct follows. This program creates an array, and asks the user to specify an element of the array to display. The user will supply a subscript number, and the program displays the corresponding array element. The statements in the `try` block will always be executed in this program, while the statements in the `catch` block will be executed only if an error occurs in the `try` block.

```
% Initialize array
a = [ 1 -3 2 5];
try
   % Try to display an element
   index = input('Enter subscript of element to display:');
   disp( ['a(' int2str(index) ') = ' num2str(a(index))] );
catch
   % If we get here an error occurred
   disp( ['Illegal subscript: ' int2str(index)] );
end
```

When this program is executed, the results are:

```
» try_catch
Enter subscript of element to display: 3
a(3) = 2
» try_catch
Enter subscript of element to display: 8
Illegal subscript: 8
```

Quiz 3.2

This quiz provides a quick check to see if you have understood the concepts introduced in Section 3.4. If you have trouble with the quiz, reread the section, ask your instructor, or discuss the material with a fellow student. The answers to this quiz are found in the back of the book.

Write MATLAB statements that perform the functions described below.

1. If x is greater than or equal to zero, assign the square root of x to variable `sqrt_x` and print out the result. Otherwise, print out an error message about the argument of the square root function, and set `sqrt_x` to zero.
2. A variable `fun` is calculated as `numerator/denominator`. If the absolute value of `denominator` is less than 1.0E-300, write "Divide by 0 error." Otherwise, calculate and print out `fun`.
3. The cost per mile for a rented vehicle is $1.00 for the first 100 miles, $0.80 for the next 200 miles, and $0.70 for all miles in excess of 300 miles. Write MATLAB statements that determine the total cost and the average cost per mile for a given number of miles (stored in variable `distance`).

Examine the following MATLAB statements. Are they correct or incorrect? If they are correct, what do they output? If they are incorrect, what is wrong with them?

4.
```
if volts > 125
   disp('WARNING: High voltage on line.');
if volts < 105
   disp('WARNING: Low voltage on line.');
else
   disp('Line voltage is within tolerances.');
end
```

5.
```
color = 'yellow';
switch ( color )
case 'red',
   disp('Stop now!');
case 'yellow',
   disp('Prepare to stop.');
case 'green',
   disp('Proceed through intersection.');
otherwise,
   disp('Illegal color encountered.');
end
```

```
6. if temperature > 37
      disp('Human body temperature exceeded.');
   elseif temperature > 100
      disp('Boiling point of water exceeded.');
   end
```

3.5 Additional Plotting Features

This section describes additional features of the simple two-dimensional plots introduced in Chapter 2. These features permit us to control the range of x and y values displayed on a plot, to lay multiple plots on top of each other, to create multiple figures, to create multiple subplots within a figure, and to provide greater control of the plotted lines and text strings. In addition, we will learn how to create polar plots.

3.5.1 Controlling x- and y-axis Plotting Limits

By default, a plot is displayed with x- and y-axis ranges wide enough to show every point in an input data set. However, it is sometimes useful to display only the subset of the data that is of particular interest. This can be done using the **axis** command/function (see the Sidebar on the next page about the relationship between MATLAB commands and functions).

Some of the forms of the `axis` command/function are shown in Table 3.5 below. The two most important forms are shown in bold type—they let a

Table 3.5 Forms of the axis Function/Command

Command	Description
`v = axis;`	This function returns a 4-element row vector containing `[xmin xmax ymin ymax]`, where `xmin`, `xmax`, `ymin`, and `ymax` are the current limits of the plot.
`axis ([xmin xmax ymin ymax]);`	This function sets the x and y limits of the plot to the specified values.
`axis equal`	This command sets the axis increments to be equal on both axes.
`axis square`	This command makes the current axis box square.
`axis normal`	This command cancels the effect of axis equal and axis square.
`axis off`	This command turns off all axis labeling, tick marks, and background.
`axis on`	This command turns on all axis labeling, tick marks, and background (default case).

Command/Function Duality

Some MATLAB commands seem to be unable to make up their minds whether they are commands or functions. For example, sometimes `axis` seems to be a command and sometimes it seems to be a function. Sometimes we treat it as a command: `axis on`, and other times we might treat it as a function: `axis([0 20 0 35])`. How is this possible?

The short answer is that MATLAB commands are really implemented by functions, and the MATLAB interpreter is smart enough to substitute the function call whenever it encounters the command. It is always possible to call the command directly as a function instead of using the command syntax. Thus the following two statements are identical:

```
axis on;
axis ('on');
```

Whenever MATLAB encounters a command, it forms a function from the command by treating each command argument as a character string and calling the equivalent function with those character strings as arguments. Thus MATLAB interprets the command

```
garbage 1 2 3
```

as the following function call:

```
garbage('1','2','3')
```

Note that *only functions with character arguments can be treated as commands*. Functions with numerical arguments must be used in function form only. This fact explains why `axis` is sometimes treated as a command and sometimes treated as a function.

programmer get the current limits of a plot and modify them. A complete list of all options can be found in the MATLAB on-line documentation.

To illustrate the use of `axis`, we will plot the function $f(x) = \sin x$ from -2π to -2π, and then restrict the axes to the region to $0 \le x \le \pi$ and $0 \le y \le 1$. The statements to create this plot are shown below, and the resulting plot is shown in Figure 3.4*a*.

```
x = -2*pi:pi/20:2*pi;
y = sin(x);
plot(x,y);
title ('Plot of sin(x) vs x');
grid on;
```

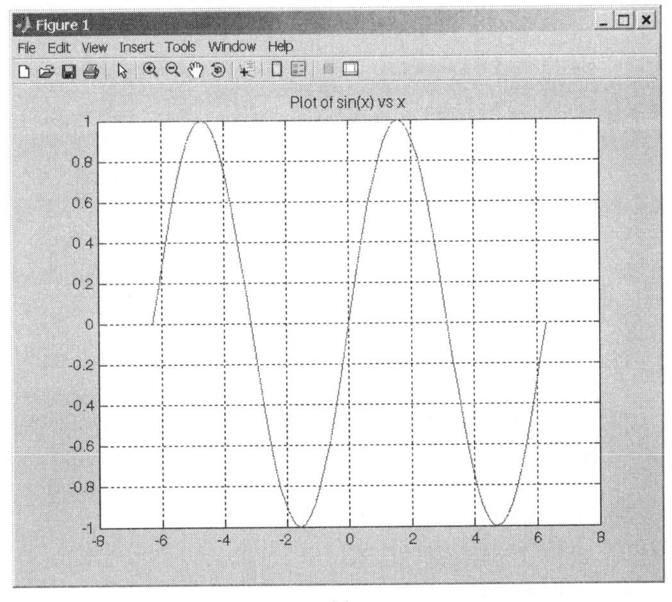

(a)

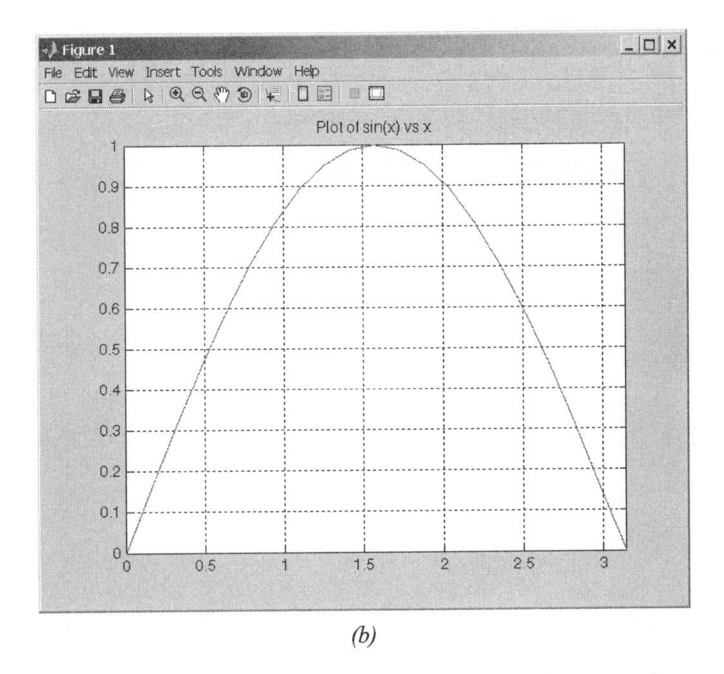

(b)

Figure 3.4 *(a)* Plot of sin x versus x. *(b)* Closeup of the region `[0 π 0 1]`.

The current limits of this plot can be determined from the basic `axis` function.

```
» limits=axis
limits =
   -8  8  -1  1
```

These limits can be modified with the function call `axis([0 pi 0 1])`. After that function is executed, the resulting plot is shown in Figure 3.4*b*.

3.5.2 Plotting Multiple Plots on the Same Axes

Normally, a new plot is created each time that a `plot` command is issued, and the previous data are lost. This behavior can be modified with the **`hold` command**. After a `hold on` command is issued, all additional plots will be laid on top of the previously existing plots. A `hold off` command switches plotting behavior back to the default situation, in which a new plot replaces the previous one.

For example, the following commands plot sin *x* and cos *x* on the same axes. The resulting plot is shown in Figure 3.5.

```
x = -pi:pi/20:pi;
y1 = sin(x);
y2 = cos(x);
plot(x,y1,'b-');
hold on;
plot(x,y2,'k--');
hold off;
legend ('sin x','cos x');
```

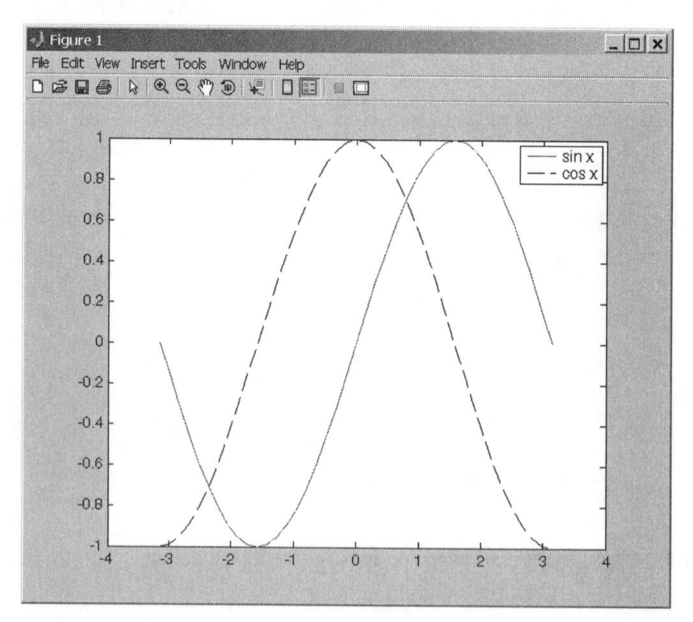

Figure 3.5 Multiple curves plotted on a single set of axes using the `hold` command.

3.5.3 Creating Multiple Figures

MATLAB can create multiple Figure Windows, with different data displayed in each window. Each Figure Window is identified by a *figure number,* which is a small positive integer. The first Figure Window is Figure 1, the second is Figure 2, etc. One of the Figure Windows will be the **current figure**, and all new plotting commands will be displayed in that window.

The current figure is selected with the **`figure` function**. This function takes the form "`figure(n)`", where `n` is a figure number. When this command is executed, Figure `n` becomes the current figure and is used for all plotting commands. The figure is automatically created if it does not already exist. The current figure may also be selected by clicking on it with the mouse.

The function `gcf` returns the number of the current figure. This function can be used by an M-file if it needs to know the current figure.

The following commands illustrate the use of the figure function. They create two figures, displaying e^x in the first figure and e^{-x} in the second one.

```
figure(1)
x = 0:0.05:2;
y1 = exp(x);
plot(x,y1);
figure(2)
y2 = exp(-x);
plot(x,y2);
```

3.5.4 Subplots

It is possible to place more than one set of axes on a single figure, creating multiple **subplots**. Subplots are created with a `subplot` command of the form

```
subplot(m,n,p)
```

This command divides the current figure into `m` × `n` equal-sized regions, arranged in `m` rows and `n` columns, and creates a set of axes at position `p` to receive all current plotting commands. The subplots are numbered from left to right and from top to bottom. For example, the command `subplot(2,3,4)` would divide the current figure into six regions arranged in two rows and three columns, and create an axis in position 4 (the lower left one) to accept new plot data (see Figure 3.6 on the next page).

If a `subplot` command creates a new set of axes that conflict with a previously existing set, the older axes are automatically deleted.

The commands below create two subplots within a single window and display the separate graphs in each subplot. The resulting figure is shown in Figure 3.7 on the next page.

```
figure(1)
subplot(2,1,1)
x = -pi:pi/20:pi;
y = sin(x);
```

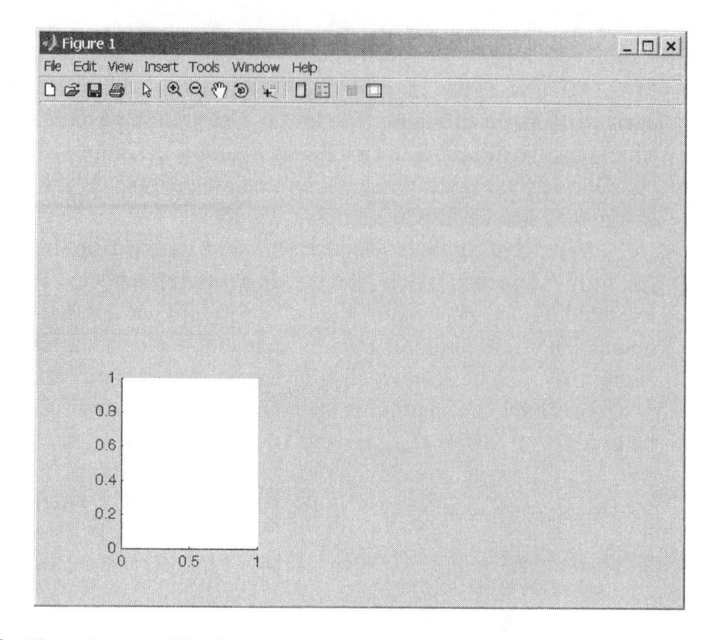

Figure 3.6 The axis created by the `subplot(2,3,4)` command.

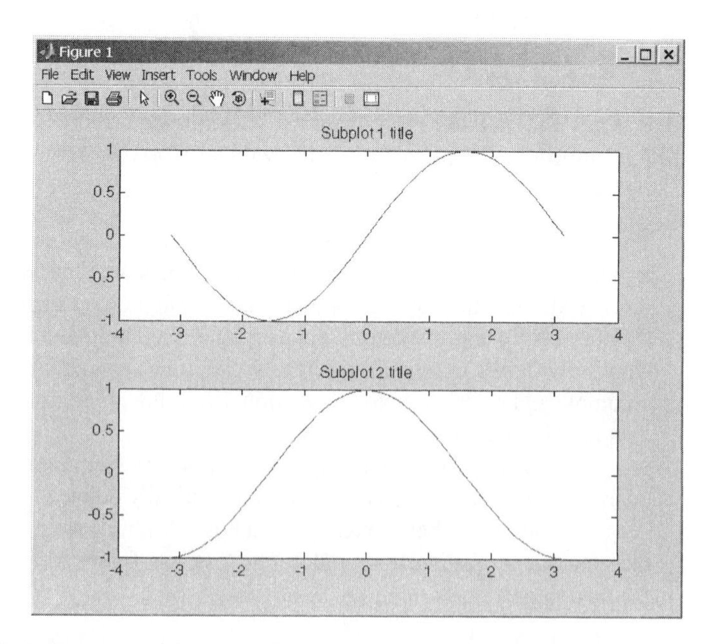

Figure 3.7 A figure containing two subplots.

```
plot(x,y);
title('Subplot 1 title');
subplot(2,1,2)
x = -pi:pi/20:pi;
y = cos(x);
plot(x,y);
title('Subplot 2 title');
```

3.5.5 Enhanced Control of Plotted Lines

In Chapter 1 we learned how to set the color, style, and marker type for a line. It is also possible to set four additional properties associated with each line:

- `LineWidth`—specifies the width of each line in points.
- `MarkerEdgeColor`—specifies the color of the marker or the edge color for filled markers.
- `MarkerFaceColor`—specifies the color of the face of filled markers.
- `MarkerSize`—specifies the size of the marker in points.

These properties are specified in the `plot` command after the data to be plotted in the following fashion:

```
plot(x,y,'PropertyName',value,...)
```

For example, the following command plots a 3-point-wide solid black line with 6-point-wide circular markers at the data points. Each marker has a red edge and a green center, as shown in Figure 3.8.

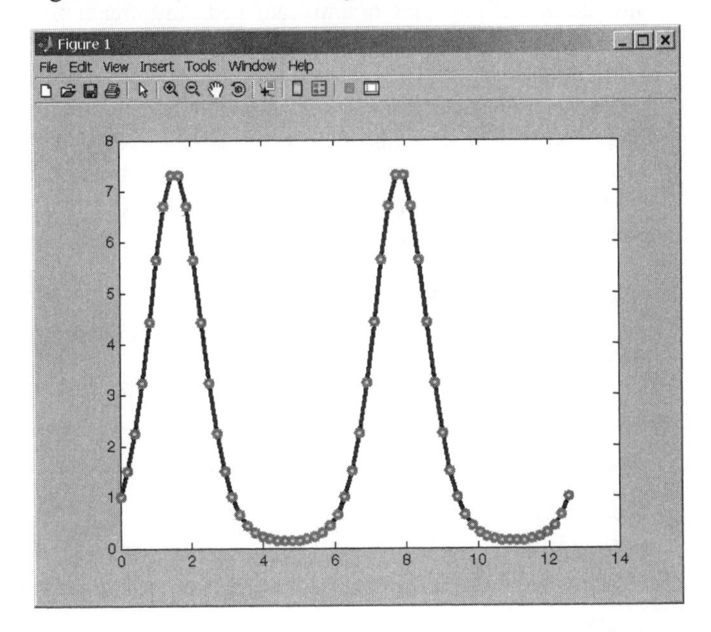

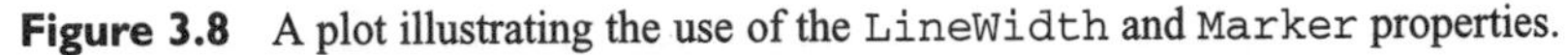

Figure 3.8 A plot illustrating the use of the `LineWidth` and `Marker` properties.

```
x = 0:pi/15:4*pi;
y = exp(2*sin(x));
plot(x,y,'-ko','LineWidth',3.0,'MarkerSize',6,...
   'MarkerEdgeColor','r','MarkerFaceColor','g')
```

3.5.6 Enhanced Control of Text Strings

It is possible to enhance plotted text strings (titles, axis labels, etc.) with formatting such as bold face, italics, etc., and with special characters such as Greek and mathematical symbols.

The font used to display the text can be modified by **stream modifiers**. A stream modifier is a special sequence of characters that tells the MATLAB interpreter to change its behavior. The most common stream modifiers are:

- `\bf`—Bold face.
- `\it`—Italics.
- `\rm`—Removes stream modifiers, restoring normal font.
- `\fontname{`*`fontname`*`}`—Specifies the font name to use.
- `\fontsize{`*`fontsize`*`}`—Specifies font size.
- `_{xxx}`—The characters inside the braces are subscripts.
- `^{xxx}`—The characters inside the braces are superscripts.

Once a stream modifier has been inserted into a text string, it will remain in effect until the end of the string or until canceled. Any stream modifier can be followed by braces {}. If a modifier is followed by braces, only the text within the braces is affected.

Special Greek and mathematical symbols may also be used in text strings. They are created by embedding *escape sequences* into the text string. These escape sequences are the same as those defined in the TeX language. A sample of the possible escape sequences is shown in Table 3.6; the full set of possibilities is included in the MATLAB on-line documentation.

If one of the special escape characters \, {, }, _, or ^ must be printed, precede it by a backslash character.

The following examples illustrate the use of stream modifiers and special characters.

String	Result
`\tau_{ind} versus \omega_{\itm}`	τ_{ind} versus ω_m
`\theta varies from 0\circ to 90\circ`	θ varies from 0° to 90°
`\bf{B}_{\itS}`	$\mathbf{B}_S$

Table 3.6 Selected Greek and Mathematical Symbols

Character Sequence	Symbol	Character Sequence	Symbol	Character Sequence	Symbol
\alpha	α			\int	$\int$
\beta	β			\cong	$\cong$
\gamma	γ	\Gamma	Γ	\sim	$\sim$
\delta	δ	\Delta	Δ	\infty	∞
\epsilon	ε			\pm	$\pm$
\eta	η			\leq	$\leq$
\theta	θ			\geq	$\geq$
\lambda	λ	\Lambda	Λ	\neq	$\neq$
\mu	μ			\propto	$\propto$
\nu	ν			\div	$\div$
\pi	π	\Pi	Π	\circ	$\circ$
\phi	ϕ			\leftrightarrow	$\leftrightarrow$
\rho	ρ			\leftarrow	$\leftarrow$
\sigma	σ	\Sigma	Σ	\rightarrow	$\rightarrow$
\tau	τ			\uparrow	$\uparrow$
\omega	ω	\Omega	Ω	\downarrow	$\downarrow$

3.5.7 Polar Plots

MATLAB includes a special function called `polar`, which plots data in polar coordinates. The basic form of this function is

```
polar(theta,r)
```

where `theta` is an array of angles in radians, and r is an array of distances. It is useful for plotting data that is intrinsically a function of angle.

Example 3.5—Cardioid Microphone

Most microphones designed for use on a stage are directional microphones, which are specifically built to enhance the signals received from the singer in the front of the microphone while suppressing the audience noise from behind the microphone. The gain of such a microphone varies as a function of angle according to the equation

$$Gain = 2g(1 + \cos\theta) \tag{3-3}$$

where g is a constant associated with a particular microphone, and θ is the angle from the axis of the microphone to the sound source. Assume that g is 0.5 for a particular microphone, and make a polar plot the gain of the microphone as a function of the direction of the sound source.

SOLUTION We must calculate the gain of the microphone versus angle and then plot it with a polar plot. The MATLAB code to do this is shown below.

```
%   Script file: microphone.m
%
%   Purpose:
%      This program plots the gain pattern of a cardioid
%      microphone.
%
%   Record of revisions:
%        Date         Programmer          Description of change
%        ====         ==========          =====================
%      01/05/05     S. J. Chapman         Original code
%
% Define variables:
%    g           -- Microphone gain constant
%    gain        -- Gain as a function of angle
%    theta       -- Angle from microphone axis(radians)

% Calculate gain versus angle
g = 0.5;
theta = 0:pi/20:2*pi;
gain = 2*g*(1+cos(theta));

% Plot gain
polar (theta,gain,'r-');
title ('\bfGain versus angle \theta');
```

The resulting plot is shown in Figure 3.9. Note that this type of microphone is called a "cardioid microphone" because its gain pattern is heart-shaped.

Example 3.6—Electrical Engineering: Frequency Response of a Low-Pass Filter

A simple low-pass filter circuit is shown in Figure 3.10. This circuit consists of a resistor and capacitor in series, and the ratio of the output voltage V_o to the input voltage V_i is given by the equation

$$\frac{V_o}{V_i} = \frac{1}{1 + j2\pi fRC} \tag{3-4}$$

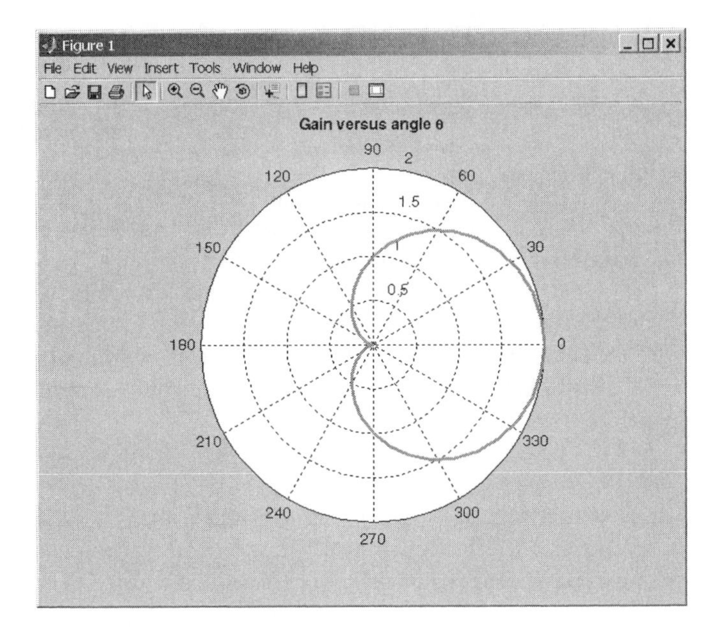

Figure 3.9 Gain of a cardioid microphone.

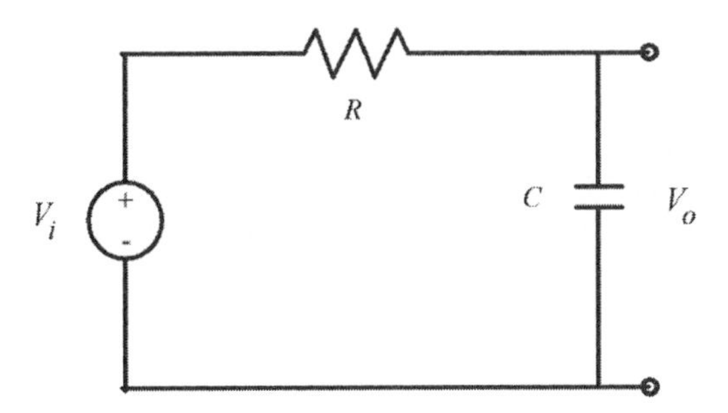

Figure 3.10 A simple low-pass filter circuit.

where V_i is a sinusoidal input voltage of frequency f, R is the resistance in ohms, C is the capacitance in farads, and j is $\sqrt{-1}$ (electrical engineers use j instead of i for $\sqrt{-1}$, because the letter i is traditionally reserved for the current in a circuit).

Assume that the resistance $R = 16$ kΩ, and capacitance $C = 1$ μF, and plot the amplitude and frequency response of this filter.

Solution The amplitude response of a filter is the ratio of the amplitude of the output voltage to the amplitude of the input voltage, and the phase response of the filter is the difference between the phase of the output voltage and the phase of the input voltage. The simplest way to calculate the amplitude and phase response of the filter is to evaluate Equation 3-4 at many different frequencies. The plot of the magnitude of Equation 3-4 versus frequency is the amplitude response of the filter, and the plot of the angle of Equation 3-4 versus frequency is the phase response of the filter.

Because the frequency and amplitude response of a filter can vary over a wide range, it is customary to plot both of these values on logarithmic scales. On the other hand, the phase varies over a very limited range, so it is customary to plot the phase of the filter on a linear scale. Therefore, we will use a `loglog` plot for the amplitude response, and a `semilogx` plot for the phase response of the filter. We will display both responses as two sub-plots within a figure.

The MATLAB code required to create and plot the responses is shown below.

```
%  Script file: plot_filter.m
%
%  Purpose:
%    This program plots the amplitude and phase responses
%    of a low-padd RC filter.
%
%  Record of revisions:
%      Date       Programmer          Description of change
%      ====       ==========          =====================
%    01/05/05    S. J. Chapman        Original code
%
% Define variables:
%   amp       -- Amplitude response
%   C         -- Capacitiance (farads)
%   f         -- Frequency of input signal (Hz)
%   phase     -- Phase response
%   R         -- Resistance (ohms)
%   res       -- Vo/Vi

% Initialize R & C
R = 16000;                  % 16 k ohms
C = 1.0E-6;                 % 1 uF

% Create array of input frequencies
f = 1:2:1000;

% Calculate response
res = 1 ./ ( 1 + j*2*pi*f*R*C );

% Calculate amplitude response
amp = abs(res);
% Calculate phase response
phase = angle(res);
```

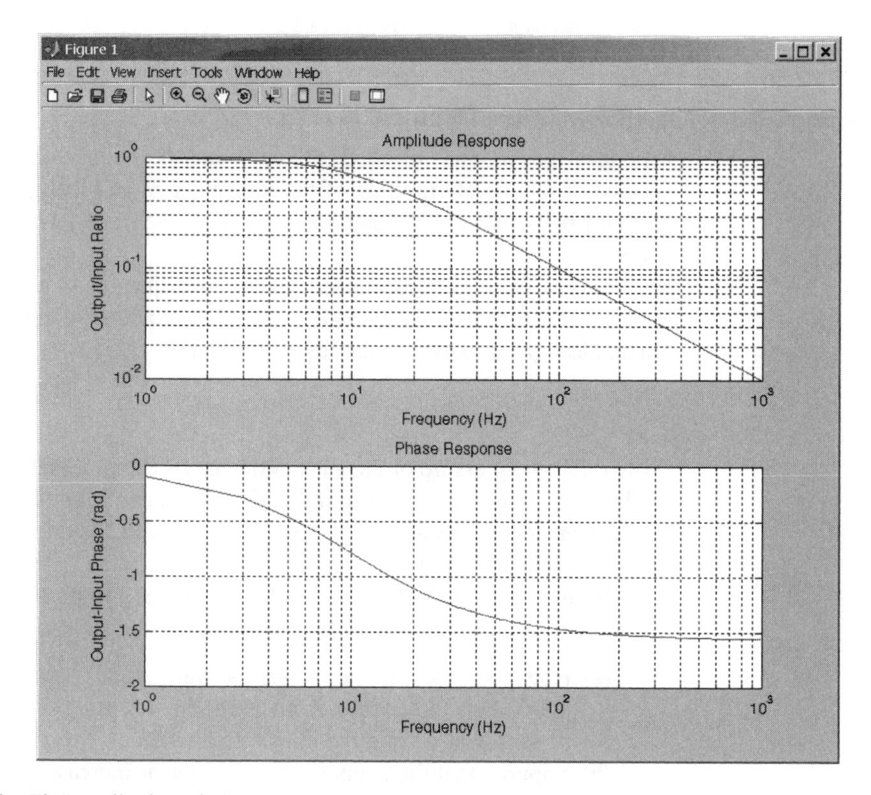

Figure 3.11 The amplitude and phase response of the low-pass filter circuit.

```
% Create plots
subplot(2,1,1);
loglog( f, amp );
title('Amplitude Response');
xlabel('Frequency (Hz)');
ylabel('Output/Input Ratio');
grid on;

subplot(2,1,2);
semilogx( f, phase );
title('Phase Response');
xlabel('Frequency (Hz)');
ylabel('Output-Input Phase (rad)');
grid on;
```

The resulting amplitude and phase responses are shown in Figure 3.11. Note that this circuit is called a low-pass filter because low frequencies are passed through with little attenuation, while high frequencies are strongly attenuated.

Example 3.7—Thermodynamics: The Ideal Gas Law

An ideal gas is one in which all collisions between molecules are perfectly elastic. It is possible to think of the molecules in an ideal gas as perfectly hard billiard balls that collide and bounce off of each other without losing kinetic energy.

Such a gas can be characterized by three quantities: absolute pressure (P), volume (V), and absolute temperature (T). The relationship among these quantities in an ideal gas is known as the Ideal Gas Law:

$$PV = nRT \tag{3-5}$$

where P is the pressure of the gas in kilopascals (kPa), V is the volume of the gas in liters (L), n is the number of molecules of the gas in units of moles (mol), R is the universal gas constant (8.314 L·kPa/mol·K), and T is the absolute temperature in kelvins (K). (Note: 1 mol = 6.02×10^{23} molecules)

Assume that a sample of an ideal gas contains 1 mole of molecules at a temperature of 273 K, and answer the following questions.

(a) How does the volume of this gas vary as its pressure varies from 1 to 1000 kPa? Plot pressure versus volume for this gas on an appropriate set of axes. Use a solid red line, with a width of 2 pixels.
(b) Suppose that the temperature of the gas is increased to 373 K. How does the volume of this gas vary with pressure now? Plot pressure versus volume for this gas on an the same set of axes as part (a). Use a dashed blue line, with a width of 2 pixels.

Include a bold-face title and x- and y-axis labels on the plot, as well as legends for each line.

SOLUTION The values that we wish to plot both vary by a factor of 1,000, so an ordinary linear plot will not produce a useful plot. Therefore, we will plot the data on a log-log scale.

Note that we must plot two curves on the same set of axes, so we must issue the command `hold on` after the first one is plotted, and `hold off` after the plot is complete. It will also be necessary to specify the color, style, and width of each line and to specify that labels be in bold face.

A program that calculates the volume of the gas as a function of pressure and creates the appropriate plot is shown below. Note that the special features controlling the style of the plot are shown in bold face.

```
%  Script file: ideal_gas.m
%
%  Purpose:
%     This program plots the pressure versus volume of an
%     ideal gas.
```

```
%
%  Record of revisions:
%       Date       Programmer          Description of change
%       ====       ==========          =====================
%     01/05/05    S. J. Chapman        Original code
%
% Define variables:
%    n          -- Number of atoms (mol)
%    P          -- Pressure (kPa)
%    R          -- Ideal gas constant (L kPa/mol K)
%    T          -- Temperature (K)
%    V          -- volume (L)

% Initialize nRT
n = 1;                 % Moles of atoms
R = 8.314;             % Ideal gas constant
T = 273;               % Temperature (K)

% Create array of input pressures. Note that this
% array must be quite dense to catch the major
% changes in volume at low pressures.
P = 1:0.1:1000;

% Calculate volumes
V = (n * R * T) ./ P;

% Create first plot
figure(1);
loglog( P, V, 'r-', 'LineWidth', 2 );
title('\bfVolume vs Pressure in an Ideal Gas');
xlabel('\bfPressure (kPa)');
ylabel('\bfVolume (L)');
grid on;
hold on;

% Now increase temperature
T = 373;               % Temperature (K)

% Calculate volumes
V = (n * R * T) ./ P;

% Add second line to plot
figure(1);
```

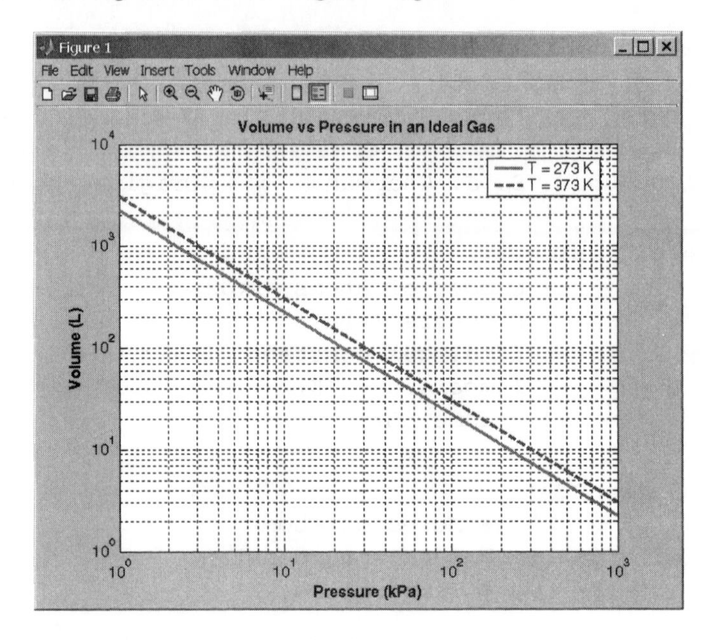

Figure 3.12 Pressure versus volume for an ideal gas.

```
loglog( P, V, 'b--', 'LineWidth', 2 );
hold off;

% Add legend
legend('T = 273 K','T = 373 k');
```

The resulting volume versus pressure plot shown in Figure 3.12.

3.5.8 Annotating and Saving Plots

Once a plot has been created by a MATLAB program, a user can edit and annotate the plot using the GUI-based tools available from the plot toolbar. Figure 3.13 shows the tools available, which allow the user to edit the properties of any objects on the plot or to add annotations to the plot. When the editing button ( ) is selected from the toolbar, the editing tools become available for use. When the button is depressed, clicking any line or text on the figure will cause it to be selected for editing, and double-clicking the line or text will open a Property Editor window that allows you to modify any or all of the characteristics of that object. Figure 3.14 shows Figure 3.12 after a user has clicked on the blue line to change it to a 3-pixel-wide dashed line.

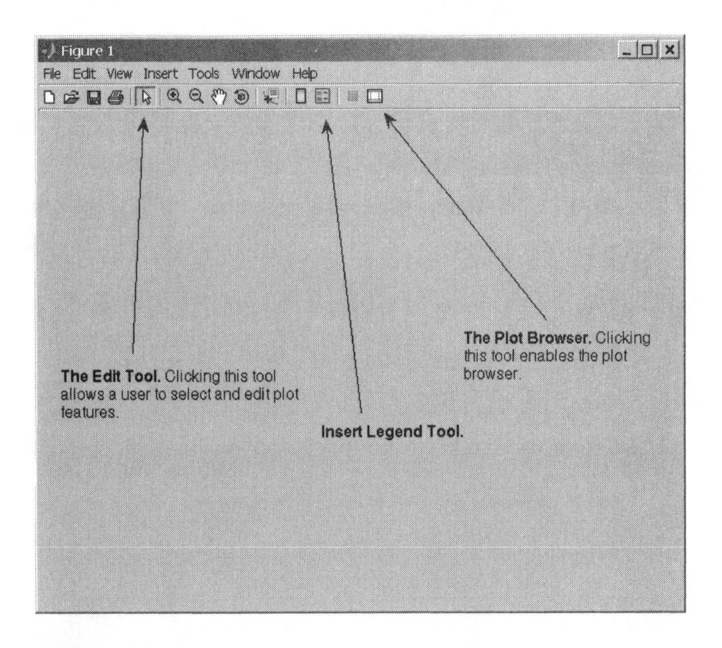

Figure 3.13 The editing tools on the figure toolbar.

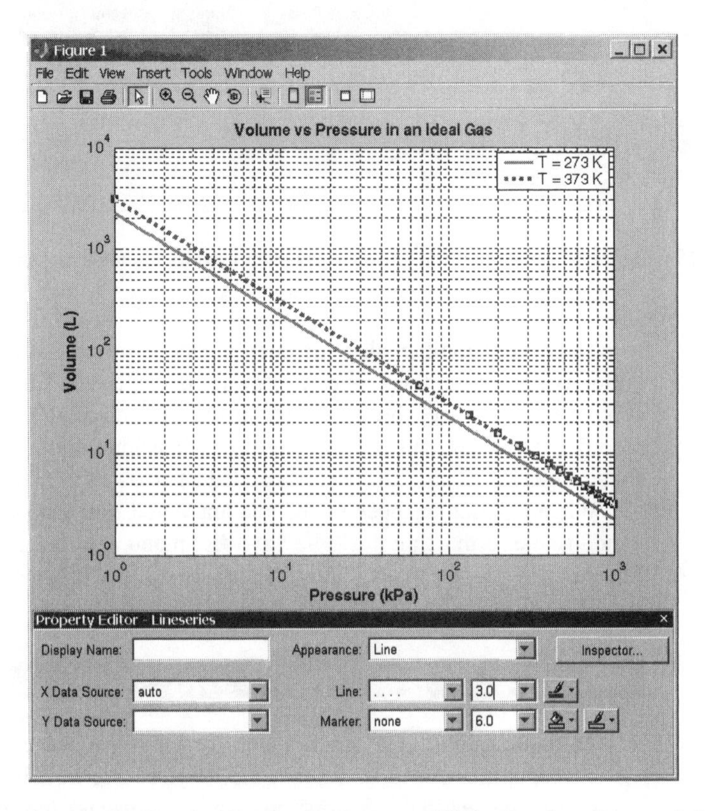

Figure 3.14 Figure 3.12 after the blue line has been modified using the editing tools built into the figure toolbar.

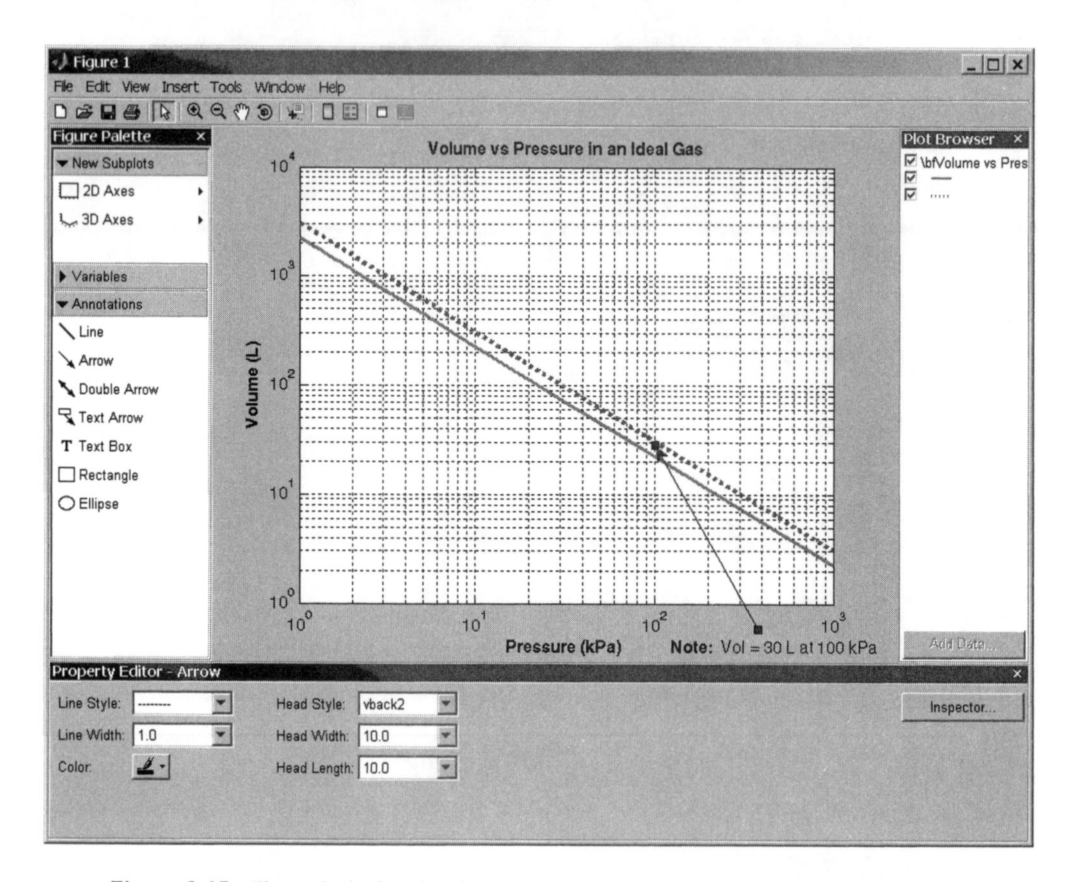

Figure 3.15 Figure 3.12 after the Plot Browser has been used to add an arrow and an annotation.

The figure toolbar also includes a Plot Browser button ( ). When this button is depressed, the Property Browser is displayed. This tool gives the user complete control over the figure. He or she can add axes, edit object properties, modify data values, and add annotations such as lines and text boxes. Figure 3.15 shows Figure 3.12 after the user has added an arrow and an annotation to the plot.

When the plot has been edited and annotated, you can save the entire plot in a modifiable form using the "File/Save As" menu item from the Figure Window. The resulting figure file (`*.fig`) contains all the information required to re-create the figure plus annotations at any time in the future.

Quiz 3.3

This quiz provides a quick check to see if you have understood the concepts introduced in Section 3.5. If you have trouble with the quiz, reread the section, ask your instructor, or discuss the material with a fellow student. The answers to this quiz are found in the back of the book.

1. Write the MATLAB statements required to plot sin x versus cos $2x$ from 0 to 2π in steps of $\pi/10$. The points should be connected by a 2-pixel-wide red line, and each point should marked with a 6-pixel-wide blue circular marker.
2. Use the Figure Editing tools to change the markers on the previous plot into black squares. Add an arrow and annotation pointing to the location $x = \pi$ on the plot.

Write the MATLAB text string that will produce the following expressions:

3. $f(x) = \sin\theta\cos 2\phi$
4. Plot of versus $\sum x^2$ versus x

Write the expression produced by the following text strings:

5. `'\tau\it_{m}'`
6. `'\bf\itx_{1}^{2} + x_{2}^{2} \rm(units: \bfm^{2} \rm)'`
7. How do you display the backslash (\) character in a text string?

3.6 More on Debugging MATLAB Programs

It is much easier to make a mistake when writing a program containing branches and loops than it is when writing simple sequential programs. Even after going through the full design process, a program of any size is almost guaranteed not to be completely correct the first time it is used. Suppose that we have built the program and tested it, only to find that the output values are in error. How do we go about finding the bugs and fixing them?

Once programs start to include loops and branches, the best way to locate an error is to use the symbolic debugger supplied with MATLAB. This debugger is integrated with the MATLAB editor.

To use the debugger, first open the file that you would like to debug using the "File/Open" menu selection in the MATLAB Command Window. When the file is opened, it is loaded into the editor and the syntax is automatically color coded. Comments in the file appear in green, variables and numbers appear in black, character strings appear in red, and language keywords appear in blue. Figure 3.16 on the next page shows an example Edit/Debug window containing the file `calc_roots.m`.

Let's say that we would like to determine what happens when the program is executed. To do this, we can set one or more **breakpoints** by right-clicking the

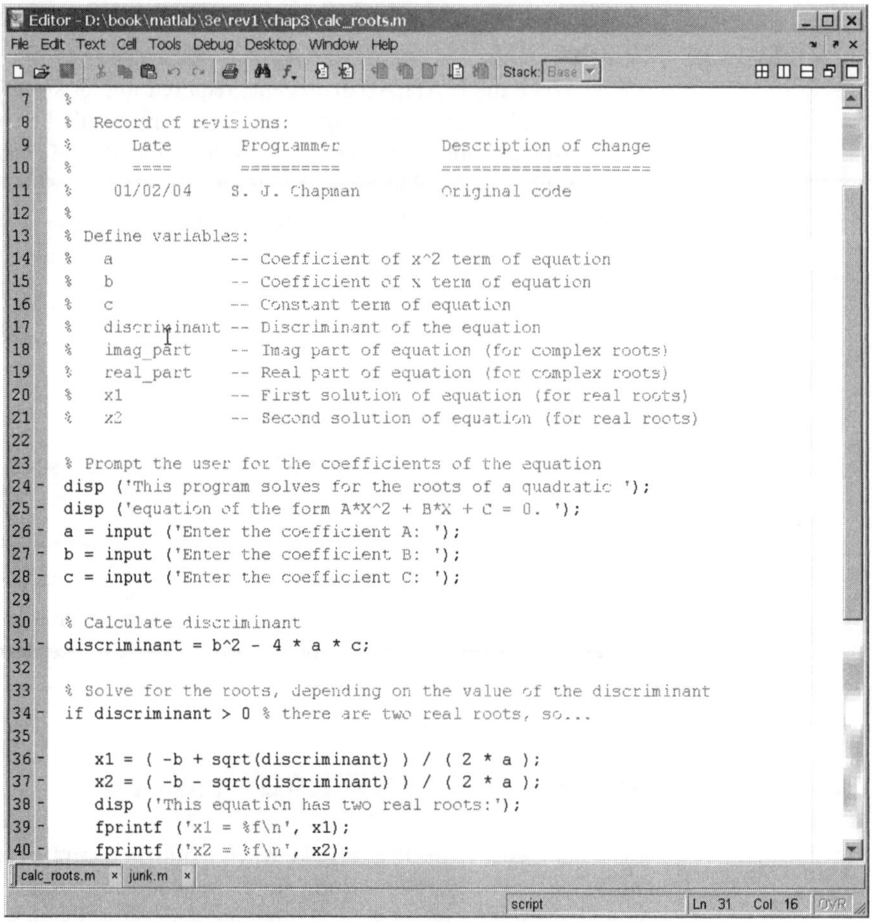

Editor - D:\book\matlab\3e\rev1\chap3\calc_roots.m

File Edit Text Cell Tools Debug Desktop Window Help

```
7     %
8     %  Record of revisions:
9     %      Date       Programmer          Description of change
10    %      ====       ==========          =====================
11    %    01/02/04    S. J. Chapman        Original code
12    %
13    % Define variables:
14    %   a             -- Coefficient of x^2 term of equation
15    %   b             -- Coefficient of x term of equation
16    %   c             -- Constant term of equation
17    %   discriminant -- Discriminant of the equation
18    %   imag_part     -- Imag part of equation (for complex roots)
19    %   real_part     -- Real part of equation (for complex roots)
20    %   x1            -- First solution of equation (for real roots)
21    %   x2            -- Second solution of equation (for real roots)
22
23    % Prompt the user for the coefficients of the equation
24 -  disp ('This program solves for the roots of a quadratic ');
25 -  disp ('equation of the form A*X^2 + B*X + C = 0. ');
26 -  a = input ('Enter the coefficient A: ');
27 -  b = input ('Enter the coefficient B: ');
28 -  c = input ('Enter the coefficient C: ');
29
30    % Calculate discriminant
31 -  discriminant = b^2 - 4 * a * c;
32
33    % Solve for the roots, depending on the value of the discriminant
34 -  if discriminant > 0 % there are two real roots, so...
35
36 -     x1 = ( -b + sqrt(discriminant) ) / ( 2 * a );
37 -     x2 = ( -b - sqrt(discriminant) ) / ( 2 * a );
38 -     disp ('This equation has two real roots:');
39 -     fprintf ('x1 = %f\n', x1);
40 -     fprintf ('x2 = %f\n', x2);
```

calc_roots.m × junk.m × | script | Ln 31 Col 16 OVR

Figure 3.16 An Edit/Debug window with a MATLAB program loaded.

mouse on the lines of interest and choosing the "Set/Clear Breakpoint" option. When a breakpoint is set, a red dot appears to the left of that line containing the breakpoint, as shown in Figure 3.17.

Once the breakpoints have been set, execute the program as usual by typing `calc_roots` in the Command Window. The program will run until it reaches the first breakpoint and stop there. A green arrow will appear by the current line during the debugging process, as shown in Figure 3.18 on page 138. When the breakpoint is reached, the programmer can examine and/or modify any variable in the workspace by typing its name in the Command Window. When the programmer is satisfied with the program at that point, he or she can either step through the program a line at a time by repeatedly pressing F10 or else run to the

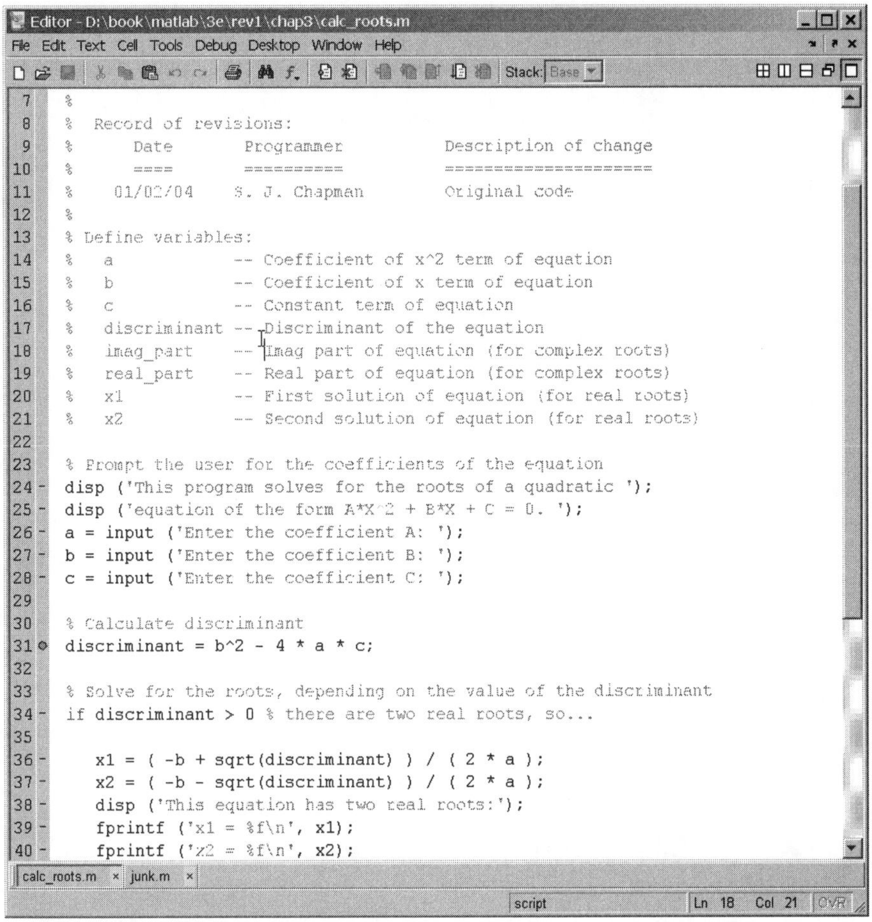

```
Editor - D:\book\matlab\3e\rev1\chap3\calc_roots.m
File Edit Text Cell Tools Debug Desktop Window Help
7    %
8    %  Record of revisions:
9    %      Date         Programmer           Description of change
10   %      ====         ==========           =====================
11   %    01/02/04     S. J. Chapman         Original code
12   %
13   % Define variables:
14   %   a             -- Coefficient of x^2 term of equation
15   %   b             -- Coefficient of x term of equation
16   %   c             -- Constant term of equation
17   %   discriminant  -- Discriminant of the equation
18   %   imag_part     -- Imag part of equation (for complex roots)
19   %   real_part     -- Real part of equation (for complex roots)
20   %   x1            -- First solution of equation (for real roots)
21   %   x2            -- Second solution of equation (for real roots)
22
23   % Prompt the user for the coefficients of the equation
24 - disp ('This program solves for the roots of a quadratic ');
25 - disp ('equation of the form A*X^2 + B*X + C = 0. ');
26 - a = input ('Enter the coefficient A: ');
27 - b = input ('Enter the coefficient B: ');
28 - c = input ('Enter the coefficient C: ');
29
30   % Calculate discriminant
31 o discriminant = b^2 - 4 * a * c;
32
33   % Solve for the roots, depending on the value of the discriminant
34 - if discriminant > 0 % there are two real roots, so...
35
36 -    x1 = ( -b + sqrt(discriminant) ) / ( 2 * a );
37 -    x2 = ( -b - sqrt(discriminant) ) / ( 2 * a );
38 -    disp ('This equation has two real roots:');
39 -    fprintf ('x1 = %f\n', x1);
40 -    fprintf ('x2 = %f\n', x2);
calc_roots.m  x   junk.m  x
script    Ln 18   Col 21   OVR
```

Figure 3.17 The window after a breakpoint has been set. Note the red dot to the left of the line with the breakpoint.

next breakpoint by pressing F5. It is always possible to examine the values of any variable at any point in the program.

When a bug is found, the programmer can use the Editor to correct the MATLAB program and save the modified version to disk. Note that all breakpoints may be lost when the program is saved to disk, so they may have to be set again before debugging can continue. This process is repeated until the program appears to be bug-free.

Two other very important features of the debugger are found on the "Breakpoints" menu. The first feature is "Set/Modify Conditional Breakpoint". A **conditional breakpoint** is a breakpoint where the code stops only if some condition is true. For example, a conditional breakpoint can be used to stop execution inside a `for` loop on its 200th execution. This can be very important if a bug

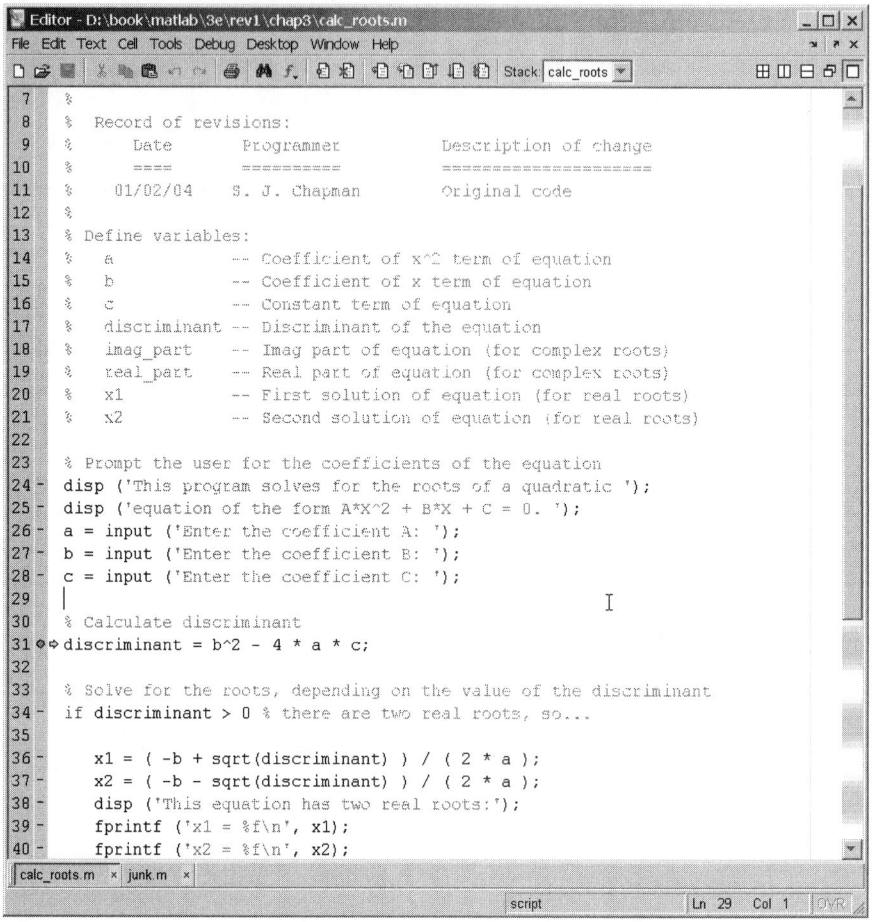

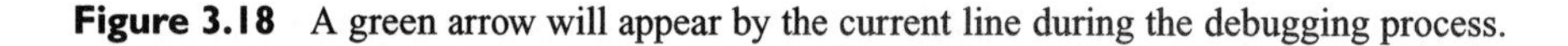
Editor - D:\book\matlab\3e\rev1\chap3\calc_roots.m
File Edit Text Cell Tools Debug Desktop Window Help
Stack: calc_roots
7 %
8 % Record of revisions:
9 % Date Programmer Description of change
10 % ==== ========== =====================
11 % 01/02/04 S. J. Chapman Original code
12 %
13 % Define variables:
14 % a -- Coefficient of x^2 term of equation
15 % b -- Coefficient of x term of equation
16 % c -- Constant term of equation
17 % discriminant -- Discriminant of the equation
18 % imag_part -- Imag part of equation (for complex roots)
19 % real_part -- Real part of equation (for complex roots)
20 % x1 -- First solution of equation (for real roots)
21 % x2 -- Second solution of equation (for real roots)
22
23 % Prompt the user for the coefficients of the equation
24 - disp ('This program solves for the roots of a quadratic ');
25 - disp ('equation of the form A*X^2 + B*X + C = 0. ');
26 - a = input ('Enter the coefficient A: ');
27 - b = input ('Enter the coefficient B: ');
28 - c = input ('Enter the coefficient C: ');
29
30 % Calculate discriminant
31 discriminant = b^2 - 4 * a * c;
32
33 % Solve for the roots, depending on the value of the discriminant
34 - if discriminant > 0 % there are two real roots, so...
35
36 - x1 = (-b + sqrt(discriminant)) / (2 * a);
37 - x2 = (-b - sqrt(discriminant)) / (2 * a);
38 - disp ('This equation has two real roots:');
39 - fprintf ('x1 = %f\n', x1);
40 - fprintf ('x2 = %f\n', x2);
calc_roots.m junk.m
script Ln 29 Col 1 OVR

Figure 3.18 A green arrow will appear by the current line during the debugging process.

appears only after a loop has been executed many times. The condition that causes the breakpoint to stop execution can be modified, and the breakpoint can be enabled or disabled during debugging.

The second feature is "Set Error Breakpoints for All Files." If an error is occurring in a program that causes it to crash or generate warning messages, the programmer can turn this item on and execute the program. It will run to the point of the error and stop there, allowing the programmer to examine the values of variables and exactly what is causing the problem.

A final critical feature is found on the "Debug" menu. It is "Check Code with M-Lint." M-Lint is a program that examines one or more M-files and reports any examples of improper or questionable usage. It is a *great* tool for locating errors, poor usage, or obsolete features in MATLAB code, including such things as variables that are defined but never used. You should always run M-Lint over your programs when they are finished as a final check that everything has been done properly.

Take some time now to become familiar with the Editor/Debugger and its supporting tools; it is a very worthwhile investment.

3.7 Summary

In Chapter 3 we have presented the basic types of MATLAB branches and the relational and logic operations used to control them. The principal type of branch is the `if` construct. This construct is very flexible. It can have as many `elseif` clauses as needed to construct any desired test. Furthermore, `if` constructs can be nested to produce more complex tests. A second type of branch is the `switch` construct. It may be used to select among mutually exclusive alternatives specified by a control expression. A third type of branch is the `try/catch` construct. It is used to trap errors that might occur during execution.

This chapter also included additional information about plots. The `axis` command allows a programmer to select the specific range of x and y data to be plotted. The `hold` command allows later plots to be plotted on top of earlier ones, so that elements can be added to a graph a piece at a time. The `figure` command allows the programmer to create and select among multiple Figure Windows, so that a program can create multiple plots in separate windows. The `subplot` command allows the programmer to create and select among multiple plots within a single Figure Window.

In addition, we learned how to control additional characteristics of our plots, such as the line width and marker color. These properties may be controlled by specifying `'PropertyName',value` pairs in the plot command after the data to be plotted.

Text strings in plots may be enhanced with stream modifiers and escape sequences. Stream modifiers allow a programmer to specify features like bold face, italic, superscripts, subscripts, font size, and font name. Escape sequences allow the programmer to include special characters such as Greek and mathematical symbols in the text string.

The MATLAB symbolic debugger and related tools such as M-Lint make debugging MATLAB code much easier. You should invest some time to become familiar with these tools.

3.7.1 Summary of Good Programming Practice

The following guidelines should be adhered to when programming with branch or loop constructs. By following them consistently, your code will contain fewer bugs, will be easier to debug, and will be more understandable to others who may need to work with it in the future.

1. Follow the steps of the program design process to produce reliable, understandable MATLAB programs.
2. Be cautious about testing for equality with numeric values, since round-off errors may cause two variables that should be equal to fail a test for equality. Instead, test to see if the variables are *nearly* equal within the roundoff error to be expected on the computer you are working with.
3. Use the `&` AND operator if it is necessary to ensure that both operands are evaluated in an expression or if the comparison is between arrays. Otherwise, use the `&&` AND operator, since the partial evaluation will make the operation faster in the cases where the first operand is `false`. The `&` operator is preferred in most practical cases.
4. Use the | inclusive OR operator if it is necessary to ensure that both operands are evaluated in an expression or if the comparison is between arrays. Otherwise, use the || operator, since the partial evaluation will make the operation faster in the cases where the first operand is `true`. The | operator is preferred in most practical cases.
5. Always indent code blocks in `if`, `switch`, and `try/catch` constructs to make them more readable.
6. For branches in which there are many mutually exclusive options, use a single `if` construct with multiple `elseif` clauses in preference to nested `if` constructs.

3.7.2 MATLAB Summary

The following summary lists all of the MATLAB commands and functions described in this chapter, along with a brief description of each one.

Commands and Functions

`axis`	(*a*) Set the *x* and *y* limits of the data to be plotted. (*b*) Get the *x* and *y* limits of the data to be plotted. (*c*) Set other axis-related properties.
`figure`	Select a Figure Window to be the current Figure Window. If the selected Figure Window does not exist, it is automatically created.
`hold`	Allows multiple plot commands to write on top of each other.
`if` construct	Selects a block of statements to execute if a specified condition is satisfied.
`ischar(a)`	Returns a 1 if `a` is a character array and a 0 otherwise.
`isempty(a)`	Returns a 1 if `a` is an empty array and a 0 otherwise.
`isinf(a)`	Returns a 1 if the value of `a` is infinite (`Inf`) and a 0 otherwise.
`isnan(a)`	Returns a 1 if the value of `a` is `NaN` (not a number) and a 0 otherwise.
`isnumeric(a)`	Returns a 1 if the `a` is a numeric array and a 0 otherwise.
`logical`	Converts numeric data to logical data, with nonzero values becoming `true` and zero values becoming `false`.

Commands and Functions

`polar`	Creates a polar plot.
`subplot`	Selects a subplot in the current Figure Window. If the selected subplot does not exist, it is automatically created. If the new subplot conflicts with a previously existing set of axes, they are automatically deleted.
`switch` construct	Selects a block of statements to execute from a set of mutually-exclusive choices based on the result of a single expression.
`try/catch` construct	A special construct used to trap errors. MATLAB executes the code in the `try` block. If an error occurs, execution of the `try` block stops immediately and the code in the `catch` block is executed.

3.8 Exercises

3.1 Evaluate the following MATLAB expressions.

(a) `5.5 >= 5`
(b) `20 > 20`
(c) `xor( 17 - pi < 15, pi < 3)`
(d) `true > false`
(e) `~~(35/17) == (35/17)`
(f) `(7 <= 8) == (3/2 == 1)`
(g) `17.5 && (3.3 > 2.)`

3.2 The tangent function is defined as $\tan\theta = \sin\theta/\cos\theta$. This expression can be evaluated to solve for the tangent as long as the magnitude of $\cos\theta$ is not too near to 0. (If $\cos\theta$ is 0, evaluating the equation for $\tan\theta$ will produce the nonnumerical value `Inf`.) Assume that θ is given in *degrees*, and write the MATLAB statements to evaluate $\tan\theta$ as long as the magnitude of $\cos\theta$ is greater than or equal to 10^{-20}. If the magnitude of $\cos\theta$ is less than 10^{-20}, write out an error message instead.

3.3 The following statements are intended to alert a user to dangerously high oral thermometer readings (values are in degrees Fahrenheit). Are they correct or incorrect? If they are incorrect, explain why and correct them.

```
if temp < 97.5
   disp('Temperature below normal');
elseif temp > 97.5
   disp('Temperature normal');
elseif temp > 99.5
   disp('Temperature slightly high');
elseif temp > 103.0
   disp('Temperature dangerously high');
end
```

3.4 The cost of sending a package by an express delivery service is $12.00 for the first two pounds, and $4.50 for each pound or fraction thereof over two pounds. If the package weighs more than 70 pounds, a $15.00 excess weight surcharge is added to the cost. No package over 100 pounds will be accepted. Write a program that accepts the weight of a package in pounds and computes the cost of mailing the package. Be sure to handle the case of overweight packages.

3.5 In Example 3.3, we wrote a program to evaluate the function $f(x, y)$ for any two user-specified values x and y, where the function $f(x, y)$ was defined as follows.

$$f(x,y) = \begin{cases} x + y & x \geq 0 \text{ and } y \geq 0 \\ x + y^2 & x \geq 0 \text{ and } y < 0 \\ x^2 + y & x < 0 \text{ and } y \geq 0 \\ x^2 + y^2 & x < 0 \text{ and } y < 0 \end{cases}$$

The problem was solved by using a single `if` construct with four code blocks to calculate $f(x, y)$ for all possible combinations of x and y. Rewrite program `funxy` to use nested `if` constructs, where the outer construct evaluates the value of x and the inner constructs evaluate the value of y.

3.6 Write a MATLAB program to evaluate the function

$$y(x) = \ln\frac{1}{1 - x}$$

for any user-specified value of x, where x is a number < 1.0 (note that ln is the natural logarithm, the logarithm to the base e). Use an `if` structure to verify that the value passed to the program is legal. If the value of x is legal, calculate $y(x)$. If not, write a suitable error message and quit.

3.7 Write a program that allows a user to enter a string containing a day of the week ('Sunday', 'Monday', 'Tuesday', etc.), and uses that a `switch` construct to convert the day to its corresponding number, where Sunday is considered the first day of the week and Saturday is considered the last day of the week. Print out the resulting day number. Also, be sure to handle the case of an illegal day name! (*Note:* Be sure to use the `'s'` option on function `input` so that the input is treated as a string.)

3.8 Suppose that a student has the option of enrolling for a single elective during a term. The student must select a course from a limited list of options: "English," "History," "Astronomy," or "Literature." Construct a fragment of MATLAB code that will prompt the student for his or her choice, read in the choice, and use the answer as the case expression for a `switch` construct. Be sure to include a default case to handle invalid inputs.

3.9 **Ideal Gas Law** The Ideal Gas Law was defined in Example 3.7. Assume that the volume of 1 mole of this gas is 10 L, and plot the pressure of the gas as a function of temperature as the temperature is changed from 250 to 400 kelvins. What sort of plot (linear, semilogx, etc.) is most appropriate for this data?

3.10 **Antenna Gain Pattern** The gain G of a certain microwave dish antenna can be expressed as a function of angle by the equation

$$G(\theta) = |\text{sinc } 4\theta| \quad \text{for } -\frac{\pi}{2} \le \theta \le \frac{\pi}{2} \tag{3-5}$$

where θ is measured in radians from the boresite of the dish, and $\text{sinc } x = \sin x/x$. Plot this gain function on a polar plot, with the title "**Antenna Gain vs θ**" in bold face.

3.11 The author of this book now lives in Australia. Australia is a great place to live, but it is also a land of high taxes. In 2002, individual citizens and residents of Australia paid the following income taxes:

Taxable Income (in A$)	Tax on This Income
$0–$6,000	Nil
$6,001–$20,000	17¢ for each $1 over $6,000
$20,001–$50,000	$2,380 plus 30¢ for each $1 over $20,000
$50,001–$60,000	$11,380 plus 42¢ for each $1 over $50,000
Over $60,000	$15,580 plus 47¢ for each $1 over $60,000

In addition, a flat 1.5% Medicare levy is charged on all income. Write a program to calculate how much income tax a person will owe based on this information. The program should accept a total income figure from the user, and calculate the income tax, Medicare levy, and total tax payable by the individual.

3.12 **Refraction** When a ray of light passes from a region with an index of refraction n_1 into a region with a different index of refraction n_2, the light ray is bent (see Figure 3.19). The angle at which the light is bent is given by Snell's Law,

$$n_1 \sin\theta_1 = n_2 \sin\theta_2 \tag{3-6}$$

where θ_1 is the angle of incidence of the light in the first region, and θ_2 is the angle of incidence of the light in the second region. Using Snell's Law, it is possible to predict the angle of incidence of a light ray in Region 2 if the angle of incidence θ_1 in Region 1 and the indices of refraction n_1 and n_2 are known. The equation to perform this calculation is

$$\theta_2 = \sin^{-1}\left(\frac{n_1}{n_2}\sin\theta_1\right) \tag{3-7}$$

Write a program to calculate the angle of incidence (in degrees) of a light ray in Region 2 given the angle of incidence θ_1 in Region 1 and the indices of refraction n_1 and n_2. (*Note*: If $n_1 > n_2$, then for some angles θ_1, Equation 3-7

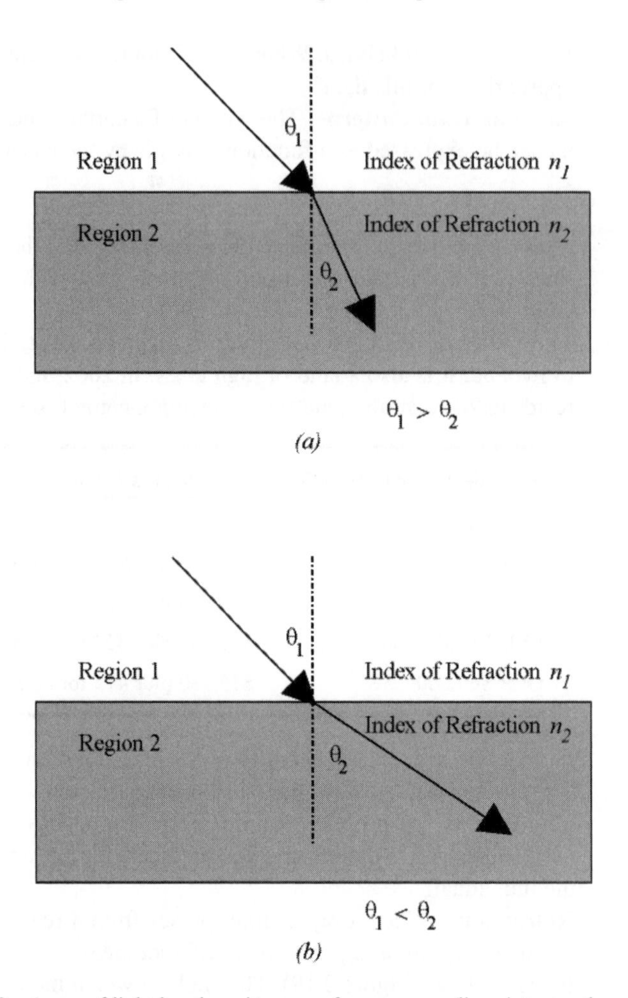

Figure 3.19 A ray of light bends as it passes from one medium into another one. *(a)* If the ray of light passes from a region with a low index of refraction into a region with a higher index of refraction, the ray of light bends more towards the vertical. *(b)* If the ray of light passes from a region with a high index of refraction into a region with a lower index of refraction, the ray of light bends away from the vertical.

will have no real solution because the absolute value of the quantity $\left(\frac{n_1}{n_2}\sin\theta_1\right)$ will be greater than 1.0. When this occurs, all light is reflected back into Region 1, and no light passes into Region 2 at all. Your program must be able to recognize and properly handle this condition.)

The program should also create a plot showing the incident ray, the boundary between the two regions, and the refracted ray on the other side of the boundary.

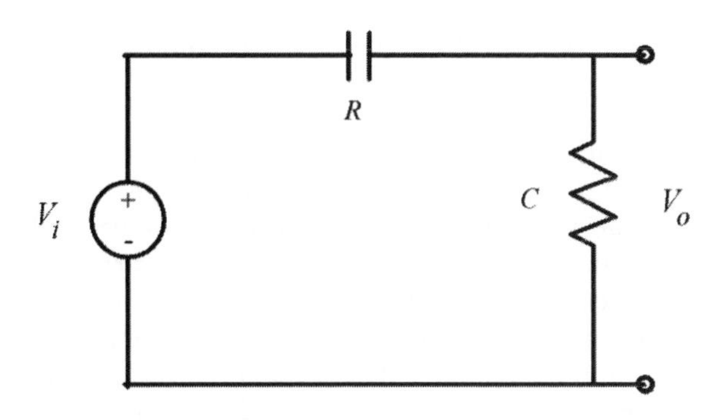

Figure 3.20 A simple high-pass filter circuit.

Test your program by running it for the following two cases: (*a*) $n_1 = 1.0$, $n_2 = 1.7$, and $\theta_1 = 45°$. (*b*) $n_1 = 1.7$, $n_2 = 1.0$, and $\theta_1 = 45°$.

3.13 Assume that the complex function $f(t)$ is defined by the equation

$$f(t) = (0.5 - 0.25i)t - 1.0$$

Plot the amplitude and phase of function f for $0 \le t \le 4$.

3.14 **High-Pass Filter** Figure 3.20 shows a simple high-pass filter consisting of a resistor and a capacitor. The ratio of the output voltage V_o to the input voltage V_i is given by the equation

$$\frac{V_o}{V_i} = \frac{j2\pi fRC}{1 + j2\pi fRC} \tag{3-8}$$

Assume that $R = 16\ \text{k}\Omega$ and $C = 1\ \mu\text{F}$. Calculate and plot the amplitude and phase response of this filter as a function of frequency.

3.15 **The Spiral of Archimedes** The spiral of Archimedes is a curve described in polar coordinates by the equation

$$r = k\theta \tag{3-9}$$

where r is the distance of a point from the origin and θ is the angle of that point in radians with respect to the origin. Plot the spiral of Archimedes for $0 \le \theta \le 6\pi$ when $k = 0.5$. Be sure to label your plot properly.

3.16 **Output Power from a Motor** The output power produced by a rotating motor is given by the equation

$$P = \tau_{\text{IND}}\,\omega_m \tag{3-10}$$

where τ_{IND} is the induced torque on the shaft in newton-meters, ω_m is the rotational speed of the shaft in radians per second, and P is in watts. Assume that the rotational speed of a particular motor shaft is given by the equation

$$\omega_m = 188.5(1 - e^{-0.2t})\ \text{rad/s}$$

and the induced torque on the shaft is given by

$$\tau_{\mathrm{IND}} = 10e^{-0.2t}\ \mathrm{N \cdot m}$$

Plot the torque, speed, and power supplied by this shaft versus time for $0 \le t \le 10$ s. Be sure to label your plot properly with the symbols τ_{IND} and ω_m where appropriate. Create two plots, one with the power displayed on a linear scale, and one with the output power displayed on a logarithmic scale. Time should always be displayed on a linear scale.

3.17 **Plotting Orbits** When a satellite orbits the Earth, the satellite's orbit will form an ellipse with the Earth located at one of the focal points of the ellipse. The satellite's orbit can be expressed in polar coordinates as

$$r = \frac{p}{1 - \varepsilon \cos\theta} \tag{3-11}$$

where r and θ are the distance and angle of the satellite from the center of the Earth, p is a parameter specifying the size of the size of the orbit, and ε is a parameter representing the eccentricity of the orbit. A circular orbit has an eccentricity ε of 0. An elliptical orbit has an eccentricity of $0 \le \varepsilon \le 1$. If $\varepsilon > 1$, the satellite follows a hyperbolic path and escapes from the Earth's gravitational field.

Consider a satellite with a size parameter $p = 1{,}000$ km. Plot the orbit of this satellite if (*a*) $\varepsilon = 0$; (*b*) $\varepsilon = 0.25$; (*c*) $\varepsilon = 0.5$. How close does each orbit come to the Earth? How far away does each orbit get from the Earth? Compare the three plots you created. Can you determine what the parameter p means from looking at the plots?

CHAPTER 4

Loops

Loops are MATLAB constructs that permit us to execute a sequence of statements more than once. There are two basic forms of loop constructs: **while loops** and **for loops**. The major difference between these two types of loops is in how the repetition is controlled. The code in a `while` loop is repeated an indefinite number of times until some user-specified condition is satisfied. By contrast, the code in a `for` loop is repeated a specified number of times, and the number of repetitions is known before the loops starts.

4.1 The while Loop

A **while loop** is a block of statements that are repeated indefinitely as long as some condition is satisfied. The general form of a while loop is

```
while expression
   ...
   ...          } Code block
   ...
end
```

The controlling expression produces a logical value. If the *expression* is `true`, the code block will be executed, and then control will return to the `while` statement. If the *expression* is still `true`, the statements will be executed again. This process will be repeated until the *expression* becomes `false`. When control returns to the `while` statement and the expression is `false`, the program will execute the first statement after the `end`.

The pseudocode corresponding to a `while` loop is

```
while expr
    ...
    ...
    ...
end
```

We will now show an example statistical analysis program that is implemented using a `while` loop.

Example 4.1—Statistical Analysis

It is very common in science and engineering to work with large sets of numbers, each of which is a measurement of some particular property that we are interested in. A simple example would be the grades on the first test in this course. Each grade would be a measurement of how much a particular student has learned in the course to date.

Much of the time, we are not interested in looking closely at every single measurement that we make. Instead, we want to summarize the results of a set of measurements with a few numbers that tell us a lot about the overall data set. Two such numbers are the *average* (or *arithmetic mean*) and the *standard deviation* of the set of measurements. The average or arithmetic mean of a set of numbers is defined as

$$\bar{x} = \frac{1}{N}\sum_{i=1}^{N} x_i \tag{4-1}$$

where x_i is sample i out of N samples. If all of the input values are available in an array, the average of a set of number may be calculated by MATLAB function `mean`. The standard deviation of a set of numbers is defined as

$$s = \sqrt{\frac{N\sum_{i=1}^{N} x_i^2 - \left(\sum_{i=1}^{N} x_i\right)^2}{N(N-1)}} \tag{4-2}$$

Standard deviation is a measure of the amount of scatter on the measurements; the greater the standard deviation, the more scattered the points in the data set are. If all of the input values are available in an array, the standard deviation of a set of number may be calculated by MATLAB function `std`.

Implement an algorithm that reads in a set of measurements and calculates the mean and the standard deviation of the input data set.

SOLUTION This program must be able to read in an arbitrary number of measurements, and then calculate the mean and standard deviation of those measurements. We will use a while loop to accumulate the input measurements before performing the calculations.

When all of the measurements have been read, we must have some way of telling the program that there is no more data to enter. For now, we will assume that all the input measurements are either positive or zero, and we will use a negative input value as a *flag* to indicate that there is no more data to read. If a negative value is entered, then the program will stop reading input values and will calculate the mean and standard deviation of the data set.

1. **State the problem**
   Since we assume that the input numbers must be positive or zero, a proper statement of this problem would be: *calculate the average and the standard deviation of a set of measurements, assuming that all of the measurements are either positive or zero, and assuming that we do not know in advance how many measurements are included in the data set. A negative input value will mark the end of the set of measurements.*

2. **Define the inputs and outputs**
   The inputs required by this program are an unknown number of positive or zero numbers. The outputs from this program are a printout of the mean and the standard deviation of the input data set. In addition, we will print out the number of data points input to the program, since this is a useful check that the input data was read correctly.

3. **Design the algorithm**
   This program can be broken down into three major steps

   ```
   Accumulate the input data
   Calculate the mean and standard deviation
   Write out the mean, standard deviation, and
     number of points
   ```

   The first major step of the program is to accumulate the input data. To do this, we will have to prompt the user to enter the desired numbers. When the numbers are entered, we will have to keep track of the number of values entered, plus the sum and the sum of the squares of those values. The pseudocode for these steps is:

   ```
   Initialize n, sum_x, and sum_x2 to 0
   Prompt user for first number
   Read in first x
   while x >= 0
      n <- n + 1
      sum_x <- sum_x + x
      sum_x2 <- sum_x2 + x^2
      Prompt user for next number
      Read in next x
   end
   ```

Note that we have to read in the first value before the while loop starts so that the while loop can have a value to test the first time it executes.

Next, we must calculate the mean and standard deviation. The pseudocode for this step is just the MATLAB versions of Equations (4-1) and (4-2).

```
x_bar <- sum_x / n
std_dev <- sqrt((n*sum_x2 - sum_x^2)/(n*(n-1)))
```

Finally, we must write out the results.

```
Write out the mean value x_bar
Write out the standard deviation std_dev
Write out the number of input data points n
```

4. **Turn the algorithm into MATLAB statements**
   The final MATLAB program is shown below:

```
%  Script file: stats_1.m
%
%  Purpose:
%    To calculate mean and the standard deviation of
%    an input data set containing an arbitrary number
%    of input values.
%
%  Record of revisions:
%      Date       Programmer          Description of change
%      ====       ==========          =====================
%    01/07/05    S. J. Chapman        Original code
%
%  Define variables:
%    n        -- The number of input samples
%    std_dev  -- The standard deviation of the input samples
%    sum_x    -- The sum of the input values
%    sum_x2   -- The sum of the squares of the input values
%    x        -- An input data value
%    xbar     -- The average of the input samples

% Initialize sums.
n = 0; sum_x = 0; sum_x2 = 0;

% Read in first value
x = input('Enter first value: ');

% While Loop to read input values.
while x >= 0
```

```
    % Accumulate sums.
    n       = n + 1;
    sum_x   = sum_x + x;
    sum_x2  = sum_x2 + x^2;

    % Read in next value
    x = input('Enter next value:  ');

end

% Calculate the mean and standard deviation
x_bar = sum_x / n;
std_dev = sqrt((n * sum_x2 - sum_x^2)/(n * (n-1)));

% Tell user.
fprintf('The mean of this data set is:    %f\n', x_bar);
fprintf('The standard deviation is:       %f\n', std_dev);
fprintf('The number of data points is:    %f\n', n);
```

5. **Test the program**
   To test this program, we will calculate the answers by hand for a simple data set, and then compare the answers to the results of the program. If we used three input values: 3, 4, and 5, the mean and standard deviation would be

$$\bar{x} = \frac{1}{N}\sum_{i=1}^{N} x_i = \frac{1}{3}(12) = 4$$

$$s = \sqrt{\frac{N\sum_{i=1}^{N} x_i^2 - \left(\sum_{i=1}^{N} x_i\right)^2}{N(N-1)}} = 1$$

When these values are fed into the program, the results are

```
» stats_1
Enter first value: 3
Enter next value:  4
Enter next value:  5
Enter next value:  -1
The mean of this data set is: 4.000000
The standard deviation is:    1.000000
The number of data points is: 3.000000
```

The program gives the correct answers for our test data set.

In the example above, we failed to follow the design process completely. This failure has left the program with a fatal flaw! Did you spot it?

We have failed because *we did not completely test the program for all possible types of inputs*. Look at the example once again. If we enter either no numbers or only one number, then we will be dividing by zero in the above equations! The division-by-zero error will cause divide-by-zero warnings to be printed, and the output values will be `NaN`. We need to modify the program to detect this problem, tell the user what the problem is, and stop gracefully.

A modified version of the program called `stats_2` is shown below. Here, we check to see if there are enough input values before performing the calculations. If not, the program will print out an intelligent error message and quit. Test the modified program for yourself.

```
%  Script file: stats_2.m
%
%  Purpose:
%    To calculate mean and the standard deviation of
%    an input data set containing an arbitrary number
%    of input values.
%
%  Record of revisions:
%      Date       Programmer        Description of change
%      ====       ==========        =====================
%    01/07/05   S. J. Chapman      Original code
% 1. 01/07/05   S. J. Chapman      Correct divide-by-0 error if
%                                  0 or 1 input values given.
%
% Define variables:
%   n           -- The number of input samples
%   std_dev     -- The standard deviation of the input samples
%   sum_x       -- The sum of the input values
%   sum_x2      -- The sum of the squares of the input values
%   x           -- An input data value`
%   xbar        -- The average of the input samples

% Initialize sums.
n = 0; sum_x = 0; sum_x2 = 0;

% Read in first value
x = input('Enter first value: ');

% While Loop to read input values.
while x >= 0

   % Accumulate sums.
   n = n + 1;
```

```
    sum_x  = sum_x + x;
    sum_x2 = sum_x2 + x^2;

    % Read in next value
    x = input('Enter next value: ');

end

% Check to see if we have enough input data.
if n < 2   % Insufficient information

   disp('At least 2 values must be entered!');

else  % There is enough information, so
      % calculate the mean and standard deviation

      x_bar = sum_x / n;
      std_dev = sqrt((n * sum_x2 - sum_x^2)/(n * (n-1)));

      % Tell user.
      fprintf('The mean of this data set is:   %f\n', x_bar);
      fprintf('The standard deviation is:      %f\n', std_dev);
      fprintf('The number of data points is:   %f\n', n);

end
```

Note that the average and standard deviation could have been calculated with the built-in MATLAB functions `mean` and `std` if all of the input values are saved in a vector and that vector is passed to these functions. You will be asked to create a version of the program that uses the standard MATLAB functions in an exercise at the end of this chapter.

4.2 The `for` Loop

The **`for` loop** is a loop that executes a block of statements a specified number of times. The `for` loop has the form

```
for index = expr
   Statement 1      }
   ...              } Body
   Statement n      }
end
```

where `index` is the loop variable (also known as the **loop index**) and *`expr`* is the loop control expression. The columns in *`expr`* are stored one at a time in the variable `index` and then the loop body is executed, so that the loop is executed once for each column in *`expr`*. The expression usually takes the form of a vector in shortcut notation `first:incr:last`.

The statements between the `for` statement and the `end` statement are known as the *body* of the loop. They are executed repeatedly during each pass of the `for` loop. The `for` loop construct functions as follows:

1. At the beginning of the loop, MATLAB generates the control expression.
2. The first time through the loop, the program assigns the first column of the expression to the loop variable `index`, and the program executes the statements within the body of the loop.
3. After the statements in the body of the loop have been executed, the program assigns the next column of the expression to the loop variable `index`, and the program executes the statements within the body of the loop again.
4. Step 3 is repeated over and over as long as there are additional columns in the control expression.

Let's look at a number of specific examples to make the operation of the `for` loop clearer. First, consider the following example:

```
for ii = 1:10
   Statement 1
   ...
   Statement n
end
```

In this case, the control expression generates a 1 × 10 array, so statements 1 through n will be executed 10 times. The loop index `ii` will be 1 on the first time, 2 on the second time, and so on. The loop index will be 10 on the last pass through the statements. When control is returned to the `for` statement after the tenth pass, there are no more columns in the control expression, so execution transfers to the first statement after the `end` statement. Note that the loop index `ii` is still set to 10 after the loop finishes executing.

Second, consider the following example:

```
for ii = 1:2:10
   Statement 1
   ...
   Statement n
end
```

In this case, the control expression generates a 1 × 5 array, so statements 1 through n will be executed five times. The loop index `ii` will be 1 the first time, 3 the second time, and so on. The loop index will be 9 on the fifth and last pass through the statements. When control is returned to the `for` statement after the fifth pass, there are no more columns in the control expression, so execution transfers to the first statement after the end statement. Note that the loop index ii is still set to 9 after the loop finishes executing.

Third, consider the following example:

```
for ii = [5 9 7]
   Statement 1
   ...
   Statement n
end
```

Here, the control expression is an explicitly-written 1×3 array, so statements 1 through n will be executed three times with the loop index set to 5 the first time, 9 the second time, and 7 the final time. The loop index ii is still set to 7 after the loop finishes executing.

Finally, consider the example:

```
for ii = [1 2 3;4 5 6]
   Statement 1
   ...
   Statement n
end
```

In this case, the control expression is a 2×3 array, so statements 1 through n will be executed three times. The loop index ii will be the column vector $\begin{bmatrix} 1 \\ 4 \end{bmatrix}$ the first time, $\begin{bmatrix} 2 \\ 5 \end{bmatrix}$ the second time, and $\begin{bmatrix} 3 \\ 6 \end{bmatrix}$ the third time. The loop index ii is still set to $\begin{bmatrix} 3 \\ 6 \end{bmatrix}$ after the loop finishes executing. This example illustrates the fact that a loop index can be a vector.

The pseudocode corresponding to a for loop looks like the loop itself:

```
for index = expression
   Statement 1
   ...
   Statement n
end
```

Example 4.2—The Factorial Function

To illustrate the operation of a for loop, we will use a for loop to calculate the factorial function. The factorial function is defined as

$$n! = \begin{cases} 1 & n = 0 \\ n \times (n-1) \times (n-2) \times \cdots \times 2 \times 1 & n > 0 \end{cases}$$

The MATLAB code to calculate N factorial for positive value of N would be

```
n_factorial = 1
for ii = 1:n
   n_factorial = n_factorial * ii;
end
```

Suppose that we wish to calculate the value of 5!. If `n` is 5, the `for` loop control expression would be the row vector `[1 2 3 4 5]`. This loop will be executed five times, with the variable `ii` taking on values of 1, 2, 3, 4, and 5 in the successive loops. The resulting value of `n_factorial` will be $1 \times 2 \times 3 \times 4 \times 5 = 120$.

◄

►

Example 4.3—Calculating the Day of Year

The *day of year* is the number of days (including the current day) which have elapsed since the beginning of a given year. It is a number in the range 1 to 365 for ordinary years, and 1 to 366 for leap years. Write a MATLAB program that accepts a day, month, and year, and calculates the day of year corresponding to that date.

SOLUTION To determine the day of year, this program will need to sum up the number of days in each month preceding the current month, plus the number of elapsed days in the current month. A `for` loop will be used to perform this sum. Since the number of days in each month varies, it is necessary to determine the correct number of days to add for each month. A `switch` construct will be used to determine the proper number of days to add for each month.

During a leap year, an extra day must be added to the day of year for any month after February. This extra day accounts for the presence of February 29 in the leap year. Therefore, to perform the day of year calculation correctly, we must determine which years are leap years. In the Gregorian calendar, leap years are determined by the following rules:

1. Years evenly divisible by 400 are leap years.
2. Years evenly divisible by 100 but *not* by 400 are not leap years.
3. All years divisible by 4 but *not* by 100 are leap years.
4. All other years are not leap years.

We will use the `mod` (for modulo) function to determine whether or not a year is evenly divisible by a given number. If the result of the `mod` function is zero, then the year was evenly divisible.

A program to calculate the day of year is shown below. Note that the program sums up the number of days in each month before the current month, and that it uses a `switch` construct to determine the number of days in each month.

```
%  Script file: doy.m
%
%  Purpose:
%    This program calculates the day of year corresponding
%    to a specified date. It illustrates the use switch and
%    for constructs.
%
%  Record of revisions:
%      Date       Programmer          Description of change
%      ====       ==========          =====================
%    01/07/05    S. J. Chapman        Original code
%
%  Define variables:
%    day            -- Day (dd)
%    day_of_year    -- Day of year
%    ii             -- Loop index
%    leap_day       -- Extra day for leap year
%    month          -- Month (mm)
%    year           -- Year (yyyy)

% Get day, month, and year to convert
disp('This program calculates the day of year given the ');
disp('current date.');
month  = input('Enter current month (1-12): ');
day    = input('Enter current day(1-31):    ');
year   = input('Enter current year(yyyy):   ');

% Check for leap year, and add extra day if necessary
if mod(year,400) == 0
   leap_day = 1;     % Years divisible by 400 are leap years
elseif mod(year,100) == 0
   leap_day = 0;     % Other centuries are not leap years
elseif mod(year,4) == 0
   leap_day = 1;     % Otherwise every 4th year is a leap year
else
   leap_day = 0;     % Other years are not leap years
end

% Calculate day of year by adding current day to the
% days in previous months.
day_of_year = day;
for ii = 1:month-1
   % Add days in months from January to last month
   switch (ii)
```

```
    case {1,3,5,7,8,10,12},
       day_of_year = day_of_year + 31;
    case {4,6,9,11},
       day_of_year = day_of_year + 30;
    case 2,
       day_of_year = day_of_year + 28 + leap_day;
    end

end

% Tell user
fprintf('The date %2d/%2d/%4d is day of year %d.\n', ...
         month, day, year, day_of_year);
```

We will use the following known results to test the program:

1. Year 1999 is not a leap year. January 1 must be day of year 1, and December 31 must be day of year 365.
2. Year 2000 is a leap year. January 1 must be day of year 1, and December 31 must be day of year 366.
3. Year 2001 is not a leap year. March 1 must be day of year 60, since January has 31 days, February has 28 days, and this is the first day of March.

If this program is executed five times with the above dates, the results are

```
» doy
This program calculates the day of year given the
current date.
Enter current month (1-12):  1
Enter current day(1-31):     1
Enter current year(yyyy):    1999
The date  1/ 1/1999 is day of year 1.
» doy
This program calculates the day of year given the
current date.
Enter current month (1-12):  12
Enter current day(1-31):     31
Enter current year(yyyy):    1999
The date 12/31/1999 is day of year 365.
» doy
This program calculates the day of year given the
current date.
Enter current month (1-12):  1
Enter current day(1-31):     1
Enter current year(yyyy):    2000
The date  1/ 1/2000 is day of year 1.
```

```
» doy
This program calculates the day of year given the
current date.
Enter current month (1-12): 12
Enter current day(1-31):    31
Enter current year(yyyy):   2000
The date 12/31/2000 is day of year 366.
» doy
This program calculates the day of year given the
current date.
Enter current month (1-12): 3
Enter current day(1-31):    1
Enter current year(yyyy):   2001
The date  3/ 1/2001 is day of year 60.
```

The program gives the correct answers for our test dates in all five test cases

Example 4.4—Statistical Analysis

Implement an algorithm that reads in a set of measurements and calculates the mean and the standard deviation of the input data set, when any value in the data set can be positive, negative, or zero.

SOLUTION This program must be able to read in an arbitrary number of measurements and then calculate the mean and standard deviation of those measurements. Each measurement can be positive, negative, or zero.

Since we cannot use a data value as a flag this time, we will ask the user for the number of input values and then use a `for` loop to read in those values. The modified program that permits the use of any input value is shown below. Verify its operation for yourself by finding the mean and standard deviation of the following five input values: 3., −1., 0., 1., and −2.

```
%  Script file: stats_3.m
%
%  Purpose:
%    To calculate mean and the standard deviation of
%    an input data set, where each input value can be
%    positive, negative, or zero.
%
%  Record of revisions:
%      Date       Programmer          Description of change
%      ====       ==========          =====================
%    01/08/05    S. J. Chapman        Original code
%
```

```
% Define variables:
%   ii        -- Loop index
%   n         -- The number of input samples
%   std_dev   -- The standard deviation of the input samples
%   sum_x     -- The sum of the input values
%   sum_x2    -- The sum of the squares of the input values
%   x         -- An input data value
%   xbar      -- The average of the input samples

% Initialize sums.
sum_x = 0; sum_x2 = 0;

% Get the number of points to input.
n = input('Enter number of points: ');

% Check to see if we have enough input data.
if n < 2          % Insufficient data

  disp ('At least 2 values must be entered.');

else % we will have enough data, so let's get it.

  % Loop to read input values.
  for ii = 1:n

     % Read in next value
     x = input('Enter value:  ');

     % Accumulate sums.
     sum_x  = sum_x + x;
     sum_x2 = sum_x2 + x^2;

  end

  % Now calculate statistics.
  x_bar = sum_x / n;
  std_dev = sqrt( (n * sum_x2 - sum_x^2) / (n * (n-1)) );

  % Tell user.
  fprintf('The mean of this data set is:    %f\n', x_bar);
  fprintf('The standard deviation is:       %f\n', std_dev);
  fprintf('The number of data points is:    %f\n', n);

end
```

4.2.1 Details of Operation

Now that we have seen examples of a for loop in operation, we must examine some important details required to use for loops properly.

1. **Indent the bodies of loops.** It is not necessary to indent the body of a for loop as we have shown above. MATLAB will recognize the loop even if every statement in it starts in column 1. However, the code is much more readable if the body of the for loop is indented, so you should always indent the bodies of loops.

✱ Good Programming Practice

Always indent the body of a for loop by two or more spaces to improve the readability of the code.

2. **Don't modify the loop index within the body of a loop.** The loop index of a for loop *should not be modified anywhere within the body of the loop*. The index variable is often used as a counter within the loop, and modifying its value can cause strange and hard-to-find errors. The example shown below is intended to initialize the elements of an array, but the statement "ii = 5" has been accidentally inserted into the body of the loop. As a result, only a(5) is initialized, and it gets the values that should have gone into a(1), a(2), etc.

```
for ii = 1:10
   ...
   ii = 5;      % Error!
   ...
   a(ii) = <calculation>
end
```

✱ Good Programming Practice

Never modify the value of a loop index within the body of the loop.

3. **Preallocating Arrays.** We learned in Chapter 2 that it is possible to extend an existing array simply by assigning a value to a higher array element. For example, the statement

```
arr = 1:4;
```

defines a 4-element array containing the values [1 2 3 4]. If the statement

```
arr(8) = 6;
```

is executed, the array will be automatically extended to eight elements, and will contain the values [1 2 3 4 0 0 0 6]. Unfortunately, each time that an array is extended, MATLAB has to (1) create a new array, (2) copy the contents of the old array to the new longer array, (3) add the new value to the array, and (4) delete the old array. This process is very time-consuming for long arrays.

When a for loop stores values in a previously undefined array, the loop forces MATLAB to go through this process each time the loop is executed. On the other hand, if the array is **preallocated** to its maximum size before the loop starts executing, no copying is required, and the code executes much faster. The code fragment shown below shows how to preallocate an array before the starting the loop.

```
square = zeros(1,100);
for ii = 1:100
   square(ii) = ii^2;
end
```

✳ Good Programming Practice

Always preallocate all arrays used in a loop before executing the loop. This practice greatly increases the execution speed of the loop.

4. **Vectorizing Arrays.** It is often possible to perform calculations with either for loops or vectors. For example, the following code fragment calculates the squares, square roots, and cube roots of all integers between 1 and 100 using a for loop.

```
for ii = 1:100
   square(ii) = ii^2;
   square_root(ii) = ii^(1/2);
   cube_root(ii) = ii^(1/3);
end
```

The following code fragment performs the same calculation with vectors.

```
ii = 1:100;
square = ii.^2;
square_root = ii.^(1/2);
cube_root(ii) = ii.^(1/3);
```

Even though these two calculations produce the same answers, they are *not* equivalent. The version with the `for` loop can be *more than 15 times slower* than the vectorized version! This happens because the statements in the `for` loop must be interpreted[1] and executed a line at a time by MATLAB during each pass of the loop. In effect, MATLAB must interpret and execute 300 separate lines of code. In contrast, MATLAB has to interpret and execute only four lines in the vectorized case. Since MATLAB is designed to implement vectorized statements in a very efficient fashion, it is much faster in that mode.

In MATLAB, the process of replacing loops by vectorized statements is known as **vectorization**. Vectorization can yield dramatic improvements in performance for many MATLAB programs.

✱ Good Programming Practice

If it is possible to implement a calculation either with a `for` loop or, using vectors, to implement the calculation with vectors. Your program will be much faster.

4.2.2 The MATLAB Just-In-Time (JIT) Compiler

A just-in-time (JIT) compiler was added to MATLAB 6.5 and later versions. The JIT compiler examines MATLAB code before it is executed, and, where possible, compiles the code before executing it. Since the MATLAB code is compiled instead of being interpreted, it runs almost as fast as vectorized code. The JIT compiler can sometimes dramatically speed up the execution of `for` loops.

The JIT compiler is a very nice tool when it works, since it speeds up the loops without any action by the programmer. However, the JIT compiler has many limitations that prevent it from speeding up all loops. A full list of JIT compiler limitations appears in the MATLAB documentation, but some of the more important limitations are:

1. The JIT accelerates only loops containing `double`, `logical`, and `char` data types (plus integer data types that are not discussed in this book). If other data types such as cell arrays or structures[2] appear in the loop, it will not be accelerated.
2. If an array in the loop has more than two dimensions, the loop will not be accelerated.
3. If the code in the loop calls external functions (other than built-in functions), it will not be accelerated.

[1]But see the next item about the MATLAB Just-In-Time compiler.

[2]We will learn about these data types in Chapter 7.

4. If the code in the loop changes the data type of a variable within a loop, the loop will not be accelerated.

Because of these limitations, a good programmer using vectorization can almost always create a faster program than one relying on the JIT compiler.

✱ Good Programming Practice

Do not rely on the JIT compiler to speed up your code. It has many limitations, and a programmer can typically do a better job with manual vectorization.

Example 4.5—Comparing Loops and Vectors

To compare the execution speeds of loops and vectors, perform and time the following four sets of calculations.

1. Calculate the squares of every integer from 1 to 10,000 in a `for` loop without initializing the array of squares first.
2. Calculate the squares of every integer from 1 to 10,000 in a `for` loop, using the zeros function to preallocate the array of squares first, but calling an external function to perform the squaring. (This will disable the JIT compiler.)
3. Calculate the squares of every integer from 1 to 10,000 in a `for` loop, using the zeros function to preallocate the array of squares first, and calculating the square of the number in-line. (This will allow the JIT compiler to function.)
4. Calculate the squares of every integer from 1 to 10,000 with vectors.

SOLUTION This program must calculate the squares of the integers from 1 to 10,000 in each of the four ways described above, timing the executions in each case. The timing can be accomplished using the MATLAB functions `tic` and `toc`. Function `tic` resets the built-in elapsed time counter, and function `toc` returns the elapsed time in seconds since the last call to function `tic`.

Since the real-time clocks in many computers have a fairly coarse granularity, it may be necessary to execute each set of instructions multiple times to get a valid average time.

A MATLAB program to compare the speeds of the four approaches is shown below:

```
% Script file: timings.m
%
% Purpose:
```

```
%   This program calculates the time required to
%   calculate the squares of all integers from 1 to
%   10,000 in four different ways:
%    1. Using a for loop with an uninitialized output
%       array.
%    2. Using a for loop with a pre-allocated output
%       array and NO JIT compiler.
%    3. Using a for loop with a pre-allocated output
%       array and the JIT compiler.
%    4. Using vectors.
%
% Record of revisions:
%     Date        Programmer     Description of change
%     ====        ==========     =====================
%   01/09/05    S. J. Chapman    Original code
%
% Define variables:
%   ii, jj     -- Loop index
%   average1   -- Average time for calculation 1
%   average2   -- Average time for calculation 2
%   average3   -- Average time for calculation 3
%   average4   -- Average time for calculation 4
%   maxcount   -- Number of times to loop calculation
%   square     -- Array of squares

% Perform calculation with an uninitialized array
% "square". This calculation is done only once
% because it is so slow.
maxcount = 1;                 % Number of repetitions
tic;                          % Start timer
for jj = 1:maxcount
   clear square               % Clear output array
   for ii = 1:10000
      square(ii) = ii^2;      % Calculate square
   end
end
average1 = (toc)/maxcount;    % Calculate average time

% Perform calculation with a pre-allocated array
% "square", calling an external function to square
% the number. This calculation is averaged over 10
% loops.
maxcount = 10;                % Number of repetitions
tic;                          % Start timer
```

```
      for jj = 1:maxcount
         clear square                  % Clear output array
         square = zeros(1,10000);      % Pre-initialize array
         for ii = 1:10000
            square(ii) = sqr(ii);      % Calculate square
         end
      end
      average2 = (toc)/maxcount;       % Calculate average time

      % Perform calculation with a pre-allocated array
      % "square". This calculation is averaged over 100
      % loops.
      maxcount = 100;                  % Number of repetitions
      tic;                             % Start timer
      for jj = 1:maxcount
         clear square                  % Clear output array
         square = zeros(1,10000);      % Pre-initialize array
         for ii = 1:10000
            square(ii) = ii^2;         % Calculate square
         end
      end
      average3 = (toc)/maxcount;       % Calculate average time

      % Perform calculation with vectors. This calculation
      % averaged over 1000 executions.
      maxcount = 1000;                 % Number of repetitions
      tic;                             % Start timer
      for jj = 1:maxcount
         clear square                  % Clear output array
         ii = 1:10000;                 % Set up vector
         square = ii.^2;               % Calculate square
      end
      average4 = (toc)/maxcount;       % Calculate average time

% Display results
fprintf('Loop / uninitialized array          = %8.4f\n', average1);
fprintf('Loop / initialized array / no JIT   = %8.4f\n', average2);
fprintf('Loop / initialized array / JIT      = %8.4f\n', average3);
fprintf('Vectorized                          = %8.4f\n', average4);
```

When this program is executed using MATLAB 7.0 on a 2.4 GHz Pentium IV computer, the results are:

```
» timings
Loop / uninitialized array          = 0.1100
Loop / initialized array / no JIT   = 0.1797
```

```
Loop / initialized array / JIT      = 0.0005
Vectorized                          = 0.0001
```

The loop with the initialized array and the loop with the initialized array but no JIT were very slow compared the loop executed with the JIT compiler or the vectorized loop. The vectorized loop was fastest way to perform the calculation, but if the JIT compiler works for your loop, you get most of the acceleration without having to do anything! As you can see, designing loops to allow the JIT compiler to function or replacing the loops with vectorized calculations can make an incredible difference in the speed of your MATLAB code!

4.2.3 The `break` and `continue` Statements

There are two additional statements that can be used to control the operation of `while` loops and `for` loops: the `break` and `continue` statements. The `break` statement terminates the execution of a loop and passes control to the next statement after the end of the loop, and the `continue` statement terminates the current pass through the loop and returns control to the top of the loop.

If a `break` statement is executed in the body of a loop, the execution of the body will stop and control will be transferred to the first executable statement after the loop. An example of the `break` statement in a `for` loop is shown below.

```
for ii = 1:5
  if ii == 3;
     break;
  end
  fprintf('ii = %d\n',ii);
end
disp(['End of loop!']);
```

When this program is executed, the output is

```
» test_break
ii = 1
ii = 2
End of loop!
```

Note that the `break` statement was executed on the iteration when `ii` was 3, and control transferred to the first executable statement after the loop without executing the `fprintf` statement.

If a `continue` statement is executed in the body of a loop, the execution of the current pass through the loop will stop and control will return to the top of the loop. The controlling variable in the `for` loop will take on its next value, and the loop will be executed again. An example of the continue statement in a for loop is shown below.

```
for ii = 1:5
   if ii == 3;
      continue;
   end
   fprintf('ii = %d\n',ii);
end
disp(['End of loop!']);
```

When this program is executed, the output is:

```
» test_continue
ii = 1
ii = 2
ii = 4
ii = 5
End of loop!
```

Note that the continue statement was executed on the iteration when ii was 3, and control transferred to the top of the loop without executing the fprintf statement.

The break and continue statements work with both while loops and for loops.

4.2.4 Nesting Loops

It is possible for one loop to be completely inside another loop. If one loop is completely inside another one, the two loops are called **nested loops**. The following example shows two nested for loops used to calculate and write out the product of two integers.

```
for ii = 1:3
   for jj = 1:3
      product = ii * jj;
      fprintf('%d * %d = %d\n',ii,jj,product);
   end
end
```

In this example, the outer for loop will assign a value of 1 to index variable ii, and then the inner for loop will be executed. The inner for loop will be executed three times with index variable jj having values 1, 2, and 3. When the entire inner for loop has been completed, the outer for loop will assign a value of 2 to index variable ii, and the inner for loop will be executed again. This process repeats until the outer for loop has executed three times, and the resulting output is

```
1 * 1 = 1
1 * 2 = 2
1 * 3 = 3
2 * 1 = 2
2 * 2 = 4
2 * 3 = 6
3 * 1 = 3
3 * 2 = 6
3 * 3 = 9
```

Note that the inner for loop executes completely before the index variable of the outer for loop is incremented.

When MATLAB encounters an end *statement, it associates that statement with the innermost currently open construct.* Therefore, the first end statement above closes the "for jj = 1:3" loop, and the second end statement above closes the "for ii = 1:3" loop. This fact can produce hard-to-find errors if an end statement is accidentally deleted somewhere within a nested loop construct.

If for *loops are nested, they should have independent loop index variables.* If they have the same index variable, the inner loop will change the value of the loop index that the outer loop just set.

If a break or continue statement appears inside a set of nested loops, that statement refers to the *innermost* of the loops containing it. For example, consider the following program

```
for ii = 1:3
   for jj = 1:3
      if jj == 3;
         break;
      end
      product = ii * jj;
      fprintf('%d * %d = %d\n',ii,jj,product);
   end
   fprintf('End of inner loop\n');
end
fprintf('End of outer loop\n');
```

If the inner loop counter jj is equal to 3, the break statement will be executed. This will cause the program to exit the innermost loop. The program will print out "End of inner loop," the index of the outer loop will be increased by 1, and execution of the innermost loop will start over. The resulting output values are

```
1 * 1 = 1
1 * 2 = 2
End of inner loop
2 * 1 = 2
2 * 2 = 4
End of inner loop
3 * 1 = 3
3 * 2 = 6
End of inner loop
End of outer loop
```

4.3 Logical Arrays and Vectorization

We learned about the `logical` data type in Chapter 3. Logical data can have one of two possible values: `true` (1) or `false` (0). Scalars and arrays of `logical` data are created as the output of relational and logic operators.

For example, consider the following statements:

```
a = [1 2 3; 4 5 6; 7 8 9];
b = a > 5;
```

These statements produced two arrays `a` and `b`. Array `a` is a `double` array containing the values $\begin{bmatrix} 1 & 2 & 3 \\ 4 & 5 & 6 \\ 7 & 8 & 9 \end{bmatrix}$, while array `b` is a `logical` array with the logical property set, containing the values $\begin{bmatrix} 0 & 0 & 0 \\ 0 & 0 & 1 \\ 1 & 1 & 1 \end{bmatrix}$. When the `whos` command is executed, the results are as follows.

```
» whos
  Name      Size        Bytes      Class

  a         3x3            72      double array
  b         3x3             9      logical array

Grand total is 18 elements using 81 bytes
```

Logical arrays have a very important special property—*they can serve as a* ***mask*** *for arithmetic operations*. A mask is an array that selects the elements of another array for use in an operation. The specified operation will be applied to the selected elements, and *not* to the remaining elements.

For example, suppose that arrays `a` and `b` are as defined above. Then the statement `a(b) = sqrt(a(b))` will take the square root of all elements for which the logical array b is true, and leave all the other elements in the array unchanged.

```
» a(b) = sqrt(a(b))
a =
    1.0000    2.0000    3.0000
    4.0000    5.0000    2.4495
    2.6458    2.8284    3.0000
```

This is a very fast and very clever way of performing an operation on a subset of an array without needing loops and branches.

The following two code fragments both take the square root of all elements in array a whose value is greater than 5, but the vectorized approach is much faster than the loop approach.

```
for ii = 1:size(a,1)
   for jj = 1:size(a,2)
      if a(ii,jj) > 5
         a(ii,jj) = sqrt(a(ii,jj));
      end
   end
end

b = a > 5;
a(b) = sqrt(a(b));
```

Example 4.6—Using Logical Arrays to Mask Operations

To compare the execution speeds of loops and branches versus vectorized code using a logical array, we will perform and time the following two sets of calculations.

1. Create a 10,000-element array containing the values, 1, 2, . . . , 10,000. Then take the square root of all elements whose value is greater than 5,000 using a for loop and an if construct.
2. Create a 10,000-element array containing the values, 1, 2, . . . , 10,000. Then take the square root of all elements whose value is greater than 5,000 using a logical array.

SOLUTION This program must create an array containing the integers from 1 to 10,000, and take the square roots of those values that are greater than 5,000 in each of the two ways previously described.

A MATLAB program to compare the speeds of the two approaches is shown below:

```
% Script file: logical1.m
%
% Purpose:
```

```
%   This program calculates the time required to
%   calculate the square roots of all elements in
%    array a whose value exceeds 5000. This is done
%    in two different ways:
%    1. Using a for loop and if construct.
%    2. Using a logical array.
%
% Record of revisions:
%      Date        Programmer       Description of change
%      ====        ==========       =====================
%   01/10/05     S. J. Chapman    Original code
%
% Define variables:
%   a           -- Array of input values
%   b           -- Logical array to serve as a mask
%   ii, jj      -- Loop index
%   average1    -- Average time for calculation 1
%   average2    -- Average time for calculation 2
%   maxcount    -- Number of times to loop calculation


% Perform calculation using loops and branches.
maxcount = 1;                  % One repetition
tic;                           % Start timer
for jj = 1:maxcount
   a = 1:10000;                % Declare array a
   for ii = 1:10000
      if a(ii) > 5000
         a(ii) = sqrt(a(ii));
      end
   end
end
average1 = (toc)/maxcount; % Calculate average time

% Perform calculation using logical arrays.
maxcount = 10;                 % One repetition
tic;                           % Start timer
for jj = 1:maxcount
   a = 1:10000;                % Declare array a
   b = a > 5000;               % Create mask
   a(b) = sqrt(a(b));          % Take square root
end
average2 = (toc)/maxcount; % Calculate average time
```

```
% Display results
fprintf('Loop / if approach =         %8.4f\n', ...
        average1);
fprintf('Logical array approach =  %8.4f\n', ...
        average2);
```

When this program is executed using MATLAB 7.0 on a 2.4 GHz Pentium IV computer, the results are:

```
» logical1
Loop / if approach =        0.1200
Logical array approach =  0.0060
```

As you can see, the use of logical arrays can speed up code execution by a factor of 20!

◄

✷ Good Programming Practice

Where possible, use logical arrays as masks to select the elements of an array for processing. If logical arrays are used instead of loops and `if` constructs, your program will be much faster.

4.3.1 Creating the Equivalent of `if/else` Constructs with Logical Arrays

Logical arrays can also be used to implement the equivalent of an `if/else` construct inside a set of `for` loops. As we saw in the last section, it is possible to apply an operation to selected elements of an array using a logical array as a mask. It is also possible to apply a different set of operations to the *unselected* elements of the array by simply adding the not operator (~) to the logical mask. For example, suppose that we wanted to take the square root of any elements in a two-dimensional array whose value is greater than 5, and to square the remaining elements in the array. The code for this operation using loops and branches is

```
for ii = 1:size(a,1)
   for jj = 1:size(a,2)
      if a(ii,jj) > 5
         a(ii,jj) = sqrt(a(ii,jj));
      else
         a(ii,jj) = a(ii,jj)^2;
      end
   end
end
```

The vectorized code for this operation is

```
b = a > 5;
a(b) = sqrt(a(b));
a(~b) = a(~b).^2;
```

The vectorized code is enormously faster than the loops-and-branches version.

Quiz 4.1

This quiz provides a quick check to see if you have understood the concepts introduced in Sections 4.1 through 4.3. If you have trouble with the quiz, reread the section, ask your instructor, or discuss the material with a fellow student. The answers to this quiz are found in the back of the book.

Examine the following `for` loops and determine how many times each loop will be executed.

1. `for index = 7:10`
2. `for jj = 7:-1:10`
3. `for index = 1:10:10`
4. `for ii = -10:3:-7`
5. `for kk = [0 5 ; 3 3]`

Examine the following loops and determine the value in `ires` at the end of each of the loops.

6.
```
ires = 0;
for index = 1:10
   ires = ires + 1;
end
```

7.
```
ires = 0;
for index = 1:10
   ires =  ires + index;
end
```

8.
```
ires = 0;
for index1 = 1:10
   for index2 = index1:10
      if index2 == 6
         break;
      end
      ires = ires + 1;
   end
end
```

9.

```
ires = 0;
for index1 = 1:10
   for index2 = index1:10
      if index2 == 6
         continue;
      end
      ires = ires + 1;
   end
end
```

10. Write the MATLAB statements to calculate the values of the function

$$f(t) = \begin{cases} \sin t & \text{for all } t \text{ where } \sin t > 0 \\ 0 & \text{elsewhere} \end{cases}$$

for $-6\pi \leq t \leq 6\pi$ at intervals of $\pi/10$. Do this twice, once using loops and branches, and once using vectorized code.

4.4 Additional Examples

Example 4.7—Fitting a Line to a Set of Noisy Measurements

The velocity of a falling object in the presence of a constant gravitational field is given by the equation

$$v(t) = at + v_0 \tag{4-3}$$

where $v(t)$ is the velocity at any time t, a is the acceleration due to gravity, and v_0 is the velocity at time 0. This equation is derived from elementary physics—it is known to every freshman physics student. If we plot velocity versus time for the falling object, our (v, t) measurement points should fall along a straight line. However, the same freshman physics student also knows that if we go out into the laboratory and attempt to *measure* the velocity versus time of an object, our measurements will *not* fall along a straight line. They may come close, but they will never line up perfectly. Why not? Because we can never make perfect measurements. There is always some *noise* included in the measurements, which distorts them.

There are many cases in science and engineering where there are noisy sets of data such as this, and we wish to estimate the straight line which "best fits" the data. This problem is called the *linear regression* problem. Given a noisy set of measurements (x, y) that appear to fall along a straight line, how can we find the equation of the line

$$y = mx + b \tag{4-4}$$

which "best fits" the measurements? If we can determine the regression coefficients m and b, we can use this equation to predict the value of y at any given x by evaluating Equation 4-4 for that value of x.

A standard method for finding the regression coefficients m and b is the *method of least squares*. This method is named "least squares" because it produces the line $y = mx + b$ for which the sum of the squares of the differences between the observed y values and the predicted y values is as small as possible. The slope of the least squares line is given by

$$m = \frac{(\sum xy) - (\sum x)\bar{y}}{(\sum x^2) - (\sum x)\bar{x}} \tag{4-5}$$

and the intercept of the least squares line is given by

$$b = \bar{y} - m\bar{x} \tag{4-6}$$

where

$\sum x$ is the sum of the x values
$\sum x^2$ is the sum of the squares of the x values
$\sum xy$ is the sum of the products of the corresponding x and y values
$\bar{x}$ is the mean (average) of the x values
$\bar{y}$ is the mean (average) of the y values

Write a program which will calculate the least-squares slope m and y-axis intercept b for a given set of noisy measured data points (x, y). The data points should be read from the keyboard, and both the individual data points and the resulting least-squares fitted line should be plotted.

SOLUTION

1. **State the problem**
   Calculate the slope m and intercept b of a least-squares line that best fits an input data set consisting of an arbitrary number of (x, y) pairs. The input (x, y) data is read from the keyboard. Plot both the input data points and the fitted line on a single plot.

2. **Define the inputs and outputs**
   The inputs required by this program are the number of points to read, plus the pairs of points (x, y).

   The outputs from this program are the slope and intercept of the least-squares fitted line, the number of points going into the fit, and a plot of the input data and the fitted line.

3. **Describe the algorithm**
   This program can be broken down into six major steps

```
Get the number of input data points
Read the input statistics
Calculate the required statistics
```

```
Calculate the slope and intercept
Write out the slope and intercept
Plot the input points and the fitted line
```

The first major step of the program is to get the number of points to read in. To do this, we will prompt the user and read his or her answer with an `input` function. Next we will read the input (x, y) pairs one pair at a time using an `input` function in a `for` loop. Each pair of input value will be placed in an array (`[x y]`), and the function will return that array to the calling program. Note that a `for` loop is appropriate because we know in advance how many times the loop will be executed.

The pseudocode for these steps is shown below:

```
Print message describing purpose of the program
n_points <- input('Enter number of [x y] pairs: ');
for ii = 1:n_points
   temp <- input('Enter [x y] pair: ');
   x(ii) <- temp(1)
   y(ii) <- temp(2)
end
```

Next, we must accumulate the statistics required for the calculation. These statistics are the sums $\sum x$, $\sum y$, $\sum x^2$, and $\sum xy$. The pseudocode for these steps is:

```
Clear the variables sum_x, sum_y, sum_x2, and sum_y2
for ii = 1:n_points
   sum_x <- sum_x + x(ii)
   sum_y <- sum_y + y(ii)
   sum_x2 <- sum_x2 + x(ii)^2
   sum_xy <- sum_xy + x(ii)*y(ii)
end
```

Next, we must calculate the slope and intercept of the least-squares line. The pseudocode for this step is just the MATLAB versions of Equations 4-4 and 4-5.

```
x_bar <- sum_x / n_points
y_bar <- sum_y / n_points
slope <- (sum_xy-sum_x * y_bar)/( sum_x2 - sum_x * x_bar)
y_int <- y_bar - slope * x_bar
```

Finally, we must write out and plot the results. The input data points should be plotted with circular markers and without a connecting line, while the fitted line should be plotted as a solid 2-pixel wide line. To do this, we will need to plot the points first, set `hold on`, plot the fitted line, and set `hold off`. We will add titles and a legend to the plot for completeness.

4. **Turn the algorithm into MATLAB statements.**
   The final MATLAB program is shown below:

```
%
% Purpose:
%   To perform a least-squares fit of an input data set
%   to a straight line, and print out the resulting slope
%   and intercept values. The input data for this fit
%   comes from a user-specified input data file.
%
% Record of revisions:
%       Date       Programmer          Description of change
%       ====       ==========          =====================
%     01/10/05    S. J. Chapman        Original code
%
% Define variables:
%   ii             -- Loop index
%   n_points       -- Number in input [x y] points
%   slope          -- Slope of the line
%   sum_x          -- Sum of all input x values
%   sum_x2         -- Sum of all input x values squared
%   sum_xy         -- Sum of all input x*y values
%   sum_y          -- Sum of all input y values
%   temp           -- Variable to read user input
%   x              -- Array of x values
%   x_bar          -- Average x value
%   y              -- Array of y values
%   y_bar          -- Average y value
%   y_int          -- y-axis intercept of the line

disp('This program performs a least-squares fit of an ');
disp('input data set to a straight line.');
n_points = input('Enter the number of input [x y] points: ');

% Read the input data
for ii = 1:n_points
   temp = input('Enter [x y] pair: ');
   x(ii) = temp(1);
   y(ii) = temp(2);
end

% Accumulate statistics
sum_x = 0;
sum_y = 0;
sum_x2 = 0;
sum_xy = 0;
```

```
for ii = 1:n_points
   sum_x  = sum_x + x(ii);
   sum_y  = sum_y + y(ii);
   sum_x2 = sum_x2 + x(ii)^2;
   sum_xy = sum_xy + x(ii) * y(ii);
end

% Now calculate the slope and intercept.
x_bar = sum_x / n_points;
y_bar = sum_y / n_points;
slope = (sum_xy - sum_x * y_bar) / ( sum_x2 - sum_x * x_bar);
y_int = y_bar - slope * x_bar;

% Tell user.
disp('Regression coefficients for the least-squares line:');
fprintf('  Slope (m)       = %8.3f\n', slope);
fprintf('  Intercept (b)   = %8.3f\n', y_int);
fprintf('  No of points    = %8d\n', n_points);

% Plot the data points as blue circles with no
% connecting lines.
plot(x,y,'bo');
hold on;

% Create the fitted line
xmin = min(x);
xmax = max(x);
ymin = slope * xmin + y_int;
ymax = slope * xmax + y_int;

% Plot a solid red line with no markers
plot([xmin xmax],[ymin ymax],'r-','LineWidth',2);
hold off;

% Add a title and legend
title ('\bfLeast-Squares Fit');
xlabel('\bf\itx');
ylabel('\bf\ity');
legend('Input data','Fitted line');
grid on
```

5. **Test the program.**
   To test this program, we will try a simple data set. For example, if every point in the input data set actually falls along a line, then the resulting slope and intercept should be exactly the slope and intercept of that line.

Thus the data set

```
[1.1 1.1]
[2.2 2.2]
[3.3 3.3]
[4.4 4.4]
[5.5 5.5]
[6.6 6.6]
[7.7 7.7]
```

should produce a slope of 1.0 and an intercept of 0.0. If we run the program with these values, the results are:

```
» lsqfit
This program performs a least-squares fit of an
input data set to a straight line.
Enter the number of input [x y] points: 7
Enter [x y] pair: [1.1 1.1]
Enter [x y] pair: [2.2 2.2]
Enter [x y] pair: [3.3 3.3]
Enter [x y] pair: [4.4 4.4]
Enter [x y] pair: [5.5 5.5]
Enter [x y] pair: [6.6 6.6]
Enter [x y] pair: [7.7 7.7]
Regression coefficients for the least-squares line:
  Slope (m)       =  1.000
  Intercept (b)   =  0.000
  No of points    =  7
```

Now let's add some noise to the measurements. The data set becomes

```
[1.1 1.01]
[2.2 2.30]
[3.3 3.05]
[4.4 4.28]
[5.5 5.75]
[6.6 6.48]
[7.7 7.84]
```

If we run the program with these values, the results are as follows:

```
» lsqfit
This program performs a least-squares fit of an
input data set to a straight line.
Enter the number of input [x y] points: 7
Enter [x y] pair: [1.1 1.01]
Enter [x y] pair: [2.2 2.30]
Enter [x y] pair: [3.3 3.05]
```

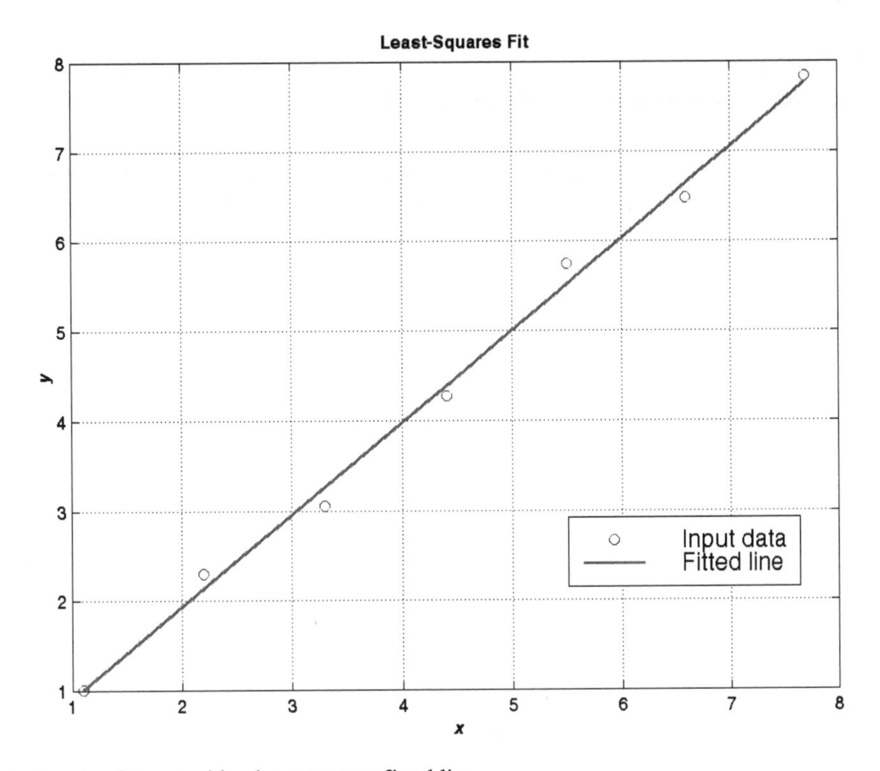

Figure 4.1 A noisy data set with a least-squares fitted line.

```
Enter [x y] pair: [4.4 4.28]
Enter [x y] pair: [5.5 5.75]
Enter [x y] pair: [6.6 6.48]
Enter [x y] pair: [7.7 7.84]
Regression coefficients for the least-squares line:
  Slope (m)       =  1.024
  Intercept (b)   = -0.120
  No of points    =  7
```

If we calculate the answer by hand, it is easy to show that the program gives the correct answers for our two test data sets. The noisy input data set and the resulting least-squares fitted line are shown in Figure 4.1.

This example uses several of the plotting capabilities that we introduced in Chapter 3. It uses the `hold` command to allow multiple plots to be placed on the same axes, the `LineWidth` property to set the width of the least-squares fitted line, and escape sequences to make the title bold face and the axis labels bold italic.

Example 4.8—Physics—The Flight of a Ball

If we assume negligible air friction and ignore the curvature of the Earth, a ball that is thrown into the air from any point on the Earth's surface will follow a parabolic flight path (see Figure 4.2*a*). The height of the ball at any time t after it is thrown is given by Equation 4-7

$$y(t) = y_0 + v_{y0}t + \frac{1}{2}gt^2 \tag{4-7}$$

where y_0 is the initial height of the object above the ground, v_{y0} is the initial vertical velocity of the object, and g is the acceleration due to the Earth's gravity. The horizontal distance (range) traveled by the ball as a function of time after it is thrown is given by Equation 4-8

$$x(t) = x_0 + v_{x0}t \tag{4-8}$$

where x_0 is the initial horizontal position of the ball on the ground, and v_{x0} is the initial horizontal velocity of the ball.

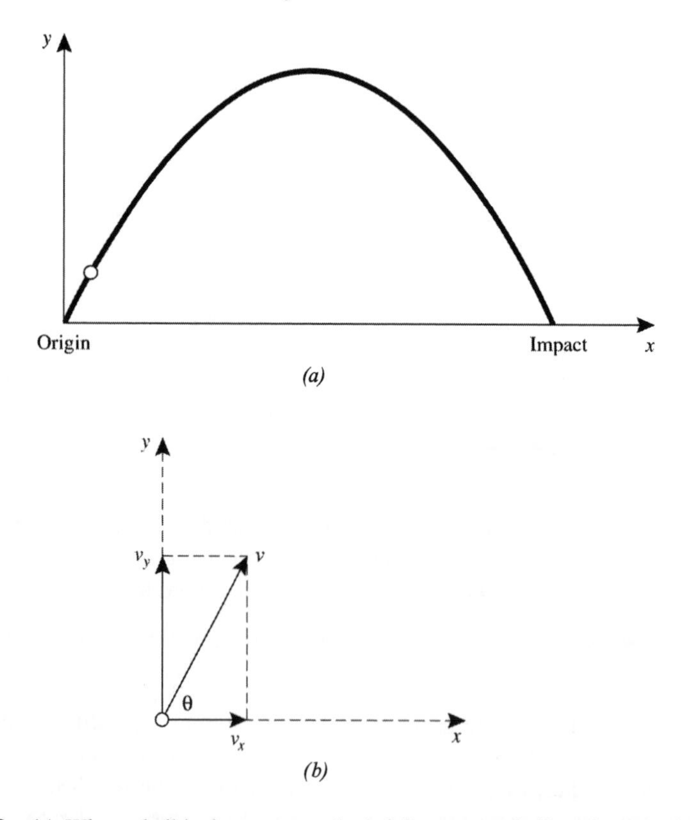

Figure 4.2 *(a)* When a ball is thrown upwards, it follows a parabolic trajectory. *(b)* The horizontal and vertical components of a velocity vector v at an angle θ with respect to the horizontal.

If the ball is thrown with some initial velocity v_0 at an angle of θ degrees with respect to the Earth's surface, then the initial horizontal and vertical components of velocity will be

$$v_{x0} = v_0 \cos \theta \tag{4-9}$$

$$v_{y0} = v_0 \sin \theta \tag{4-10}$$

Assume that the ball is initially thrown from position $(x_0, y_0) = (0, 0)$ with an initial velocity v_0 of 20 meters per second at an initial angle of θ degrees. Write a program that will plot the trajectory of the ball and also determine the horizontal distance traveled before it touches the ground again. The program should plot the trajectories of the ball for all angles θ from 5 to 85° in 10° steps, and should determine the horizontal distance traveled for all angles θ from 0 to 90° in 1° steps. Finally, it should determine the angle θ that maximizes the range of the ball, and plot that particular trajectory in a different color with a thicker line.

SOLUTION To solve this problem, we must determine an equation for the time that the ball returns to the ground. Then, we can calculate the (x, y) position of the ball using Equations 4-7 through 4-10. If we do this for many times between 0 and the time that the ball returns to the ground, we can use those points to plot the ball's trajectory.

The time that the ball will remain in the air after it is thrown may be calculated from Equation 4-7. The ball will touch the ground at the time t for which $y(t) = 0$. Remembering that the ball will start from ground level ($y(0) = 0$), and solving for t, we get:

$$y(t) = y_0 + v_{y0}t + \frac{1}{2}gt^2 \tag{4-7}$$

$$0 = 0 + v_{y0}t + \frac{1}{2}gt^2$$

$$0 = \left(v_{y0} + \frac{1}{2}gt\right)t$$

so the ball will be at ground level at time $t_1 = 0$ (when we threw it), and at time

$$t_2 = -\frac{2v_{y0}}{g} \tag{4-11}$$

From the problem statement, we know that the initial velocity v_0 is 20 meters per second, and that the ball will be thrown at all angles from 0° to 90° in 1° steps. Finally, any elementary physics textbook will tell us that the acceleration due to the earth's gravity is -9.81 meters per second squared.

Now let's apply our design technique to this problem.

1. **State the problem**
   A proper statement of this problem would be: *Calculate the range that a ball would travel when it is thrown with an initial velocity of v_0 of 20 m/s*

at an initial angle θ. Calculate this range for all angles between 0° and 90°, in 1° steps. Determine the angle θ that will result in the maximum range for the ball. Plot the trajectory of the ball for angles between 5° and 85°, in 10° increments. Plot the maximum-range trajectory in a different color and with a thicker line. Assume that there is no air friction.

2. **Define the inputs and outputs**
   As the problem is defined above, no inputs are required. We know from the problem statement what v_0 and θ will be, so there is no need to input them. The outputs from this program will be a table showing the range of the ball for each angle θ, the angle θ for which the range is maximum, and a plot of the specified trajectories.

3. **Design the algorithm**
   This program can be broken down into the following major steps:

```
Calculate the range of the ball for θ between 0 and 90°
Write a table of ranges
Determine the maximum range and write it out
Plot the trajectories for θ between 5 and 85°
Plot the maximum-range trajectory
```

Since we know the exact number of times that the loops will be repeated, `for` loops are appropriate for this algorithm. We will now refine the pseudocode for each of the preceding major steps.

To calculate the maximum range of the ball for each angle, we will first calculate the initial horizontal and vertical velocity from Equations 4-9 and 4-10. Then we will determine the time when the ball returns to Earth from Equation 4-11. Finally, we will calculate the range at that time from Equation 4-7. The detailed pseudocode for these steps is shown below. Note that we must convert all angles to radians before using the trig functions!

```
Create and initialize an array to hold ranges
for ii = 1:91
   theta <- ii - 1
   vxo <- vo * cos(theta*conv)
   vyo <- vo * sin(theta*conv)
   max_time <- -2 * vyo / g
   range(ii) <- vxo * max_time
end
```

Next, we must write a table of ranges. The pseudocode for this step is:

```
Write heading
for ii = 1:91
   theta <- ii - 1
   print theta and range
end
```

The maximum range can be found with the `max` function. Recall that this function returns both the maximum value and its location. The pseudocode for this step is:

```
[maxrange index] <- max(range)
Print out maximum range and angle (=index-1)
```

We will use nested `for` loops to calculate and plot the trajectories. To get all of the plots to appear on the screen, we must plot the first trajectory and then set `hold on` before plotting any other trajectories. After plotting the last trajectory, we must set `hold off`. To perform this calculation, we will divide each trajectory into 21 time steps and find the x and y positions of the ball for each time step. Then, we will plot those (x, y) positions. The pseudocode for this step is as follows:

```
for ii = 5:10:85

   % Get velocities and max time for this angle
   theta <- ii - 1
   vxo <- vo * cos(theta*conv)
   vyo <- vo * sin(theta*conv)
   max_time <- -2 * vyo / g

   Initialize x and y arrays
   for jj = 1:21
      time <- (jj-1) * max_time/20
      x(time) <- vxo * time
      y(time) <- vyo * time + 0.5 * g * time^2
   end
   plot(x,y) with thin green lines
   Set "hold on" after first plot
end
Add titles and axis labels
```

Finally, we must plot the maximum range trajectory in a different color and with a thicker line.

```
vxo <- vo * cos(max_angle*conv)
vyo <- vo * sin(max_angle*conv)
max_time <- -2 * vyo / g

Initialize x and y arrays
for jj = 1:21
   time <- (jj-1) * max_time/20
   x(jj) <- vxo * time
   y(jj) <- vyo * time + 0.5 * g * time^2
end
```

```
plot(x,y) with a thick red line
hold off
```

4. **Turn the algorithm into MATLAB statements.**
   The final MATLAB program is shown below:

```
% Script file: ball.m
%
% Purpose:
%   This program calculates the distance traveled by a
%   ball thrown at a specified angle "theta" and a
%   specified velocity "vo" from a point on the surface of
%   the Earth, ignoring air friction and the Earth's
%   curvature. It calculates the angle yielding maximum
%   range, and also plots selected trajectories.
%
% Record of revisions:
%     Date       Programmer          Description of change
%     ====       ==========          =====================
%   01/10/05    S. J. Chapman        Original code
%
% Define variables:
%   conv         -- Degrees to radians conv factor
%   gravity      -- Accel. due to gravity (m/s^2)
%   ii, jj       -- Loop index
%   index        -- Location of maximum range in array
%   maxangle     -- Angle that gives maximum range (deg)
%   maxrange     -- Maximum range (m)
%   range        -- Range for a particular angle (m)
%   time         -- Time (s)
%   theta        -- Initial angle (deg)
%   traj_time    -- Total trajectory time (s)
%   vo           -- Initial velocity (m/s)
%   vxo          -- X-component of initial velocity (m/s)
%   vyo          -- Y-component of initial velocity (m/s)
%   x            -- X-position of ball (m)
%   y            -- Y-position of ball (m)

% Constants
conv = pi / 180;  % Degrees-to-radians conversion factor
g = -9.81;        % Accel. due to gravity
vo = 20;          % Initial velocity

%Create an array to hold ranges
range = zeros(1,91);
```

```
% Calculate maximum ranges
for ii = 1:91
   theta = ii - 1;
   vxo = vo * cos(theta*conv);
   vyo = vo * sin(theta*conv);
   max_time = -2 * vyo / g;
   range(ii) = vxo * max_time;
end

% Write out table of ranges
fprintf ('Range versus angle theta:\n');
for ii = 1:91
   theta = ii - 1;
   fprintf('  %2d    %8.4f\n',theta, range(ii));
end

% Calculate the maximum range and angle
[maxrange index] = max(range);
maxangle = index - 1;
fprintf ('\nMax range is %8.4f at %2d degrees.\n',...
         maxrange, maxangle);

% Now plot the trajectories
for ii = 5:10:85

   % Get velocities and max time for this angle
   theta = ii;
   vxo = vo * cos(theta*conv);
   vyo = vo * sin(theta*conv);
   max_time = -2 * vyo / g;

   % Calculate the (x,y) positions
   x = zeros(1,21);
   y = zeros(1,21);
   for jj = 1:21
      time = (jj-1) * max_time/20;
      x(jj) = vxo * time;
      y(jj) = vyo * time + 0.5 * g * time^2;
   end
   plot(x,y,'b');
   if ii == 5
      hold on;
   end
end

% Add titles and axis lables
title ('\bfTrajectory of Ball vs Initial Angle \theta');
```

```
xlabel ('\bf\itx \rm\bf(meters)');
ylabel ('\bf\ity \rm\bf(meters)');
axis ([0 45 0 25]);
grid on;

% Now plot the max range trajectory
vxo = vo * cos(maxangle*conv);
vyo = vo * sin(maxangle*conv);
max_time = -2 * vyo / g;

% Calculate the (x,y) positions
x = zeros(1,21);
y = zeros(1,21);
for jj = 1:21
   time = (jj-1) * max_time/20;
   x(jj) = vxo * time;
   y(jj) = vyo * time + 0.5 * g * time^2;
end
plot(x,y,'r','LineWidth',3.0);
hold off
```

The acceleration due to gravity at sea level can be found in any physics text. It is about 9.81 m/sec^2, directed downward.

5. **Test the program.**
To test this program, we will calculate the answers by hand for a few of the angles, and compare the results with the output of the program.

θ	$v_{x0} = v_0 \cos\theta$	$v_{y0} = v_0 \sin\theta$	$t_2 = -\frac{2v_{y0}}{g}$	$x = v_{x0}t_2$
0°	20 m/s	0 m/s	0 s	0 m
5°	19.92 m/s	1.74 m/s	0.355 s	7.08 m
40°	15.32 m/s	12.86 m/s	2.621 s	40.15 m
45°	14.14 m/s	14.14 m/s	2.883 s	40.77 m

When program `ball` is executed, a 91-line table of angles and ranges is produced. To save space, only a portion of the table is reproduced here.

```
» ball
Range versus angle theta:
   0     0.0000
   1     1.4230
```

```
        2      2.8443
        3      4.2621
        4      5.6747
        5      7.0805
...
       40     40.1553
       41     40.3779
       42     40.5514
       43     40.6754
       44     40.7499
       45     40.7747
       46     40.7499
       47     40.6754
       48     40.5514
       49     40.3779
       50     40.1553
...

       85      7.0805
       86      5.6747
       87      4.2621
       88      2.8443
       89      1.4230
       90      0.0000

Max range is  40.7747 at 45 degrees.
```

The resulting plot is shown in Figure 4.3 on the next page. The program output matches our hand calculation for the angles calculated above to the 4-digit accuracy of the hand calculation. Note that the maximum range occurred at an angle of 45°.

◄

This example uses several of the plotting capabilities that we introduced in Chapter 3. It uses the `axis` command to set the range of data to display, the `hold` command to allow multiple plots to be placed on the same axes, the `LineWidth` property to set the width of the line corresponding to the maximum-range trajectory, and escape sequences to create the desired title and x- and y-axis labels.

However, this program is not written in the most efficient manner, since there are a number of loops that could have been better replaced by vectorized statements. You will be asked to rewrite and improve `ball.m` in Exercise 4.11 at the end of this chapter.

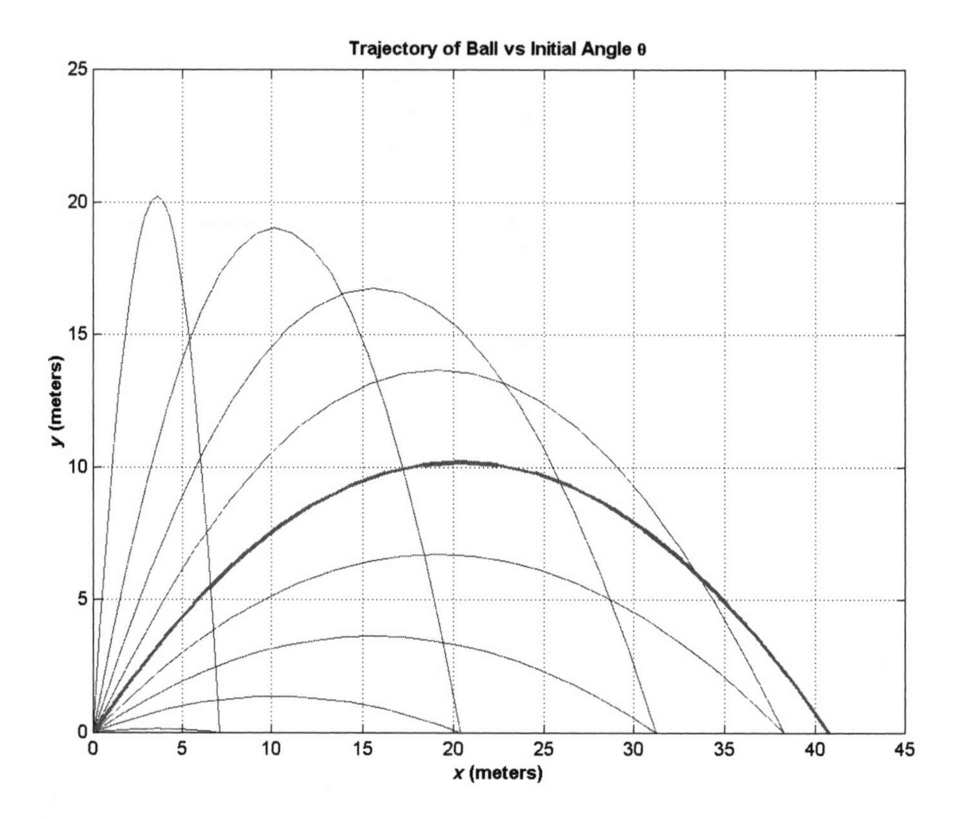

Figure 4.3 Possible trajectories for the ball.

4.5 Summary

There are two basic types of loops in MATLAB, the `while` loop and the `for` loop. The `while` loop is used to repeat a section of code in cases where we do not know in advance how many times the loop must be repeated. The `for` loop is used to repeat a section of code in cases where we know in advance how many times the loop should be repeated. It is possible to exit from either type of loop at any time using the `break` statement.

4.5.1 Summary of Good Programming Practice

The following guidelines should be adhered to when programming with loop constructs. By following them consistently, your code will contain fewer bugs, will be easier to debug, and will be more understandable to others who may need to work with it in the future.

1. Always indent code blocks in `while` and `for` constructs to make them more readable.
2. Use a `while` loop to repeat sections of code when you don't know in advance how often the loop will be executed.

3. Use a `for` loop to repeat sections of code when you know in advance how often the loop will be executed.
4. Never modify the values of a `for` loop index while inside the loop.
5. Always preallocate all arrays used in a loop before executing the loop. This practice greatly increases the execution speed of the loop.
6. If it is possible to implement a calculation either with a `for` loop or using vectors, implement the calculation with vectors. Your program will be much faster.
7. Do not rely on the JIT compiler to speed up your code. It has many limitations, and a programmer can typically do a better job with manual vectorization.
8. Where possible, use logical arrays as masks to select the elements of an array for processing. If logical arrays are used instead of loops and `if` constructs, your program will be much faster.

4.5.2 MATLAB Summary

The following summary lists all of the MATLAB commands and functions described in this chapter, along with a brief description of each one.

Commands and Functions

`break`	Stop the execution of a loop, and transfer control to the first statement after the end of the loop.
`continue`	Stop the execution of a loop, and transfer control to the top of the loop for the next iteration.
`for` loop	Loops over a block of statements a specified number of times.
`tic`	Resets elapsed time counter.
`toc`	Returns elapsed time since last call to `tic`.
`while` loop	Loops over a block of statements until a test condition becomes 0 (false).

4.6 Exercises

4.1 Write the MATLAB statements required to calculate $y(t)$ from the equation

$$y(t) = \begin{cases} -3t^2 + 5 & t \geq 0 \\ 3t^2 + 5 & t < 0 \end{cases}$$

for values of t between -9 and 9 in steps of 0.5. Use loops and branches to perform this calculation.

4.2 Rewrite the statements required to solve Exercise 4.1 using vectorization.

4.3 Write the MATLAB statements required to calculate and print out the squares of all the even integers between 0 and 50. Create a table consisting of each integer and its square, with appropriate labels over each column.

4.4 Write an M-file to evaluate the equation $y(x) = x^2 - 3x + 2$ for all values of x between -1 and 3, in steps of 0.1. Do this twice, once with a `for` loop and once with vectors. Plot the resulting function using a 3-point thick dashed red line.

4.5 Write an M-file to calculate the factorial function N!, as defined in Example 4.2. Be sure to handle the special case of 0! Also, be sure to report an error if N is negative or not an integer.

4.6 Examine the following `for` statements and determine how many times each loop will be executed.

(a) `for ii = -32768:32767`
(b) `for ii = 32768:32767`
(c) `for kk = 2:4:3`
(d) `for jj = ones(5,5)`

4.7 Examine the following `for` loops and determine the value of `ires` at the end of each of the loops, and also the number of times each loop executes.

(a)
```
ires = 0;
for index = -10:10
   ires = ires + 1;
end
```

(b)
```
ires = 0;
for index = 10:-2:4
   if index == 6
      continue;
   end
   ires = ires + index;
end
```

(c)
```
ires = 0;
for index = 10:-2:4
   if index == 6
      break;
   end
   ires = ires + index;
end
```

(d)
```
ires = 0;
for index1 = 10:-2:4
   for index2 = 2:2:index1
      if index2 == 6
         break
      end
      ires = ires + index2;
   end
end
```

4.8 Examine the following `while` loops and determine the value of `ires` at the end of each of the loops, and the number of times each loop executes.

(a)
```
ires = 1;
while mod(ires,10) ~= 0
   ires = ires + 1;
end
```

(b)
```
ires = 2;
while ires <= 200
   ires = ires^2;
end
```

(c)
```
ires = 2;
while ires > 200
   ires = ires^2;
end
```

4.9 What is contained in array `arr1` after each of the following sets of statements are executed?

(a)
```
arr1 = [1 2 3 4; 5 6 7 8; 9 10 11 12];
mask = mod(arr1,2) == 0;
arr1(mask) = -arr1(mask);
```

(b)
```
arr1 = [1 2 3 4; 5 6 7 8; 9 10 11 12];
arr2 = arr1 <= 5;
arr1(arr2) = 0;
arr1(~arr2) = arr1(~arr2).^2;
```

4.10 How can a logical array be made to behave as a logical mask for vector operations?

4.11 Modify program `ball` from Example 4.8 by replacing the inner `for` loops with vectorized calculations.

4.12 Modify program `ball` from Example 4.8 to read in the acceleration due to gravity at a particular location, and to calculate the maximum range of the ball for that acceleration. After modifying the program, run it with accelerations of -9.8 m/sec^2, -9.7 m/sec^2, and -9.6 m/sec^2. What effect does the reduction in gravitational attraction have on the range of the ball? What effect does the reduction in gravitational attraction have on the best angle θ at which to throw the ball?

4.13 Modify program `ball` from Example 4.8 to read in the initial velocity with which the ball is thrown. After modifying the program, run it with initial velocities of 10 m/sec, 20 m/sec, and 30 m/sec. What effect does changing the initial velocity v_0 have on the range of the ball? What effect does it have on the best angle θ at which to throw the ball?

4.14 Program `lsqfit` from Example 4.7 required the user to specify the number of input data points before entering the values. Modify the program so that it reads an arbitrary number of data values using a `while` loop, and stops reading input values when the user pressed the Enter key without typing any values. Test your program using the same two data sets that were used in Example 4.7. (*Hint:* The `input` function returns an empty array (`[]`) if a user presses Enter without supplying any data. You can use function `isempty` to test for an empty array, and stop reading data when one is detected.)

4.15 Modify program `lsqfit` from Example 4.7 to read its input values from an ASCII file named `input1.dat`. The data in the file will be organized in rows, with one pair of (x, y) values on each row, as follows:

```
1.1   2.2
2.2   3.3
...
```

Test your program using the same two data sets that were used in Example 4.6. (*Hint:* Use the `load` command to read the data into an array named `input1`, and then store the first column of `input1` into array `x` and the second column of `input1` into array `y`.)

4.16 **MATLAB Least-Squares Fit Function** MATLAB includes a standard function that performs a least-squares fit to a polynomial. Function `polyfit` calculates the least-squares fit of a data set to a polynomial of order N:

$$p(x) = a_n x^n + a_{n-1} x^{n-1} + \cdots + a_1 x + a_0 \tag{4-12}$$

where N can be any value greater than or equal to 1. Note that for $N = 1$, this polynomial is a linear equation, with the slope being the coefficient a_1 and the y-intercept being the coefficient a_0. The form of this function is

```
p = polyfit(x,y,n)
```

where `x` and `y` are vectors of x and y components, and `n` is the order of the fit.

Write a program that calculates the least-squares fit of a data set to a straight line using `polyfit`. Plot the input data points and the resulting fitted line. Compare the produced by the program using polyfit with the result produced by `lsqfit` for the input data set in Example 4.6.

4.17 Program `doy` in Example 4.3 calculates the day of year associated with any given month, day, and year. As written, this program does not check to see if the data entered by the user is valid. It will accept nonsense values for months and days, and do calculations with them to produce meaningless results. Modify the program so that it checks the input values for validity before using them. If the inputs are invalid, the program should tell the user what is wrong, and quit. The year should be a number greater than zero, the month should be a number between 1 and 12, and the day should be a number between 1 and a maximum that depends on the month. Use a `switch` construct to implement the bounds checking performed on the day.

4.18 Write a MATLAB program to evaluate the function

$$y(x) = \ln\frac{1}{1 - x}$$

for any user-specified value of x, where ln is the natural logarithm (logarithm to the base e). Write the program with a `while` loop, so that the program repeats the calculation for each legal value of x entered into the program. When an illegal value of x is entered, terminate the program. (Any $x \geq 1$ is considered an illegal value.)

4.19 **Fibonacci Numbers** The nth Fibonacci number is defined by the following recursive equations:

$$\begin{aligned} f(1) &= 1 \\ f(2) &= 2 \\ f(n) &= f(n-1) + f(n-2) \end{aligned}$$

Therefore, $f(3) = f(2) + f(1) = 2 + 1 = 3$, and so forth for higher numbers. Write an M-file to calculate and write out the nth Fibonacci number for $n > 2$, where n is input by the user. Use a `while` loop to perform the calculation.

4.20 **Current Through a Diode** The current flowing through the semiconductor diode shown in Figure 4.4 is given by the equation

$$i_D = I_0\left(e^{\frac{qv_D}{kT}} - 1\right) \tag{4-13}$$

where i_D = the voltage across the diode, in volts
v_D = the current flow through the diode, in amps
I_0 = the leakage current of the diode, in amps
q = the charge on an electron, 1.602×10^{-19} coulombs
k = Boltzmann's constant, 1.38×10^{-23} joule/K
T = temperature, in kelvins (K)

The leakage current I_0 of the diode is 2.0 μA. Write a program to calculate the current flowing through this diode for all voltages from -1.0 V to $+0.6$ V, in 0.1 V steps. Repeat this process for the following temperatures: 75°F and 100°F, and 125°F. Create a plot of the current as a function of

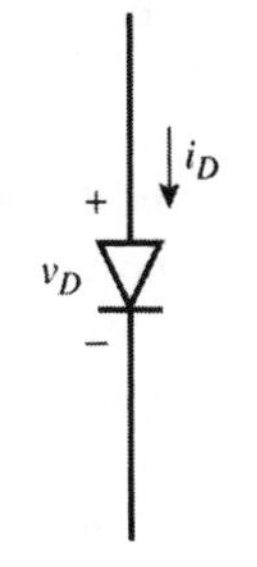

Figure 4.4 A semiconductor diode.

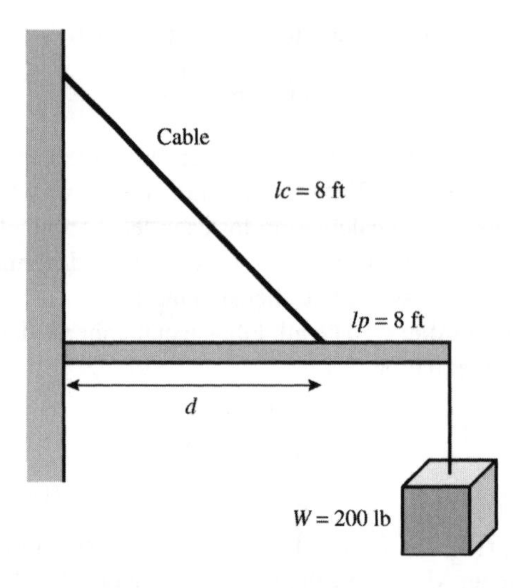

Figure 4.5 A 200-pound weight suspended from a rigid bar supported by a cable.

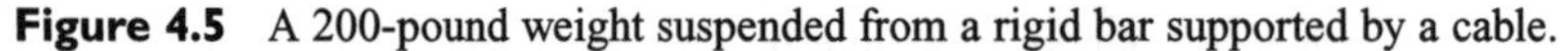

applied voltage, with the curves for the three different temperatures appearing as different colors.

4.21 Tension on a Cable A 200-pound object is to be hung from the end of a rigid 8-foot horizontal pole of negligible weight, as shown in Figure 4.5. The pole is attached to a wall by a pivot and is supported by an 8-foot cable that is attached to the wall at a higher point. The tension on this cable is given by the equation

$$T = \frac{W \cdot lc \cdot lp}{d\sqrt{lp^2 - d^2}} \tag{4-14}$$

where T is the tension on the cable, W is the weight of the object, lc is the length of the cable, lp is the length of the pole, and d is the distance along the pole at which the cable is attached. Write a program to determine the distance d at which to attach the cable to the pole in order to minimize the tension on the cable. To do this, the program should calculate the tension on the cable at regular one-foot intervals from $d = 1$ foot to $d = 7$ feet, and should locate the position d that produces the minimum tension. Also, the program should plot the tension on the cable as a function of d, with appropriate titles and axis labels.

4.22 Bacterial Growth Suppose that a biologist performs an experiment in which he or she measures the rate at which a specific type of bacterium reproduces asexually in different culture media. The experiment shows that in Medium A the bacteria reproduce once every 60 minutes, and in Medium B the bacteria reproduce once every 90 minutes. Assume that a single bacterium is placed on each culture medium at the beginning of the experiment. Write a program that calculates and plots the number of bacteria present in each culture at intervals of three hours from the beginning of the experiment until 24 hours have elapsed. Make two plots, one a linear *xy* plot and the other a linear-log (`semilogy`) plot. How do the numbers of bacteria compare on the two media after 24 hours?

4.23 Decibels Engineers often measure the ratio of two power measurements in *decibels*, or dB. The equation for the ratio of two power measurements in decibels is

$$\mathrm{dB} = 10 \log_{10} \frac{P_2}{P_1} \tag{4-15}$$

where P_2 is the power level being measured, and P_1 is some reference power level. Assume that the reference power level P_1 is 1 watt, and write a program that calculates the decibel level corresponding to power levels between 1 and 20 watts, in 0.5 W steps. Plot the dB-versus-power curve on a log-linear scale.

4.24 Geometric Mean The *geometric mean* of a set of numbers x_1 through x_n is defined as the *n*th root of the product of the numbers:

$$\text{geometric mean} = \sqrt[n]{x_1 x_2 x_3 \ldots x_n} \tag{4-16}$$

Write a MATLAB program that will accept an arbitrary number of positive input values and calculate both the arithmetic mean (*i.e.*, the average) and the geometric mean of the numbers. Use a `while` loop to get the input values, and terminate the inputs a user enters a negative number. Test your program by calculating the average and geometric mean of the four numbers 10, 5, 2, and 5.

4.25 RMS Average The *root-mean-square* (rms) *average* is another way of calculating a mean for a set of numbers. The rms average of a series of numbers is the square root of the arithmetic mean of the squares of the numbers:

$$\text{rms average} = \sqrt{\frac{1}{N}\sum_{i=1}^{N} x_i^2} \tag{4-17}$$

Write a MATLAB program that will accept an arbitrary number of positive input values and calculate the rms average of the numbers. Prompt the user for the number of values to be entered, and use a `for` loop to read in the numbers. Test your program by calculating the rms average of the four numbers 10, 5, 2, and 5.

4.26 Harmonic Mean The *harmonic mean* is yet another way of calculating a mean for a set of numbers. The harmonic mean of a set of numbers is given by the equation:

$$\text{harmonic mean} = \frac{N}{\dfrac{1}{x_1} + \dfrac{1}{x_2} + \cdots + \dfrac{1}{x_N}} \tag{4-18}$$

Write a MATLAB program that will read in an arbitrary number of positive input values and calculate the harmonic mean of the numbers. Use any method that you desire to read in the input values. Test your program by calculating the harmonic mean of the four numbers 10, 5, 2, and 5.

4.27 Write a single program that calculates the arithmetic mean (average), rms average, geometric mean, and harmonic mean for a set of positive numbers. Use any method that you desire to read in the input values. Compare these values for each of the following sets of numbers:

(a) 4, 4, 4, 4, 4, 4, 4
(b) 4, 3, 4, 5, 4, 3, 5
(c) 4, 1, 4, 7, 4, 1, 7
(d) 1, 2, 3, 4, 5, 6, 7

4.28 **Mean Time Between Failure Calculations** The reliability of a piece of electronic equipment is usually measured in terms of Mean Time Between Failures (MTBF), where MTBF is the average time that the piece of equipment can operate before a failure occurs in it. For large systems containing many pieces of electronic equipment, it is customary to determine the MTBFs of each component, and to calculate the overall MTBF of the system from the failure rates of the individual components. If the system is structured like the one shown in Figure 4.6, every component must work in order for the whole system to work, and the overall system MTBF can be calculated as

$$\text{MTBF}_{\text{sys}} = \frac{1}{\dfrac{1}{\text{MTBF}_1} + \dfrac{1}{\text{MTBF}_2} + \cdots + \dfrac{1}{\text{MTBF}_n}} \tag{4-19}$$

Write a program that reads in the number of series components in a system and the MTBFs for each component, and then calculates the overall MTBF for the system. To test your program, determine the MTBF for a radar system consisting of an antenna subsystem with an MTBF of 2,000 hours, a transmitter with an MTBF of 800 hours, a receiver with an MTBF of 3,000 hours, and a computer with an MTBF of 5,000 hours.

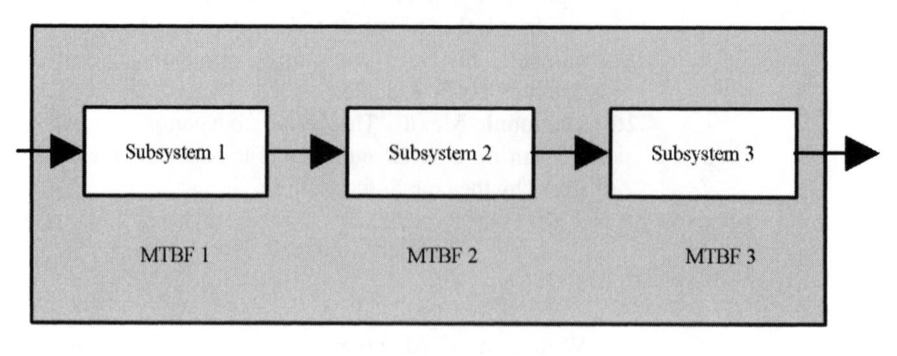

Figure 4.6 An electronic system containing three subsystems with known MTBFs.

CHAPTER 5

User-Defined Functions

In Chapter 3, we learned the importance of good program design. The basic technique that we employed is **top-down design**. In top-down design, the programmer starts with a statement of the problem to be solved and the required inputs and outputs. Next, he or she describes the algorithm to be implemented by the program in broad outline, and applies *decomposition* to break the algorithm down into logical subdivisions called sub-tasks. Then, the programmer breaks down each sub-task until he or she winds up with many small pieces, each of which does a simple, clearly understandable job. Finally, the individual pieces are turned into MATLAB code.

Although we have followed this design process in our examples, the results have been somewhat restricted, because we have had to combine the final MATLAB code generated for each sub-task into a single large program. There has been no way to code, verify, and test each sub-task independently before combining them into the final program.

Fortunately, MATLAB has a special mechanism designed to make sub-tasks easy to develop and debug independently before building the final program. It is possible to code each sub-task as a separate **function**, and each function can be tested and debugged independently of all of the other sub-tasks in the program.

Well-designed functions enormously reduce the effort required on a large programming project. Their benefits include:

1. **Independent testing of sub-tasks.** Each sub-task can be written as an independent unit. The sub-task can be tested separately to ensure that it performs properly by itself before combining it into the larger program. This step is know as **unit testing**. It eliminates a major source of problems before the final program is even built.

2. **Reusable code.** In many cases, the same basic sub-task is needed in many parts of a program. For example, it may be necessary to sort a list of values into ascending order many different times within a program, or even in other programs. It is possible to design, code, test, and debug a *single* function to do the sorting, and then to reuse that function whenever sorting is required. This reusable code has two major advantages: it reduces the total programming effort required, and it simplifies debugging, since the sorting function only needs to be debugged once.
3. **Isolation from unintended side effects.** Functions receive input data from the program that invokes them through a list of variables called an **input argument list**, and return results to the program through an **output argument list**. Each function has its own workspace with its own variables, independent of all other functions and of the calling program. *The only variables in the calling program that can be seen by the function are those in the input argument list, and the only variables in the function that can be seen by the calling program are those in the output argument list.* This is very important, since accidental programming mistakes within a function can only affect the variables within function in which the mistake occurred.

Once a large program is written and released, it has to be *maintained.* Program maintenance involves fixing bugs and modifying the program to handle new and unforeseen circumstances. The programmer who modifies a program during maintenance is often not the person who originally wrote it. In poorly written programs, it is common for the programmer modifying the program to make a change in one region of the code, and to have that change cause unintended side effects in a totally different part of the program. This happens because variable names are re-used in different portions of the program. When the programmer changes the values left behind in some of the variables, those values are accidentally picked up and used in other portions of the code.

The use of well-designed functions minimizes this problem by **data hiding**. The variables in the main program are not visible to the function (except for those in the input argument list), and the variables in the main program cannot be accidentally modified by anything occurring in the function. Therefore, mistakes or changes in the function's variables cannot accidentally cause unintended side effects in the other parts of the program.

✷ Good Programming Practice

Break large program tasks into functions whenever practical to achieve the important benefits of independent component testing, reusability, and isolation from undesired side effects.

5.1 Introduction to MATLAB Functions

All of the M-files that we have seen so far have been **script files**. Script files are just collections of MATLAB statements that are stored in a file. When a script file is executed, the result is the same as it would be if all of the commands had been typed directly into the Command Window. Script files share the Command Window's workspace, so any variables that were defined before the script file starts are visible to the script file, and any variables created by the script file remain in the workspace after the script file finishes executing. A script file has no input arguments and returns no results, but script files can communicate with other script files through the data left behind in the workspace.

In contrast, a **MATLAB function** is a special type of M-file that runs in its own independent workspace. It receives input data through an **input argument list**, and returns results to the caller through an **output argument list**. The general form of a MATLAB function is

```
function [outarg1, outarg2, ...] = fname(inarg1, inarg2, ...)
% H1 comment line
% Other comment lines
...
(Executable code)
...
(return)
(end)
```

The `function` statement marks the beginning of the function. It specifies the name of the function and the input and output argument lists. The input argument list appears in parentheses after the function name, and the output argument list appears in brackets to the left of the equal sign. (If there is only one output argument, the brackets can be dropped.)

Each MATLAB ordinary function should be placed in a file with the same name (including capitalization) as the function, and the file extent "`.m`". For example, if a function is named `My_fun`, then that function should be placed in a file named `My_fun.m`.

The input argument list is a list of names representing values that will be passed from the caller to the function. These names are called **dummy arguments**. They are just placeholders for actual values that are passed from the caller when the function is invoked. Similarly, the output argument list contains a list of dummy arguments that are placeholders for the values returned to the caller when the function finishes executing.

A function is invoked by naming it in an expression together with a list of **actual arguments**. A function can be invoked by typing its name directly in the Command Window, or by including it in a script file or another function. The name in the calling program must *exactly match* the function name (including capitalization).[1] When the function is invoked, the value of the first actual argument is used in place of the first dummy argument, and so forth for each other actual argument/dummy argument pair.

Execution begins at the top of the function, and ends when either a `return` statement, an `end` statement, or the end of the function is reached. Because execution stops at the end of a function anyway, the `return` statement is not actually required in most functions, and is rarely used. Each item in the output argument list must appear on the left side of a least one assignment statement in the function. When the function returns, the values stored in the output argument list are returned to the caller, and may be used in further calculations.

The use of an `end` statement to terminate a function is a new feature of MATLAB 7.0. In earlier versions of MATLAB, the `end` statement was only used to terminate structures such as `if`, `for`, `while`, etc. It is optional in MATLAB 7 unless a file includes nested functions, which are a special feature not covered in this book.

The initial comment lines in a function serve a special purpose. The first comment line after the function statement is called the **H1 comment line**. It should always contain a one-line summary of the purpose of the function. The special significance of this line is that it is searched and displayed by the `lookfor` command. The remaining comment lines from the H1 line until the first blank line or the first executable statement are displayed by the `help` command. They should contain a brief summary of how to use the function.

A simple example of a user-defined function is shown below. Function `dist2` calculates the distance between points (x_1, y_1) and (x_2, y_2) in a Cartesian coordinate system.

```
function distance = dist2 (x1, y1, x2, y2)
%DIST2 Calculate the distance between two points
% Function DIST2 calculates the distance between
% two points (x1,y1) and (x2,y2) in a Cartesian
% coordinate system.
%
% Calling sequence:
%   distance = dist2(x1, y1, x2, y2)
```

[1]For example, suppose that a function has been declared with the name `My_Fun`, and placed in file `My_Fun.m`. Then this function should be called with the name `My_Fun`, not `my_fun` or `MY_FUN`. If the capitalization fails to match, this will produce an error on Linux, Unix, and Macintosh computers, and a warning on Windows-based computers.

```
% Define variables:
%   x1       -- x-position of point 1
%   y1       -- y-position of point 1
%   x2       -- x-position of point 2
%   y2       -- y-position of point 2
%   distance -- Distance between points

%  Record of revisions:
%     Date         Programmer          Description of change
%     ====         ==========          =====================
%   01/12/05     S. J. Chapman         Original code

% Calculate distance.
distance = sqrt((x2-x1).^2 + (y2-y1).^2);
```

This function has four input arguments and one output argument. A simple script file using this function is shown below.

```
% Script file: test_dist2.m
%
% Purpose:
%   This program tests function dist2.
%
% Record of revisions:
%     Date         Programmer          Description of change
%     ====         ==========          =====================
%   01/12/05     S. J. Chapman         Original code
%
% Define variables:
%   ax       -- x-position of point a
%   ay       -- y-position of point a
%   bx       -- x-position of point b
%   by       -- y-position of point b
%   result   -- Distance between the points

% Get input data.
disp('Calculate the distance between two points:');
ax = input('Enter x value of point a: ');
ay = input('Enter y value of point a: ');
bx = input('Enter x value of point b: ');
by = input('Enter y value of point b: ');

% Evaluate function
result = dist2 (ax, ay, bx, by);
```

```
% Write out result.
fprintf('The distance between points a and b is %f\n', result);
```

When this script file is executed, the results are:

```
» test_dist2
Calculate the distance between two points:
Enter x value of point a: 1
Enter y value of point a: 1
Enter x value of point b: 4
Enter y value of point b: 5
The distance between points a and b is 5.000000
```

These results are correct, as we can verify from simple hand calculations.

Function `dist2` also supports the MATLAB help subsystem. If we type "`help dist2`," the results are:

```
» help dist2
DIST2 Calculate the distance between two points
 Function DIST2 calculates the distance between
 two points (x1,y1) and (x2,y2) in a Cartesian
 coordinate system.

 Calling sequence:
    res = dist2(x1, y1, x2, y2)
```

Similarly, "`lookfor distance`" produces the result

```
» lookfor distance
DIST2 Calculate the distance between two points
MAHAL Mahalanobis distance.
DIST Distances between vectors.
NBDIST Neighborhood matrix using vector distance.
NBGRID Neighborhood matrix using grid distance.
NBMAN Neighborhood matrix using Manhattan-distance.
```

To observe the behavior of the MATLAB workspace before, during, and after the function is executed, we will load function `dist2` and the script file `test_dist2` into the MATLAB debugger, and set breakpoints before, during and after the function call (see Figure 5.1). When the program stops at the breakpoint *before* the function call, the workspace is as shown in Figure 5.2(*a*) on page 206. Note that variables `ax`, `ay`, `bx`, and `by` are defined in the workspace,

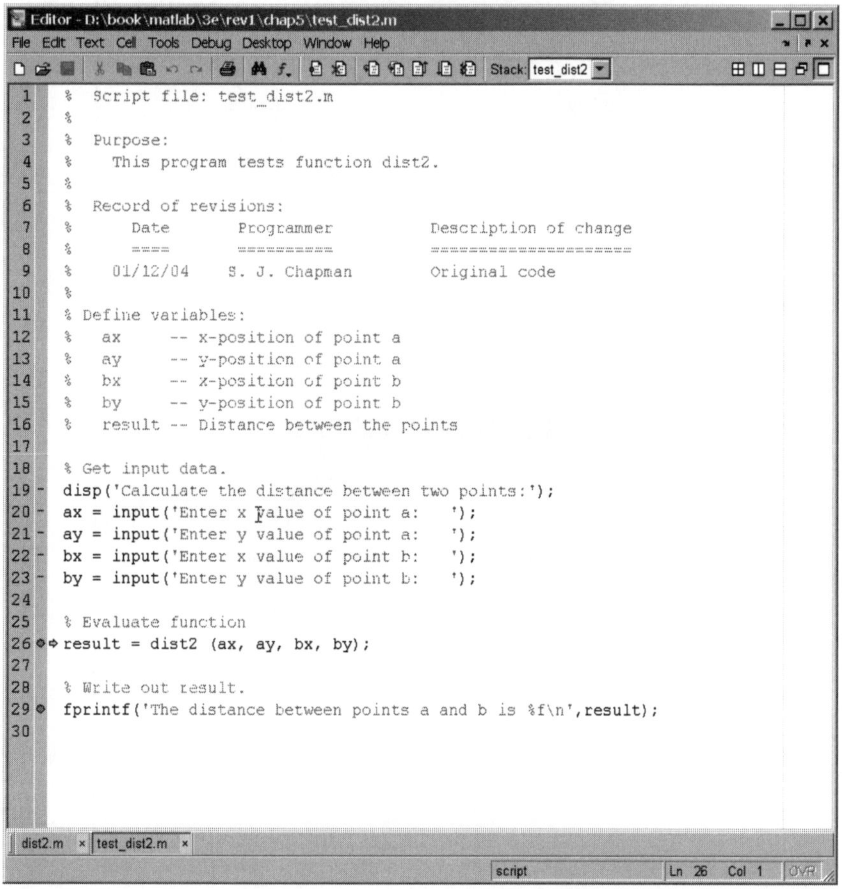

Editor - D:\book\matlab\3e\rev1\chap5\test_dist2.m

File Edit Text Cell Tools Debug Desktop Window Help

Stack: test_dist2

```
1    %  Script file: test_dist2.m
2    %
3    %  Purpose:
4    %    This program tests function dist2.
5    %
6    %  Record of revisions:
7    %      Date        Programmer          Description of change
8    %      ====        ==========          =====================
9    %    01/12/04    S. J. Chapman         Original code
10   %
11   % Define variables:
12   %   ax      -- x-position of point a
13   %   ay      -- y-position of point a
14   %   bx      -- x-position of point b
15   %   by      -- y-position of point b
16   %   result -- Distance between the points
17
18   % Get input data.
19 - disp('Calculate the distance between two points:');
20 - ax = input('Enter x value of point a:   ');
21 - ay = input('Enter y value of point a:   ');
22 - bx = input('Enter x value of point b:   ');
23 - by = input('Enter y value of point b:   ');
24
25   % Evaluate function
26 ●⇨ result = dist2 (ax, ay, bx, by);
27
28   % Write out result.
29 ● fprintf('The distance between points a and b is %f\n',result);
30
```

dist2.m × test_dist2.m ×

script Ln 26 Col 1 OVR

Figure 5.1 M-file `test_dist2` and function `dist2` are loaded into the debugger, with breakpoints set before, during, and after the function call.

with the values that we have entered. When the program stops at the breakpoint *within* the function call, the function's workspace is active. It is as shown in Figure 5.2(*b*) on page 206. Note that variables `x1`, `x2`, `y1`, `y2`, and `distance` are defined in the function's workspace, and the variables defined in the calling M-file not present. When the program stops in the calling program at the breakpoint *after* the function call, the workspace is as shown in Figure 5.2(*c*) on page 206. Now the original variables are back, with the variable `result` added to contain the value returned by the function. These figures show that the workspace of the function is different than the workspace of the calling M-file.

Name	Value	Class
ax	1	double
ay	1	double
bx	4	double
by	5	double

(a)

Name	Value	Class
distance	5	double
x1	1	double
x2	4	double
y1	1	double
y2	5	double

(b)

Name	Value	Class
ax	1	double
ay	1	double
bx	4	double
by	5	double
result	5	double

(c)

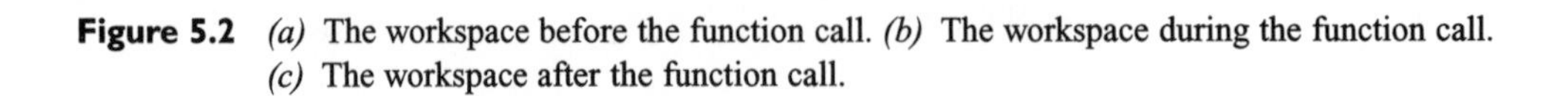

Figure 5.2 *(a)* The workspace before the function call. *(b)* The workspace during the function call. *(c)* The workspace after the function call.

5.2 Variable Passing in MATLAB: The Pass-By-Value Scheme

MATLAB programs communicate with their functions using a **pass-by-value** scheme. When a function call occurs, MATLAB makes a *copy* of the actual arguments and passes them to the function. This copying is very significant, because it means that even if the function modifies the input arguments, it won't affect the original data in the caller. This feature helps to prevent unintended side effects, in which an error in the function might unintentionally modify variables in the calling program.

This behavior is illustrated in the function shown below. This function has two input arguments: a and b. During its calculations, it modifies both input arguments.

```
function out = sample(a, b)
fprintf('In     sample: a = %f, b = %f %f\n',a,b);
a = b(1) + 2*a;
b = a .* b;
out = a + b(1);
fprintf('In     sample: a = %f, b = %f %f\n',a,b);
```

A simple test program to call this function is shown below.

```
a = 2; b = [6 4];
fprintf('Before sample: a = %f, b = %f %f\n',a,b);
out = sample(a,b);
fprintf('After  sample: a = %f, b = %f %f\n',a,b);
fprintf('After  sample: out = %f\n',out);
```

When this program is executed, the results are:

```
» test_sample
Before sample: a = 2.000000, b = 6.000000 4.000000
In     sample: a = 2.000000, b = 6.000000 4.000000
In     sample: a = 10.000000, b = 60.000000 40.000000
After  sample: a = 2.000000, b = 6.000000 4.000000
After  sample: out = 70.000000
```

Note that a and b were both changed inside function sample, but those changes had *no effect on the values in the calling program.*

Users of the C language will be familiar with the pass-by-value scheme, since C uses it for scalar values passed to functions. However C does *not* use the pass-by-value scheme when passing arrays, so an unintended modification to a dummy array in a C function can cause side effects in the calling program.

MATLAB improves on this by using the pass-by-value scheme for both scalars and arrays.[2]

Example 5-1—Rectangular-to-Polar Conversion

The location of a point in a Cartesian plane can be expressed in either the rectangular coordinates (x, y) or the polar coordinates (r, θ), as shown in Figure 5.3. The relationships among these two sets of coordinates are given by the following equations:

$$x = r\cos\theta \tag{5-1}$$

$$y = r\sin\theta \tag{5-2}$$

$$r = \sqrt{x^2 + y^2} \tag{5-3}$$

$$\theta = \tan^{-1}\frac{y}{x} \tag{5-4}$$

Write two functions `rect2polar` and `polar2rect` that convert coordinates from rectangular to polar form, and vice versa, where the angle θ is expressed in degrees.

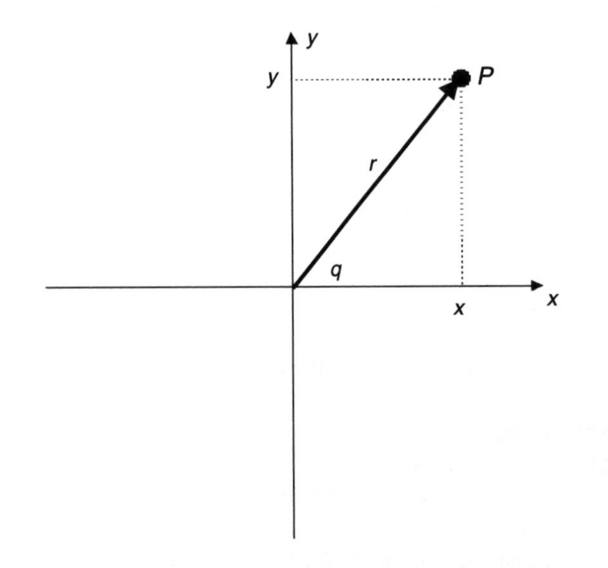

Figure 5.3 A point P in a Cartesian plane can be located by either the rectangular coordinates (x, y) or the polar coordinates (r, θ).

[2]The implementation of argument passing in MATLAB is actually more sophisticated than this discussion indicates. As pointed out previously, the copying associated with pass-by-value takes up a lot of time, but it provides protection against unintended side effects. MATLAB actually uses the best of both approaches: it analyzes each argument of each function, and determines whether or not the function modifies that argument. If the function modifies the argument, then MATLAB makes a copy of it. If it does not modify the argument, then MATLAB simply points to the existing value in the calling program. This practice increases speed while still providing protection against side effects!

SOLUTION We will apply our standard problem-solving approach to creating these functions. Note that MATLAB's trigonometric functions work in radians, so we must convert from degrees to radians and vice versa when solving this problem. The basic relationship between degrees and radians is

$$180° = \pi \text{ radians} \quad (5\text{-}5)$$

1. **State the problem**

   A succinct statement of the problem is:

   > Write a function that converts a location on a Cartesian plane expressed in rectangular coordinates into the corresponding polar coordinates, with the angle θ is expressed in degrees. Also, write a function that converts a location on a Cartesian plane expressed in polar coordinates with the angle θ is expressed in degrees into the corresponding rectangular coordinates.

2. **Define the inputs and outputs**

   The inputs to function `rect2polar` are the rectangular (x, y) location of a point. The outputs of the function are the polar (r, θ) location of the point. The inputs to function `polar2rect` are the polar (r, θ) location of a point. The outputs of the function are the rectangular (x, y) location of the point.

3. **Describe the algorithm**

   These functions are very simple, so we can directly write the final pseudocode for them. The pseudocode for function `polar2rect` is:

   ```
   x <- r * cos(theta * pi/180)
   y <- r * sin(theta * pi/180)
   ```

   The pseudocode for function `rect2polar` will use the function `atan2`, because that function works over all four quadrants of the Cartesian plane. (Look that function up in the MATLAB Help Browser!)

   ```
   r <- sqrt( x.^2 + y .^2 )
   theta <- 180/pi * atan2(y,x)
   ```

4. **Turn the algorithm into MATLAB statements.**

   The MATLAB code for the selection `polar2rect` function is shown below.

```
function [x, y] = polar2rect(r,theta)
%POLAR2RECT Convert rectangular to polar coordinates
% Function POLAR2RECT accepts the polar coordinates
% (r,theta), where theta is expressed in degrees,
% and converts them into the rectangular coordinates
% (x,y).
%
```

```
% Calling sequence:
%   [x, y] = polar2rect(r,theta)

% Define variables:
%   r              -- Length of polar vector
%   theta          -- Angle of vector in degrees
%   x              -- x-position of point
%   y              -- y-position of point

% Record of revisions:
%     Date           Programmer          Description of change
%     ====           ==========          =====================
%   01/12/05       S. J. Chapman       Original code

x = r * cos(theta * pi/180);
y = r * sin(theta * pi/180);
```

The MATLAB code for the selection `rect2polar` function is shown below.

```
function [r, theta] = rect2polar(x,y)
%RECT2POLAR Convert rectangular to polar coordinates
% Function RECT2POLAR accepts the rectangular coordinates
% (x,y) and converts them into the polar coordinates
% (r,theta), where theta is expressed in degrees.
%
% Calling sequence:
%   [r, theta] = rect2polar(x,y)

% Define variables:
%   r         -- Length of polar vector
%   theta     -- Angle of vector in degrees
%   x         -- x-position of point
%   y         -- y-position of point

% Record of revisions:
%     Date           Programmer          Description of change
%     ====           ==========          =====================
%   01/12/05       S. J. Chapman       Original code

r = sqrt( x.^2 + y .^2 );
theta = 180/pi * atan2(y,x);
```

Note that these functions both include help information, so they will work properly with MATLAB's help subsystem and with the `lookfor` command.

5. **Test the program.**
   To test these functions, we will execute them directly in the MATLAB Command Window. We will test the functions using the 3-4-5 triangle, which is familiar to most people from secondary school. The smaller angle within a 3-4-5 triangle is approximately 36.87°. We will also test the function in all four quadrants of the Cartesian plane to ensure that the conversion are correct everywhere.

```
» [r, theta] = rect2polar(4,3)
r =
     5
theta =
   36.8699
» [r, theta] = rect2polar(-4,3)
r =
     5
theta =
   143.1301
» [r, theta] = rect2polar(-4,-3)
r =
     5
theta =
   -143.1301
» [r, theta] = rect2polar(4,-3)
r =
     5
theta =
   -36.8699
» [x, y] = polar2rect(5,36.8699)
x =
   4.0000
y =
   3.0000
» [x, y] = polar2rect(5,143.1301)
x =
  -4.0000
y =
   3.0000
» [x, y] = polar2rect(5,-143.1301)
x =
  -4.0000
y =
  -3.0000
» [x, y] = polar2rect(5,-36.8699)
x =
   4.0000
```

```
y =
   -3.0000
»
```

These functions appear to be working correctly in all quadrants of the Cartesian plane.

Example 5.2—Sorting Data

In many scientific and engineering applications, it is necessary to take a random input data set and to sort it so that the numbers in the data set are either all in *ascending order* (lowest-to-highest) or all in *descending order* (highest-to-lowest). For example, suppose that you were a zoologist studying a large population of animals, and that you wanted to identify the largest 5% of the animals in the population. The most straightforward way to approach this problem would be to sort the sizes of all of the animals in the population into ascending order, and take the top 5% of the values.

Sorting data into ascending or descending order seems to be an easy job. After all, we do it all the time. It is simple matter for us to sort the data (10, 3, 6, 4, 9) into the order (3, 4, 6, 9, 10). How do we do it? We first scan the input data list (10, 3, 6, 4, 9) to find the smallest value in the list (3), and then scan the remaining input data (10, 6, 4, 9) to find the next smallest value (4), *etc.* until the complete list is sorted.

In fact, sorting can be a very difficult job. As the number of values to be sorted increases, the time required to perform the simple sort described above increases rapidly, since we must scan the input data set once for each value sorted. For very large data sets, this technique just takes too long to be practical. Even worse, how would we sort the data if there were too many numbers to fit into the main memory of the computer? The development of efficient sorting techniques for large data sets is an active area of research, and is the subject of whole courses all by itself.

In this example, we will confine ourselves to the simplest possible algorithm to illustrate the concept of sorting. This simplest algorithm is called the **selection sort**. It is just a computer implementation of the mental math described above. The basic algorithm for the selection sort is:

1. Scan the list of numbers to be sorted to locate the smallest value in the list. Place that value at the front of the list by swapping it with the value currently at the front of the list. If the value at the front of the list is already the smallest value, then do nothing.
2. Scan the list of numbers from position 2 to the end to locate the next smallest value in the list. Place that value in position 2 of the list by swapping it with the value currently at that position. If the value in position 2 is already the next smallest value, then do nothing.
3. Scan the list of numbers from position 3 to the end to locate the third smallest value in the list. Place that value in position 3 of the list

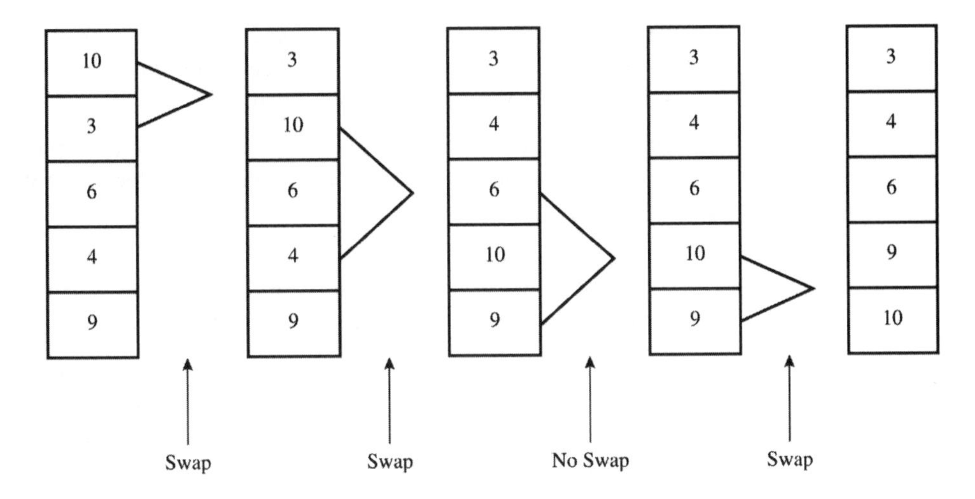

Figure 5.4 An example problem demonstrating the selection sort algorithm.

by swapping it with the value currently at that position. If the value in position 3 is already the third smallest value, then do nothing.

4. Repeat this process until the next-to-last position in the list is reached. After the next-to-last position in the list has been processed, the sort is complete.

Note that if we are sorting N values, this sorting algorithm requires N-1 scans through the data to accomplish the sort.

This process is illustrated in Figure 5.4. Since there are five values in the data set to be sorted, we will make four scans through the data. During the first pass through the entire data set, the minimum value is 3, so the 3 is swapped with the 10 which was in position 1. Pass 2 searches for the minimum value in positions 2 through 5. That minimum is 4, so the 4 is swapped with the 10 in position 2. Pass 3 searches for the minimum value in positions 3 through 5. That minimum is 6, which is already in position 3, so no swapping is required. Finally, pass four searches for the minimum value in positions 4 through 5. That minimum is 9, so the 9 is swapped with the 10 in position 4, and the sort is completed. ◀

Programming Pitfalls

The selection sort algorithm is the easiest sorting algorithm to understand, but it is computationally inefficient. *It should never be applied to sort large data sets* (say, sets with more than 1000 elements). Over the years, computer scientists have developed much more efficient sorting algorithms. The `sort` and `sortrows` functions built into MATLAB are extremely efficient and should be used for all real work.

We will now develop a program to read in a data set from the Command Window, sort it into ascending order, and display the sorted data set. The sorting will be done by a separate user-defined function.

SOLUTION This program must be able to ask the user for the input data, sort the data, and write out the sorted data. The design process for this problem is given below.

1. **State the problem**
   We have not yet specified the type of data to be sorted. If the data is numeric, then the problem may be stated as follows:

   > Develop a program to read an arbitrary number of numeric input values from the Command Window, sort the data into ascending order using a separate sorting function, and write the sorted data to the Command Window.

2. **Define the inputs and outputs**
   The inputs to this program are the numeric values typed in the Command Window by the user. The outputs from this program are the sorted data values written to the Command Window.

3. **Describe the algorithm**
   This program can be broken down into three major steps

   ```
   Read the input data into an array
   Sort the data in ascending order
   Write the sorted data
   ```

   The first major step is to read in the data. We must prompt the user for the number of input data values, and then read in the data. Sine we will know how many input values there are to read, a `for` loop is appropriate for reading in the data. The detailed pseudocode is shown below:

   ```
   Prompt user for the number of data values
   Read the number of data values
   Preallocate an input array
   for ii = 1:number of values
      Prompt for next value
      Read value
   end
   ```

   Next we have to sort the data in a separate function. We will need to make **`nvals-1`** passes through the data, finding the smallest remaining value each time. We will use a pointer to locate the smallest value in each pass. Once the smallest value is found, it will be swapped to the top of the list of it is not already there. The detailed pseudocode is shown below:

```
for ii = 1:nvals-1
   % Find the minimum value in a(ii) through a(nvals)
```

```
    iptr <- ii
    for jj == ii+1 to nvals
       if a(jj) < a(iptr)
          iptr <- jj
       end
    end

    % iptr now points to the min value, so swap a(iptr)
    % with a(ii) if iptr ~= ii.
    if i ~= iptr
       temp <- a(i)
       a(i) <- a(iptr)
       a(iptr) <- temp
    end
end
```

The final step is writing out the sorted values. No refinement of the pseudocode is required for that step. The final pseudocode is the combination of the reading, sorting and writing steps.

4. **Turn the algorithm into MATLAB statements.**
   The MATLAB code for the selection sort function is shown below.

```
function out = ssort(a)
%SSORT Selection sort data in ascending order
% Function SSORT sorts a numeric data set into
% ascending order. Note that the selection sort
% is relatively inefficient. DO NOT USE THIS
% FUNCTION FOR LARGE DATA SETS. Use MATLAB's
% "sort" function instead.

% Define variables:
%   a           -- Input array to sort
%   ii          -- Index variable
%   iptr        -- Pointer to min value
%   jj          -- Index variable
%   nvals       -- Number of values in "a"
%   out         -- Sorted output array
%   temp        -- Temp variable for swapping

% Record of revisions:
%     Date          Programmer          Description of change
%     ====          ==========          =====================
%   01/12/05      S. J. Chapman        Original code

% Get the length of the array to sort
nvals = length(a);

% Sort the input array
```

```
for ii = 1:nvals-1
   % Find the minimum value in a(ii) through a(n)
   iptr = ii;
   for jj = ii+1:nvals
      if a(jj) < a(iptr)
         iptr = jj;
      end
   end

   % iptr now points to the minimum value, so swap a(iptr)
   % with a(ii) if ii ~= iptr.
   if ii ~= iptr
      temp    = a(ii);
      a(ii)   = a(iptr);
      a(iptr) = temp;

   end
end

% Pass data back to caller
out = a;
```

The program to invoke the selection sort function is shown below.

```
% Script file: test_ssort.m
%
% Purpose:
%   To read in an input data set, sort it into ascending
%   order using the selection sort algorithm, and to
%   write the sorted data to the Command Window. This
%   program calls function "ssort" to do the actual
%   sorting.
%
% Record of revisions:
%     Date        Programmer          Description of change
%     ====        ==========          =====================
%   01/12/05    S. J. Chapman       Original code
%
% Define variables:
%   array     -- Input data array
%   ii        -- Index variable
%   nvals     -- Number of input values
%   sorted    -- Sorted data array

% Prompt for the number of values in the data set
nvals = input('Enter number of values to sort: ');
```

```
% Preallocate array
array = zeros(1,nvals);

% Get input values
for ii = 1:nvals

   % Prompt for next value
   string = ['Enter value ' int2str(ii) ': '];
   array(ii) = input(string);

end

% Now sort the data
sorted = ssort(array);

% Display the sorted result.
fprintf('\nSorted data:\n');
for ii = 1:nvals
   fprintf(' %8.4f\n',sorted(ii));
end
```

5. **Test the program.**
   To test this program, we will create an input data set and run the program with it. The data set should contain a mixture of positive and negative numbers as well as at least one duplicated value to see if the program works properly under those conditions.

```
» test_ssort
Enter number of values to sort: 6
Enter value 1: -5
Enter value 2: 4
Enter value 3: -2
Enter value 4: 3
Enter value 5: -2
Enter value 6: 0

Sorted data:
  -5.0000
  -2.0000
  -2.0000
   0.0000
   3.0000
   4.0000
```

   The program gives the correct answers for our test data set. Note that it works for both positive and negative numbers as well as for repeated numbers.

◄

5.3 Optional Arguments

Many MATLAB functions support optional input arguments and output arguments. For example, we have seen calls to the `plot` function with as few as two or as many as seven input arguments. On the other hand, the function `max` supports either one or two output arguments. If there is only one output argument, `max` returns the maximum value of an array. If there are two output arguments, `max` returns both the maximum value and the location of the maximum value in an array. How do MATLAB functions know how many input and output arguments are present, and how do they adjust their behavior accordingly?

There are eight special functions that can be used by MATLAB functions to get information about their optional arguments, and to report errors in those arguments. Six of these functions are introduced here, and the remaining two will be introduced in Chapter 7 after we learn about the cell array data type. The functions introduced now are:

- `nargin`—This function returns the number of actual input arguments that were used to call the function.
- `nargout`—This function returns the number of actual output arguments that were used to call the function.
- `nargchk`—This function returns a standard error message if a function is called with too few or too many arguments.
- `error`—Display error message and abort the function producing the error. This function is used if the argument errors are fatal.
- `warning`—Display warning message and continue function execution. This function is used if the argument errors are not fatal, and execution can continue.
- `inputname`—This function returns the actual name of the variable that corresponds to a particular argument number.

When functions `nargin` and `nargout` are called within a user-defined function, these functions return the number of actual input arguments and the number of actual output arguments that were used to when the user-defined function was called.

Function `nargchk` generates a string containing a standard error message if a function is called with too few or too many arguments. The syntax of this function is

```
message = nargchk(min_args,max_args,num_args);
```

where `min_args` is the minimum number of arguments, `max_args` is the maximum number of arguments, and `num_args` is the actual number of arguments. If the number of arguments is outside the acceptable limits, a standard error message is produced. If the number of arguments is within acceptable limits, then an empty string is returned.

Function `error` is a standard way to display an error message and abort the user-defined function causing the error. The syntax of this function is `error('msg')`, where `msg` is a character string containing an error message. When `error` is executed, it halts the current function and returns to the keyboard, displaying the error message in the Command Window. If the message string is empty, `error` does nothing, and execution continues. This function works well with `nargchk`, which produces a message string when an error occurs and an empty string when there is no error.

Function `warning` is a standard way to display a warning message that includes the function and line number where the problem occurred, but let execution continue. The syntax of this function is `warning('msg')`, where `msg` is a character string containing a warning message. When `warning` is executed, it displays the warning message in the Command Window, and lists the function name and line number where the warning came from. If the message string is empty, `warning` does nothing. In either case, execution of the function continues.

Function `inputname` returns the name of the actual argument used when a function is called. The syntax of this function is

```
name = inputname(argno);
```

where `argno` is the number of the argument. If argument is a variable, then its name is returned. If the argument is an expression, then this function will return an empty string. For example, consider the function

```
function myfun(x,y,z)
name = inputname(2);
disp(['The second argument is named ' name]);
```

When this function is called, the results are

```
» myfun(dog,cat)
The second argument is named cat
» myfun(1,2+cat)
The second argument is named
```

Function `inputname` is useful for displaying argument names in warning and error messages.

Example 5.3—Using Optional Arguments

We will illustrate the use optional arguments by creating a function that accepts an (x, y) value in rectangular coordinates, and produces the equivalent polar representation consisting of a magnitude and an angle in degrees. The function will be designed to support two input arguments, x and y. However, if only one argument is supplied, the function will assume that the y value is zero and proceed with the calculation. The function will normally return both the magnitude and the angle in degrees, but if only one output argument is present, it will return only the magnitude. This function is shown below:

```
function [mag, angle] = polar_value(x,y)
%POLAR_VALUE Converts (x,y) to (r,theta)
```

```
% Function POLAR_VALUE converts an input (x,y)
% value into (r,theta), with theta in degrees.
% It illustrates the use of optional arguments.

% Define variables:
%   angle    -- Angle in degrees
%   msg      -- Error message
%   mag      -- Magnitude
%   x        -- Input x value
%   y        -- Input y value (optional)

% Record of revisions:
%     Date        Programmer         Description of change
%     ====        ==========         =====================
%   01/12/05     S. J. Chapman       Original code

% Check for a legal number of input arguments.
msg = nargchk(1,2,nargin);
error(msg);

% If the y argument is missing, set it to 0.
if nargin < 2
   y = 0;
end

% Check for (0,0) input arguments, and print out
% a warning message.
if x == 0 & y == 0
   msg = 'Both x any y are zero: angle is meaningless!';
   warning(msg);
end

% Now calculate the magnitude.
mag = sqrt(x.^2 + y.^2);

% If the second output argument is present, calcuate
% angle in degrees.
if nargout == 2
   angle = atan2(y,x) * 180/pi;
end

end % function polar_value
```

We will test this function by calling it repeatedly from the Command Window. First, we will try to call the function with too few or too many arguments.

```
» [mag angle] = polar_value
??? Error using ==> polar_value
Not enough input arguments.
```

```
» [mag angle] = polar_value(1,-1,1)
??? Error using ==> polar_value
Too many input arguments.
```

The function provides proper error messages in both cases. Next, we will try to call the function with one or two input arguments.

```
» [mag angle] = polar_value(1)
mag =
     1
angle =
     0
» [mag angle] = polar_value(1,-1)
mag =
         1.4142
angle =
   -45
```

The function provides the correct answer in both cases. Next, we will try to call the function with one or two output arguments.

```
» mag = polar_value(1,-1)
mag =
     1.4142
» [mag angle] = polar_value(1,-1)
mag =
     1.4142
angle =
     -45
```

The function provides the correct answer in both cases. Finally, we will try to call the function with both *x* and *y* equal to zero.

```
» [mag angle] = polar_value(0,0)

Warning: Both x any y are zero: angle is meaningless!
> In d:\book\matlab\chap5\polar_value.m at line 32
mag =
     0
angle =
     0
```

In this case, the function displays the warning message, but execution continues. ◄

Note that a MATLAB function may be declared to have more output arguments than are actually used, and this is *not* an error. The function does not actually have to check nargout to determine if an output argument is present. For example, consider the following function:

```
function [z1, z2] = junk(x,y)
z1 = x + y;
z2 = x - y;
end % function junk
```

This function can be called successfully with one or two output arguments.

```
» a = junk(2,1)
a =
   3
» [a b] = junk(2,1)
a =
   3
b =
   1
```

The reason for checking nargout in a function is to prevent useless work. If a result is going to be thrown away anyway, why bother to calculate it in the first place? A programmer can speed up the operation of a program by not bothering with useless calculations.

Quiz 5.1

This quiz provides a quick check to see if you have understood the concepts introduced in Sections 5.1 through 5.3. If you have trouble with the quiz, reread the section, ask your instructor, or discuss the material with a fellow student. The answers to this quiz are found in the back of the book.

1. What are the differences between a script file and a function?
2. How does the help command work with user-defined functions?
3. What is the significance of the H1 comment line in a function?
4. What is the pass-by-value scheme? How does it contribute to good program design?
5. How can a MATLAB function be designed to have optional arguments?

For questions 6 and 7, determine whether the function calls are correct or not. If they are in error, specify what is wrong with them.

6. out = test1(6);

```
function res = test1(x,y)
res = sqrt(x.^2 + y.^2);
```

7.
```
out = test2(12);
function res = test2(x,y)
error(nargchk(1,2,nargin));
if nargin == 2
   res = sqrt(x.^2 + y.^2);
else
   res = x;
end
```

5.4 Sharing Data Using Global Memory

We have seen that programs exchange data with the functions they call through a argument lists. When a function is called, each actual argument is copied, and the copy is used by the function.

In addition to the argument list, MATLAB functions can exchange data with each other and with the base workspace through global memory. **Global memory** is a special type of memory that can be accessed from any workspace. If a variable is declared to be global in a function, then it will be placed in the global memory instead of the local workspace. If the same variable is declared to be global in another function, then that variable will refer to the *same memory location* as the variable in the first function. Each script file or function that declares the global variable will have access the same data values, so *global memory provide a way to share data between functions.*

A global variable is declared with the **global statement**. The form of a `global` statement is

```
global var1 var2 var3 ...
```

where *var1*, *var2*, *var3*, etc. are the variables to be placed in global memory. By convention, global variables are declared in all capital letters, but this is not actually a requirement.

✷ Good Programming Practice

Declare global variables in all capital letters to make them easy to distinguish from local variables.

Each global variable must be declared to be global before it is used for the first time in a function—it is an error to declare a variable to be global after it has already been created in the local workspace.[3] To avoid this error, it is customary to declare global variables immediately after the initial comments and before the first executable statement in a function.

Good Programming Practice

Declare global variables immediately after the initial comments and before the first executable statement each function that uses them.

Global variables are especially useful for sharing very large volumes of data among many functions, because the entire data set does not have to be copied each time that a function is called. The downside of using global memory to exchange data among functions is that the functions will only work for that specific data set. A function that exchanges data through input arguments can be reused by simply calling it with different arguments, but a function that exchanges data through global memory must actually be modified to allow it to work with a different data set.

Global variables are also useful for sharing hidden data among a group of related functions while keeping it invisible from the invoking program unit.

Good Programming Practice

You may use global memory to pass large amounts of data among functions within a program.

Example 5.4—Random Number Generator

It is impossible to make perfect measurements in the real world. There will always be some *measurement noise* associated with each measurement. This fact is an important consideration in the design of systems to control the operation of such real-world devices as airplanes, refineries, etc. A good engineering design must take these measurement errors into account, so that the noise in the measurements will not lead to unstable behavior (no plane crashes or refinery explosions!).

Most engineering designs are tested by running *simulations* of the operation of the system before it is ever built. These simulations involve creating

[3]If a variable is declared `global` after it has already been defined in a function, MATLAB will issue a warning message and then change the local value to match the global value. You should never rely on this capability, though, because future versions of MATLAB will not allow it.

mathematical models of the behavior of the system, and feeding the models a realistic string of input data. If the models respond correctly to the simulated input data, then we can have reasonable confidence that the real-world system will respond correctly to the real-world input data.

The simulated input data supplied to the models must be corrupted by a simulated measurement noise, which is just a string of random numbers added to the ideal input data. The simulated noise is usually produced by a *random number generator*.

A random number generator is a function that will return a different and apparently random number each time it is called. Since the numbers are in fact generated by a deterministic algorithm, they only appear to be random.[4] However, if the algorithm used to generate them is complex enough, the numbers will be random enough to use in the simulation.

One simple random number generator algorithm is described below.[5] It relies on the unpredictability of the modulo function when applied to large numbers. Consider the following equation:

$$n_{i+1} = \operatorname{mod}(8121\, n_i + 28411, 134456) \tag{5-6}$$

Assume that n_i is a non-negative integer. Then because of the modulo function, n_{i+1} will be a number between 0 and 134455 inclusive. Next, n_{i+1} can be fed into the equation to produce a number n_{i+2} that is also between 0 and 134455. This process can be repeated forever to produce a series of numbers in the range [0,134455]. If we didn't know the numbers 8121, 28411, and 134456 in advance, it would be impossible to guess the order in which the values of n would be produced. Furthermore, it turns out that there is an equal (or uniform) probability that any given number will appear in the sequence. Because of these properties, Equation 5-6 can serve as the basis for a simple random number generator with a uniform distribution.

We will now use Equation 5-6 to design a random number generator whose output is a real number in the range [0.0, 1.0).[6]

SOLUTION We will write a function that generates one random number in the range $0 \leq ran < 1.0$ each time that it is called. The random number will be based on the equation

$$ran_i = \frac{n_i}{134456} \tag{5-7}$$

where n_i is a number in the range 0 to 134455 produced by Equation 5-7.

The particular sequence produced by Equations 5-6 and 5-7 will depend on the initial value of n_0 (called the *seed*) of the sequence. We must provide a

[4]For this reason, some people refer to these functions as *pseudorandom number generators*.

[5]This algorithm is adapted from the discussion found in Chapter 7 of *Numerical Recipes: The Art of Scientific Programming*, by Press, Flannery, Teukolsky, and Vetterling, Cambridge University Press, 1986.

[6]The notation [0.0, 1.0) implies that the range of the random numbers is between 0.0 and 1.0, including the number 0.0, but excluding the number 1.0.

way for the user to specify n_0 so that the sequence may be varied from run to run.

1. **State the problem**
   Write a function `random0` that will generate and return an array `ran` containing one or more numbers with a uniform probability distribution in the range 0 ≤ `ran` < 1.0, based on the sequence specified by Equations 5-6 and 5-7. The function should have one or two input arguments (`n` and `m`) specifying the size of the array to return. If there is one argument, the function should generate square array of size `n` × `n`. If there are two arguments, the function should generate an array of size `n` × `m`. The initial value of the seed n_0 will be specified by a call to a function called `seed`.

2. **Define the inputs and outputs**
   There are two functions in this problem: `seed` and `random0`. The input to function `seed` is an integer to serve as the starting point of the sequence. There is no output from this function. The input to function `random0` is one or two integers specifying the size of the array of random numbers to be generated. If only argument `m` is supplied, the function should generate a square array of size `n` × `n`. If both arguments `m` and `n` are supplied, the function should generate an array of size `n` × `m`. The output from the function is the array of random values in the range [0.0, 1.0).

3. **Describe the algorithm**
   The pseudocode for function `random0` is:

```
function ran = random0 (n, m)
Check for valid arguments
Set m <- n if not supplied
Create output array with "zeros" function
for ii = 1:number of rows
   for jj = 1:number of columns
      ISEED <- mod (8121 * ISEED + 28411, 134456)
      ran(ii,jj) <- ISEED / 134456
   end
end
```

   where the value of `ISEED` is placed in global memory so that it is saved between calls to the function. The pseudocode for function `seed` is trivial:

```
function seed (new_seed)
new_seed <- round(new_seed)
ISEED <- abs(new_seed)
```

   The `round` function is used in case the user fails to supply an integer, and the absolute value function is used in case the user supplies a negative seed. The user will not have to know in advance that only positive integers are legal seeds.

The variable ISEED will be placed in global memory so that it may be accessed by both functions.

4. **Turn the algorithm into MATLAB statements.**
Function random0 is shown below.

```
function ran = random0(n,m)
%RANDOM0 Generate uniform random numbers in [0,1)
% Function RANDOM0 generates an array of uniform
% random numbers in the range [0,1). The usage
% is:
%
% random0(n)     -- Generate an n x n array
% random0(n,m)   -- Generate an n x m array

% Define variables:
%   ii        -- Index variable
%   ISEED     -- Random number seed (global)
%   jj        -- Index variable
%   m         -- Number of columns
%   msg       -- Error message
%   n         -- Number of rows
%   ran       -- Output array

% Record of revisions:
%     Date          Programmer         Description of change
%     ====          ==========         =====================
%   01/12/05       S. J. Chapman      Original code
%
% Declare globl values
global ISEED      % Seed for random number generator

% Check for a legal number of input arguments.
msg = nargchk(1,2,nargin);
error(msg);

% If the m argument is missing, set it to n.
if nargin < 2
   m = n;
end

% Initialize the output array
ran = zeros(n,m);

% Now calculate random values
for ii = 1:n
```

```
   for jj = 1:m
      ISEED = mod(8121*ISEED + 28411, 134456 );
      ran(ii,jj) = ISEED / 134456;
   end
end
```

Function seed is shown below.

```
function seed(new_seed)
%SEED Set new seed for function RANDOM0
% Function SEED sets a new seed for function
% RANDOM0. The new seed should be a positive
% integer.

% Define variables:
%   ISEED     -- Random number seed (global)
%   new_seed -- New seed

% Record of revisions:
%      Date          Programmer          Description of change
%      ====          ==========          =====================
%    01/12/05       S. J. Chapman        Original code
%
% Declare globl values
global ISEED       % Seed for random number generator

% Check for a legal number of input arguments.
msg = nargchk(1,1,nargin);
error(msg);

% Save seed
new_seed = round(new_seed);
ISEED = abs(new_seed);
```

5. **Test the resulting MATLAB programs.**
   If the numbers generated by these functions are truly uniformly distributed random numbers in the range $0 \le$ `ran` < 1.0, then the average of many numbers should be close to 0.5 and the standard deviation of the numbers should be close to $\frac{1}{\sqrt{12}}$.

   Furthermore, the if the range between 0 and 1 is divided into a number of equal-size bins, the number of random values falling in each bin should be about the same. A **histogram** is a plot of the number of values falling in each bin. MATLAB function `hist` will create and plot a histogram from an input data set, so we will use it to verify the distribution of random number generated by `random0`.

   To test the results of these functions, we will perform the following tests:

1. Call `seed` with `new_seed` set to 1024.

2. Call `random0(4)` to see that the results appear random.
3. Call `random0(4)` to verify that the results differ from call to call.
4. Call `seed` again with `new_seed` set to 1024.
5. Call `random0(4)` to see that the results are the same as in (2) above. This verifies that the seed is properly being reset.
6. Call `random0(2,3)` to verify that both input arguments are being used correctly.
7. Call `random0(1,20000)` and calculate the average and standard deviation of the resulting data set using MATLAB functions `mean` and `std`. Compare the results to 0.5 and $\frac{1}{\sqrt{12}}$.
8. Create a histogram of the data from (7) to see if approximately equal numbers of values fall in each bin.

We will perform these tests interactively, checking the results as we go.

```
» seed(1024)
» random0(4)
ans =
    0.0598    1.0000    0.0905    0.2060
    0.2620    0.6432    0.6325    0.8392
    0.6278    0.5463    0.7551    0.4554
    0.3177    0.9105    0.1289    0.6230
» random0(4)
ans =
    0.2266    0.3858    0.5876    0.7880
    0.8415    0.9287    0.9855    0.1314
    0.0982    0.6585    0.0543    0.4256
    0.2387    0.7153    0.2606    0.8922
» seed(1024)
» random0(4)
ans =
    0.0598    1.0000    0.0905    0.2060
    0.2620    0.6432    0.6325    0.8392
    0.6278    0.5463    0.7551    0.4554
    0.3177    0.9105    0.1289    0.6230
» random0(2,3)
ans =
    0.2266    0.3858    0.5876
    0.7880    0.8415    0.9287
» arr = random0(1,20000);
» mean(arr)
ans =
    0.5020
» std(arr)
ans =
    0.2881
```

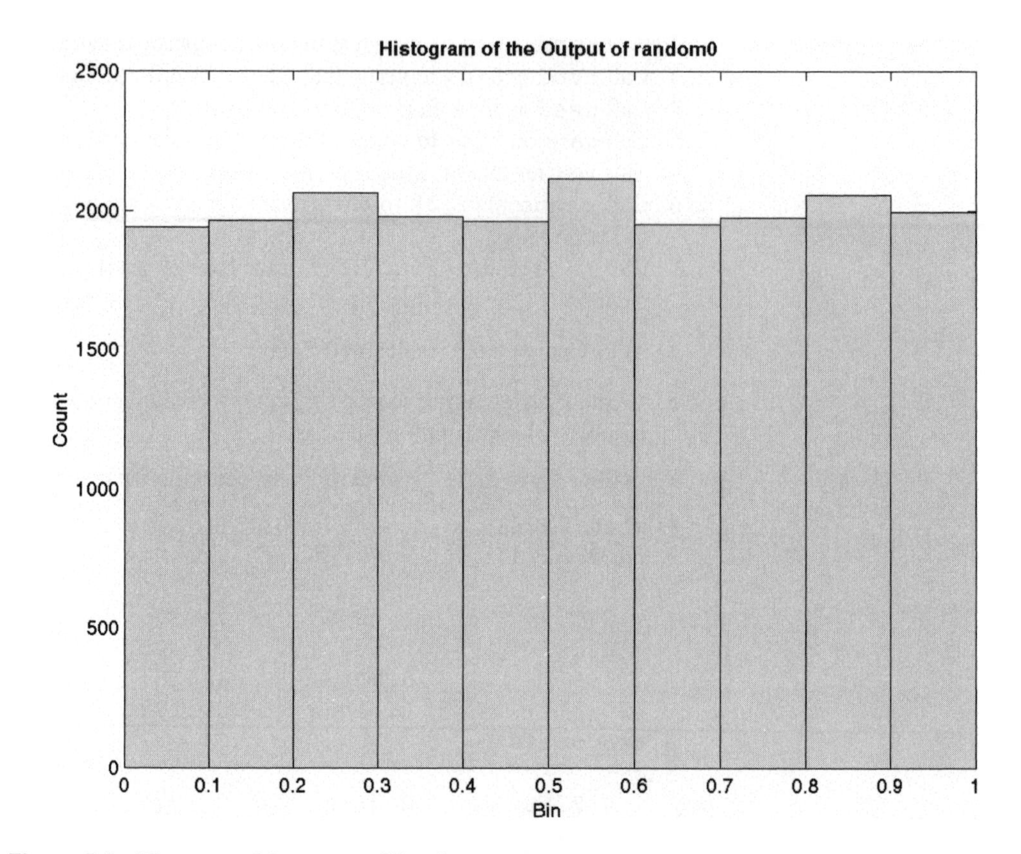

Figure 5.5 Histogram of the output of function `random0`.

```
» hist(arr,10);
» title('\bfHistogram of the Output of random0');
» xlabel('Bin')
» ylabel('Count')
```

The results of these tests look reasonable, so the function appears to be working. The average of the data set was 0.5020, which is quite close to the theoretical value of 0.5000, and the standard deviation of the data set was 0.2881, which is quite close to the theoretical value of 0.2887. The histogram is shown in Figure 5.5, and the distribution of the random values is roughly even across all of the bins.

◄

MATLAB includes two standard functions that generate random values from different distributions. They are

- `rand`—Generates random values from a uniform distribution on the range [0,1)
- `randn`—Generates random values from a normal distribution

Both of them are much faster and much more "random" than the simple function that we have created. If you really need random numbers in your programs, use one of these functions.

Functions `rand` and `randn` have the following calling sequences:

- `rand()`—Generates a single random value
- `rand(n)`—Generates an $n \times n$ array of random values
- `rand(n,m)`—Generates an $n \times m$ array of random values

5.5 Preserving Data Between Calls to a Function

When a function finishes executing, the special workspace created for that function is destroyed, so the contents of all local variables within the function will disappear. The next time that the function is called, a new workspace will be created, and all of the local variables will be returned to their default values. This behavior is usually desirable, since it ensures that MATLAB functions behave in a repeatable fashion every time they are called.

However, it is sometimes useful to preserve some local information within a function between calls to the function. For example, we might which to create a counter to count the number of times that the function has been called. If such a counter were destroyed every time the function exited, the count would never exceed 1!

MATLAB includes a special mechanism to allow local variables to be preserved between calls to a function. **Persistent memory** is a special type of memory that can only be accessed from within the function, but is preserved unchanged between calls to the function.

A persistent variable is declared with the **`persistent` statement**. The form of a global statement is

```
persistent var1 var2 var3 ...
```

where *var1*, *var2*, *var3*, etc. are the variables to be placed in persistent memory.

Good Programming Practice

Use persistent memory to preserve the values of local variables within a function between calls to the function.

Example 5.5—Running Averages

It is sometimes desirable to calculate running statistics on a data set on-the-fly as the values are being entered. The built-in MATLAB functions `mean` and `std` could perform this function, but we would have to pass the entire data set to them for re-calculation after each new data value is entered. A better result can be achieved by writing a special function that keeps tracks of the appropriate running sums between calls, and only needs the latest value to calculate the current average and standard deviation.

The average or arithmetic mean of a set of numbers is defined as

$$\bar{x} = \frac{1}{N}\sum_{i=1}^{N} x_i \tag{5-8}$$

where x_i is sample i out of N samples. The standard deviation of a set of numbers is defined as

$$s = \sqrt{\frac{N\sum_{i=1}^{N} x_i^2 - \left(\sum_{i=1}^{N} x_i\right)^2}{N(N-1)}} \tag{5-9}$$

Standard deviation is a measure of the amount of scatter on the measurements; the greater the standard deviation, the more scattered the points in the data set are. If we can keep track of the number of values N, the sum of the values Σx, and the sum of the squares of the values Σx^2, then we can calculate the average and standard deviation at any time from Equations 5-8 and 5-9.

Write a function to calculate the running average and standard deviation of a data set as it is being entered.

SOLUTION This function must be able to accept input values one at a time and keep running sums of N, Σx, and Σx^2, which will be used to calculate the current average and standard deviation. It must store the running sums in global memory so that they are preserved between calls. Finally, there must be a mechanism to reset the running sums.

1. **State the problem**

   Create a function to calculate the running average and standard deviation of a data set as new values are entered. The function must also include a feature to reset the running sums when desired.

2. **Define the inputs and outputs**

   There are two types of inputs required by this function:

   1. The character string `'reset'` to reset running sums to zero.
   2. The numeric values from the input data set, present one value per function call.

   The outputs from this function are the mean and standard deviation of the data supplied to the function so far.

3. **Design the algorithm**

   This function can be broken down into four major steps

```
Check for a legal number of arguments
Check for a 'reset', and reset sums if present
```

```
Otherwise, add current value to running sums
Calculate and return running average and std dev
   if enough data is available. Return zeros if
   not enough data is available.
```

The detailed pseudocode for these steps is:

```
Check for a legal number of arguments
if x == 'reset'
   n <- 0
   sum_x <- 0
   sum_x2 <- 0
else
   n <- n + 1
   sum_x <- sum_x + x
   sum_x2 <- sum_x2 + x^2
end

% Calculate ave and sd
if n == 0
   ave <- 0
   std <- 0
elseif n == 1
   ave <- sum_x
   std <- 0
else
   ave <- sum_x / n
   std <- sqrt((n*sum_x2 - sum_x^2)/(n*(n-1)))
end
```

4. **Turn the algorithm into MATLAB statements.**
   The final MATLAB function is shown below.

```
function [ave, std] = runstats(x)
%RUNSTATS Generate running ave / std deviation
% Function RUNSTATS generates a running average
% and standard deviation of a data set. The
% values x must be passed to this function one
% at a time. A call to RUNSTATS with the argument
% 'reset' will reset tue running sums.

% Define variables:
%   ave      -- Running average
%   msg      -- Error message
%   n        -- Number of data values
%   std      -- Running standard deviation
%   sum_x    -- Running sum of data values
%   sum_x2   -- Running sum of data values squared
```

```
%   x         -- Input value
%
% Record of revisions:
%     Date          Programmer           Description of change
%     ====          ==========           =====================
%   01/13/05      S. J. Chapman        Original code

% Declare persistent values
persistent n                % Number of input values
persistent sum_x            % Running sum of values
persistent sum_x2           % Running sum of values squared

% Check for a legal number of input arguments.
msg = nargchk(1,1,nargin);
error(msg);

% If the argument is 'reset', reset the running sums.
if x == 'reset'
   n = 0;
   sum_x = 0;
   sum_x2 = 0;
else
   n = n + 1;
   sum_x = sum_x + x;
   sum_x2 = sum_x2 + x^2;
end

% Calculate ave and sd
if n == 0
   ave = 0;
   std = 0;
elseif n == 1
   ave = sum_x;
   std = 0;
else
   ave = sum_x / n;
   std = sqrt((n*sum_x2 - sum_x^2) / (n*(n-1)));
end
```

5. **Test the program.**
   To test this function, we must create a script file that resets `runstats`, reads input values, calls `runstats`, and displays the running statistics. An appropriate script file is shown below:

```
% Script file: test_runstats.m
%
```

```
% Purpose:
%   To read in an input data set and calculate the
%   running statistics on the data set as the values
%   are read in. The running stats will be written
%   to the Command Window.
%
% Record of revisions:
%     Date          Programmer         Description of change
%     ====          ==========         =====================
%   01/13/05      S. J. Chapman      Original code
%
% Define variables:
%   array  -- Input data array
%   ave    -- Running average
%   std    -- Running standard deviation
%   ii     -- Index variable
%   nvals  -- Number of input values
%   std    -- Running standard deviation

% First reset running sums
[ave std] = runstats('reset');

% Prompt for the number of values in the data set
nvals = input('Enter number of values in data set: ');

% Get input values
for ii = 1:nvals

   % Prompt for next value
   string = ['Enter value ' int2str(ii) ': '];
   x = input(string);

   % Get running statistics
   [ave std] = runstats(x);

   % Display running statistics
   fprintf('Average = %8.4f; Std dev = %8.4f\n',ave, std);

end
```

To test this function, we will calculate running statistics by hand for a set of five numbers, and compare the hand calculations to the results from the program. If a data set is created with the following five input values

3., 2., 3., 4., 2.8

then the running statistics calculated by hand would be:

Value	n	$\sum x$	$\sum x^2$	Average	Std_dev
3.0	1	3.0	9.0	3.00	0.000
2.0	2	5.0	13.0	2.50	0.707
3.0	3	8.0	22.0	2.67	0.577
4.0	4	12.0	38.0	3.00	0.816
2.8	5	14.8	45.84	2.96	0.713

The output of the test program for the same data set is:

```
» test_runstats
Enter number of values in data set: 5
Enter value 1:  3
Average =   3.0000; Std dev =   0.0000
Enter value 2:  2
Average =   2.5000; Std dev =   0.7071
Enter value 3:  3
Average =   2.6667; Std dev =   0.5774
Enter value 4:  4
Average =   3.0000; Std dev =   0.8165
Enter value 5:  2.8
Average =   2.9600; Std dev =   0.7127
```

so the results check to the accuracy shown in the hand calculations.

5.6 Function Functions

Function functions are functions whose input arguments include the names of other functions. The functions whose names are passed to the function function are normally used during the function's execution.

For example, MATLAB contains a function function called `fzero`. This function locates a zero of the function that is passed to it. For example, the statement `fzero('cos',[0 pi])` locates a zero of the function `cos` between 0 and π, and `fzero('exp(x)-2',[0 1])` locates a zero of the function `exp(x)-2` between 0 and 1. When these statements are executed, the result is:

```
» fzero('cos',[0 pi])
ans =
   1.5708
» fzero('exp(x)-2',[0 1])
ans =
   0.6931
```

The keys to the operation of function functions are two special MATLAB functions, `eval` and `feval`. Function `eval` *evaluates a character string* as though it had been typed in the Command Window, while function `feval` *evaluates a named function* at a specific input value.

Function `eval` evaluates a character string as though it has been typed in the Command Window. This function gives MATLAB functions a chance to construct executable statements during execution. The form of the `eval` function is

```
eval(string)
```

For example, the statement `x = eval('sin(pi/4)')` produces the result

```
» x = eval('sin(pi/4)')
x =
    0.7071
```

An example where the a character string is constructed and evaluated using the `eval` function is shown below:

```
x = 1;
str = ['exp(' num2str(x) ') -1'];
res = eval(str);
```

In this case, `str` contains the character string `'exp(1)  -1'`, which `eval` evaluates to get the result 1.7183.

Function `feval` evaluates a *named function* defined by an M-file at a specified input value. The general form of the `feval` function is

```
feval(fun,value)
```

For example, the statement `x = feval('sin',pi/4)` produces the result

```
» x = feval('sin',pi/4)
x =
    0.7071
```

Some of the more common MATLAB function functions are listed in Table 5.1. Type `help fun_name` to learn how to use each of these functions.

Table 5.1 Common MATLAB Function Functions

Function Name	Description
`fminbnd`	Minimize a function of one variable.
`fzero`	Find a zero of a function of one variable.
`quad`	Numerically integrate a function.
`ezplot`	Easy to use function plotter.
`fplot`	Plot a function by name.

Example 5.6—Creating a Function Function

Create a function function that will plot any MATLAB function of a single variable between specified starting and ending values.

SOLUTION This function have two input arguments, the first one containing the name of the function to plot and the second one containing a two-element vector with the range of values to plot.

1. **State the problem**
   Create a function to plot any MATLAB function of a single variable between two user-specified limits.

2. **Define the inputs and outputs**
   There are two inputs required by this function:

   1. A character string containing the name of a function.
   2. A two-element vector containing the first and last values to plot.

   The output from this function is a plot of the function specified in the first input argument.

3. **Design the algorithm**
   This function can be broken down into four major steps

   ```
   Check for a legal number of arguments
   Check that the second argument has two elements
   Calculate the value of the function between the
      start and stop points
   Plot and label the function
   ```

   The detailed pseudocode for the evaluation and plotting steps is:

   ```
   n_steps <- 100
   step_size <- (xlim(2) - xlim(1)) / n_steps
   x <- xlim(1):step_size:xlim(2)
   y <- feval(fun,x)
   plot(x,y)
   title(['\bfPlot of function ' fun '(x)'])
   xlabel('\bfx')
   ylabel(['\bf' fun '(x)'])
   ```

4. **Turn the algorithm into MATLAB statements.**
   The final MATLAB function is shown below.

```
function quickplot(fun,xlim)
%QUICKPLOT Generate quick plot of a function
% Function QUICKPLOT generates a quick plot
```

```
% of a function contained in a external M-file,
% between user-specified x limits.

% Define variables:
%   fun       -- Function to plot
%   msg       -- Error message
%   n_steps   -- Number of steps to plot
%   step_size -- Step size
%   x         -- X-values to plot
%   y         -- Y-values to plot
%   xlim      -- Plot x limits
%
% Record of revisions:
%     Date          Programmer          Description of change
%     ====          ==========          =====================
%   01/13/05      S. J. Chapman       Original code

% Check for a legal number of input arguments.
msg = nargchk(2,2,nargin);
error(msg);

% Check the second argument to see if it has two
% elements. Note that this double test allows the
% argument to be either a row or a column vector.
if ( size(xlim,1) == 1 & size(xlim,2) == 2 ) | ...
   ( size(xlim,1) == 2 & size(xlim,2) == 1 )

   % Ok--continue processing.
   n_steps = 100;
   step_size = (xlim(2) - xlim(1)) / n_steps;
   x = xlim(1):step_size:xlim(2);
   y = feval(fun,x);
   plot(x,y);
   title(['\bfPlot of function ' fun '(x)']);
   xlabel('\bfx');
   ylabel(['\bf' fun '(x)']);

else
   % Else wrong number of elements in xlim.
   error('Incorrect number of elements in xlim.');
end
```

5. **Test the program.**

   To test this function, we must call it with correct and incorrect input arguments, verifying that it handles both correct inputs and errors properly. The results are shown below:

```
» quickplot('sin')
??? Error using ==> quickplot
Not enough input arguments.
```

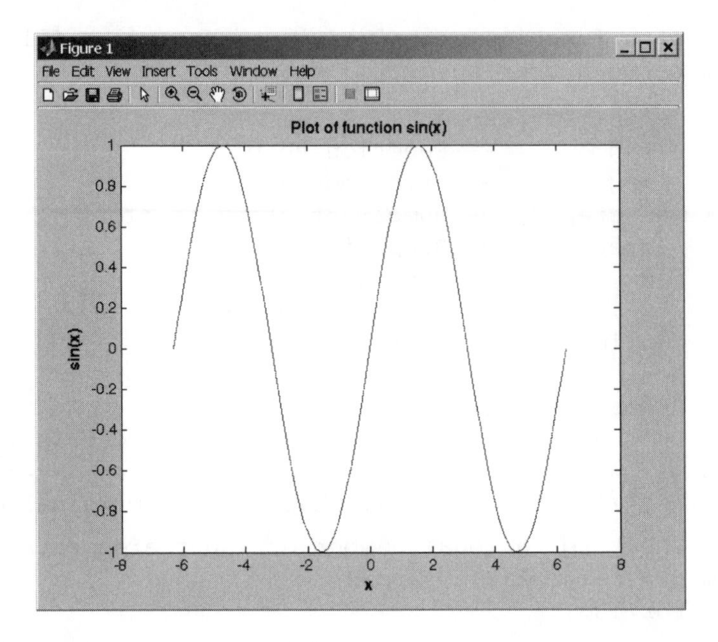

Figure 5.6 Plot of sin x versus x generated by function `quickplot`.

```
» quickplot('sin',[-2*pi 2*pi],3)
??? Error using ==> quickplot
Too many input arguments.

» quickplot('sin',-2*pi)
??? Error using ==> quickplot
Incorrect number of elements in xlim.

» quickplot('sin',[-2*pi 2*pi])
```

The last call was correct, and it produced the plot shown in Figure 5.6.

5.7 Subfunctions and Private Functions

MATLAB includes several special types of functions that behave differently than the ordinary functions we have used so far. Ordinary functions can be called by any other function, as long as they are in the same directory or in any directory on the MATLAB path.

The **scope** of a function is defined as the locations within MATLAB from which the function can be accessed. The scope of an ordinary MATLAB function is the current working directory. If the function lies in a directory on the

MATLAB path, then the scope extends to all MATLAB functions in a program, because they all check the path when trying to find a function with a given name.

In contrast, the scope of the other function types that we will discuss in the rest of this chapter is more limited in one way or another.

5.7.1 Subfunctions

It is possible to place more than one function in a single file. If more than one function is present in a file, the top function is a normal or **primary function**, while the ones below it are **subfunctions**. The primary function should have the same name as the file it appears in. Subfunctions look just like ordinary functions, but they are only accessible to the other functions within the same file. In other words, the scope of a subfunction is the other functions within the same file (see Figure 5.7).

Subfunctions are often used to implement "utility" calculations for a main function. For example, the file `mystats.m` shown below contains a primary function `mystats` and two subfunctions `mean` and `median`. Function `mystats` is a normal MATLAB function, so it can be called by any other MATLAB function in the same directory. If this file is in a directory included in the MATLAB search path, it can be called by any other MATLAB function, even if the other function is not in the same directory. By contrast, the scope of functions `mean` and `median` is restricted to other functions within the same file.

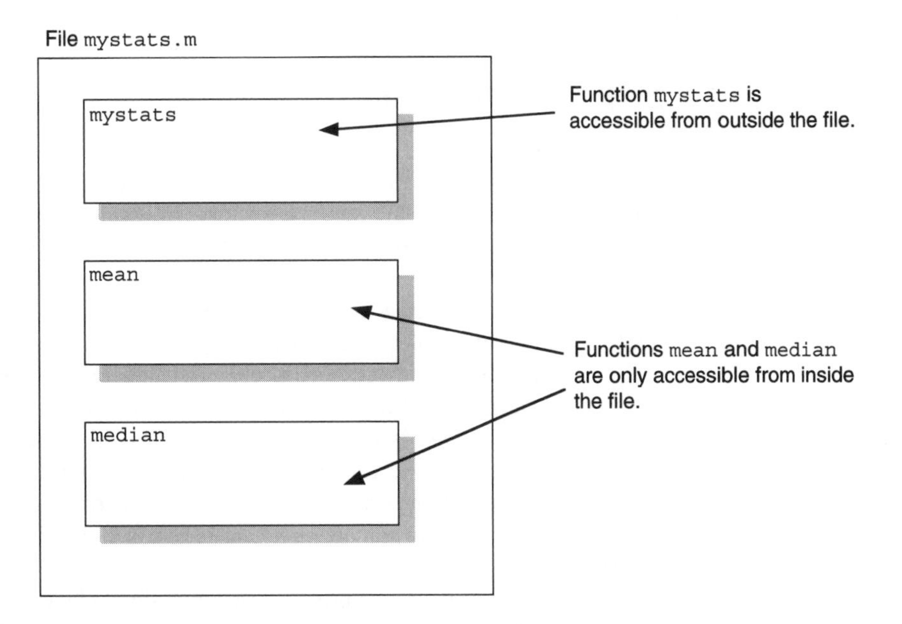

Figure 5.7 The first function in a file is called the primary function. It should have the same name as the file it appears in, and it is accessible from outside the file. The remaining functions in the file are subfunctions; they are accessible only from within the file.

Function `mystats` can call them and they can call each other, but a function outside of the file cannot. They are "utility" functions that perform a part the job of the main function `mystats`.

```
function [avg, med] = mystats(u)
% MYSTATS Find mean and median with internal functions.
% Function MYSTATS calculates the average and median
% of a data set using subfunctions.

n = length(u);
avg = mean(u,n);
med = median(u,n);

function a = mean(v,n)
% Subfunction to calculate average.
a = sum(v)/n;

function m = median(v,n)
% Subfunction to calculate median.
w = sort(v);
if rem(n,2) == 1
   m = w((n + 1)/2);
else
   m = (w(n/2)+ w(n/2 + 1))/2;
end
```

5.7.2 Private Functions

Private functions are functions that reside in subdirectories with the special name `private`. They are only visible to other functions in the `private` directory, or to functions in the parent directory. In other words, the scope of these functions is restricted to the private directory and to the parent directory that contains it.

For example, assume the directory `testing` is on the MATLAB search path. A subdirectory of `testing` called `private` can contain functions that only the functions in `testing` can call. Because private functions are invisible outside of the parent directory, they can use the same names as functions in other directories. This is useful if you want to create your own version of a particular function while retaining the original in another directory. Because MATLAB looks for private functions before standard M-file functions, it will find a private function named `test.m` before a non-private function named `test.m`.

You can create your own private directories simply by creating a subdirectory called `private` under the directory containing your functions. Do not place these private directories on your search path.

When a function is called from within an M-file, MATLAB first checks the file to see if the function is a subfunction defined in the same file. If not, it checks for a private function with that name. If it is not a private function, MATLAB checks current directory for the function name. If it is not in the current directory, MATLAB checks the standard search path for the function.

If you have special-purpose MATLAB functions that should only be used by other functions and never be called directly by the user, consider hiding them as subfunctions or private functions. Hiding the functions will prevent their accidental use, and will also prevent conflicts with other public functions of the same name.

5.7.3 Order of Function Evaluation

In a large program, there could possibly be multiple functions (subfunctions, private functions, nested functions, and public functions) with the same name. When a function with a given name is called, how do we know which copy of the function will be executed?

The answer this question is that MATLAB locates functions in a specific order as follows:

1. MATLAB checks to see if there is a subfunction with the specified name. If so, it is executed.
2. MATLAB checks for a private function with the specified name. If so, it is executed.
3. MATLAB checks for a function with the specified name in the current directory. If so, it is executed.
4. MATLAB checks for a function with the specified name on the MATLAB path. MATLAB will stop searching and execute the first function with the right name found on the path.

5.8 Summary

In Chapter 5, we presented an introduction to user-defined functions. Functions are special types of M-files that receive data through input arguments and return results through output arguments. Each function has its own independent workspace. Each normal function (one that is not a subfunction) should appear in a separate file with the same name as the function, *including capitalization*.

Functions are called by naming them in the Command Window or another M-file. The names used should match the function name exactly, including capitalization. Arguments are passed to functions using a pass-by-value scheme, meaning that MATLAB copies each argument and passes the copy to the function. This copying is important, because the function can freely modify its input arguments without affecting the actual arguments in the calling program.

MATLAB functions can support varying numbers of input and output arguments. Function `nargin` reports the number of actual input arguments used in a function call, and function `nargout` reports the number of actual output arguments used in a function call.

Data can also be shared between MATLAB functions by placing the data in global memory. Global variables are declared using the `global` statement. Global variables may be shared by all functions that declare them. By convention, global variable names are written in all capital letters.

Internal data within a function can be preserved between calls to that function by placing the data in persistent memory. Persistent variables are declared using the `persistent` statement.

Function functions are MATLAB functions whose input arguments include the names of other functions. The functions whose names are passed to the function function are normally used during that function's execution. Examples are some root-solving and plotting functions.

Subfunctions are additional functions placed within a single file. Subfunctions are only accessible from other functions within the same file. Private functions are functions placed in a special subdirectory called `private`. They are only accessible to functions in the parent directory. Subfunctions and private functions can be used to restrict access to MATLAB functions.

5.8.1 Summary of Good Programming Practice

The following guidelines should be adhered to when working with MATLAB functions.

1. Break large program tasks into smaller, more understandable functions whenever possible.
2. Declare global variables in all capital letters to make them easy to distinguish from local variables.
3. Declare global variables immediately after the initial comments and before the first executable statement each function that uses them.
4. You may use global memory to pass large amounts of data among functions within a program.
5. Use persistent memory to preserve the values of local variables within a function between calls to the function.
6. Use subfunctions or private functions to hide special-purpose calculations that should not be generally accessible to other functions. Hiding the functions will prevent their accidental use, and will also prevent conflicts with other public functions of the same name.

5.8.2 MATLAB Summary

The following summary lists all of the MATLAB commands and functions described in this chapter, along with a brief description of each one.

Commands and Functions

`error`	Displays error message and aborts the function producing the error. This function is used if the argument errors are fatal.
`eval`	Evaluates a character string as though it had been typed in the Command Window.
`ezplot`	Easy-to-use function plotter.
`feval`	Calculates the value of a function $f(x)$ defined by an M-file at a specific x.
`fmin`	Minimize a function of one variable.
`fplot`	Plot a function by name.
`fzero`	Find a zero of a function of one variable.
`global`	Declares global variables.
`hist`	Calculate and plot a histogram of a data set.
`inputname`	Returns the actual name of the variable that corresponds to a particular argument number.
`nargchk`	Returns a standard error message if a function is called with too few or too many arguments.
`nargin`	Returns the number of actual input arguments that were used to call the function.
`nargout`	Returns the number of actual output arguments that were used to call the function.
`persistent`	Declares persistent variables.
`quad`	Numerically integrate a function.
`rand`	Generates random values from a uniform distribution.
`randn`	Generates random values from a normal distribution.
`return`	Stop executing a function and return to caller.
`warning`	Displays a warning message and continues function execution. This function is used if the argument errors are not fatal, and execution can continue.

5.9 Exercises

5.1 What is the difference between a script file and a function?

5.2 When a function is called, how is data passed from the caller to the function, and how are the results of the function returned to the caller?

5.3 What are the advantages and disadvantages of the pass-by-value scheme used in MATLAB?

5.4 Modify the selection sort function developed in this chapter so that it accepts a second optional argument, which may be either `'up'` or `'down'`. If the argument is `'up'`, sort the data in ascending order. If the argument is `'down'`, sort the data in descending order. If the argument is missing, the default case is to sort the data in ascending order. (Be sure to handle the case of invalid arguments, and be sure to include the proper help information in your function.)

5.5 Modify function `random0` so that it can accept 0, 1, or 2 calling arguments. If it has no calling arguments, it should return a single random value. If it has 1 or 2 calling arguments, it should behave as it currently does.

5.6 As function `random0` is currently written, it will fail if function `seed` is not called first. Modify function `random0` so that it will function properly with some default seed even if function `seed` is never called.

5.7 Write a function that uses function `random0` to generate a random value in the range [−1.0, 1.0). Make `random0` a subfunction of your new function.

5.8 Write a function that uses function `random0` to generate a random value in the range [`low`, `high`), where `low` and `high` are passed as calling arguments. Make `random0` a subfunction called by your new function.

5.9 **Dice Simulation** It is often useful to be able to simulate the throw of a fair die. Write a MATLAB function `dice` that simulates the throw of a fair die by returning some random integer between 1 and 6 every time that it is called. (*Hint:* Call `random0` to generate a random number. Divide the possible values out of `random0` into six equal intervals, and return the number of the interval that a given random value falls into.)

5.10 **Road Traffic Density** Function `random0` produces a number with a *uniform* probability distribution in the range [0.0, 1.0). This function is suitable for simulating random events if each outcome has an equal probability of occurring. However, in many events, the probability of occurrence is *not* equal for every event, and a uniform probability distribution is not suitable for simulating such events.

For example, when traffic engineers studied the number of cars passing a given location in a time interval of length t, they discovered that the probability of k cars passing during the interval is given by the equation

$$P(k, t) = e^{-\lambda t}\frac{(\lambda t)^k}{k!} \quad \text{for } t \geq 0,\ \lambda > 0,\ \text{ and } k = 0, 1, 2, \ldots \qquad (5\text{-}10)$$

This probability distribution is known as the *Poisson distribution;* it occurs in many applications in science and engineering. For example, the number of calls k to a telephone switchboard in time interval t, the number of bacteria k in a specified volume t of liquid, and the number of failures k of a complicated system in time interval t all have Poisson distributions.

Write a function to evaluate the Poisson distribution for any k, t, and λ. Test your function by calculating the probability of 0, 1, 2, . . . , 5 cars passing a particular point on a highway in 1 minute, given that λ is 1.6 per minute for that highway. Plot the Poisson distribution for $t = 1$ and $\lambda = 1.6$.

5.11 Write three MATLAB functions to calculate the hyperbolic sine, cosine, and tangent functions:

$$\sinh(x) = \frac{e^x - e^{-x}}{2} \qquad \cosh(x) = \frac{e^x + e^{-x}}{2} \qquad \tanh(x) = \frac{e^x - e^{-x}}{e^x + e^{-x}}$$

Use your functions to plot the shapes of the hyperbolic sine, cosine, and tangent functions.

5.12 Write a single MATLAB function `hyperbolic` to calculate the hyperbolic sine, cosine, and tangent functions as defined in the previous problem. The function should have two arguments. The first argument will be a string containing the function names `'sinh'`, `'cosh'`, or `'tanh'`, and the second argument will be the value of x at which to evaluate the function. The file should also contain three subfunctions `sinh1`, `cosh1`, and `tanh1` to perform the actual calculations, and the primary function should call the proper subfunction depending on the value in the string. [**Note:** Be sure to handle the case of an incorrect number of arguments, and also the case of an invalid string. In either case, the function should generate an error.]

5.13 **Cross Product** Write a function to calculate the cross product of two vectors $\mathbf{V}_1$ and $\mathbf{V}_2$:

$$\mathbf{V}_1 \times \mathbf{V}_2 = (V_{y1}V_{z2} - V_{y2}V_{z1})\,\mathbf{i} + (V_{z1}V_{x2} - V_{z2}V_{x1})\,\mathbf{j} + (V_{x1}V_{y2} - V_{x2}V_{y1})\,\mathbf{k}$$

where $\mathbf{V}_1 = V_{x1}\,\mathbf{i} + V_{y1}\,\mathbf{j} + V_{z1}\,\mathbf{k}$ and $\mathbf{V}_2 = V_{x2}\,\mathbf{i} + V_{y2}\,\mathbf{j} + V_{z2}\,\mathbf{k}$. Note that this function will return a real array as its result. Use the function to calculate the cross product of the two vectors $\mathbf{V}_1 = [-2, 4, 0.5]$ and $\mathbf{V}_2 = [0.5, 3, 2]$.

5.14 **Sort with Carry** It is often useful to sort an array `arr1` into ascending order, while simultaneously carrying along a second array `arr2`. In such a sort, each time an element of array `arr1` is exchanged with another element of `arr1`, the corresponding elements of array `arr2` are also swapped. When the sort is over, the elements of array `arr1` are in ascending order, while the elements of array `arr2` that were associated with particular elements of array `arr1` are still associated with them. For example, suppose we have the following two arrays:

Element	arr1	arr2
1.	6.	1.
2.	1.	0.
3.	2.	10.

After sorting array `arr1` while carrying along array `arr2`, the contents of the two arrays will be:

Element	arr1	arr2
1.	1.	0.
2.	2.	10.
3.	6.	1.

Write a function to sort one real array into ascending order while carrying along a second one. Test the function with the following two 9-element arrays:

```
a = [1, 11, -6, 17, -23, 0, 5, 1, -1];
b = [31, 101, 36, -17, 0, 10, -8, -1, -1];
```

5.15 Use the Help Browser to look up information about the standard MATLAB function `sortrows`, and compare the performance of `sortrows` with the sort-with-carry function created in the previous exercise. To do this, create two copies of a 1,000 × 2 element array containing random values, and sort column 1 of each array while carrying along column 2 using both functions. Determine the execution times of each sort function using `tic` and `toc`. How does the speed of your function compare with the speed of the standard function `sortrows`?

5.16 Figure 5.8 shows two ships steaming on the ocean. Ship 1 is at position (x_1, y_1) and steaming on heading θ_1. Ship 2 is at position (x_2, y_2) and steaming on heading θ_2. Suppose that Ship 1 makes radar contact with an object at range r_1 and bearing ϕ_1. Write a MATLAB function that will calculate the range r_2 and bearing ϕ_2 at which Ship 2 should see the object.

5.17 **Minima and Maxima of a Function** Write a function that attempts to locate the maximum and minimum values of an arbitrary function $f(x)$ over a certain range. The function being evaluated should be passed to the function as a calling argument. The function should have the following input arguments:

`first_value`—The first value of x to search
`last_value`—The last value of x to search
`num_steps`—The number of steps to include in the search
`func`—The name of the function to search

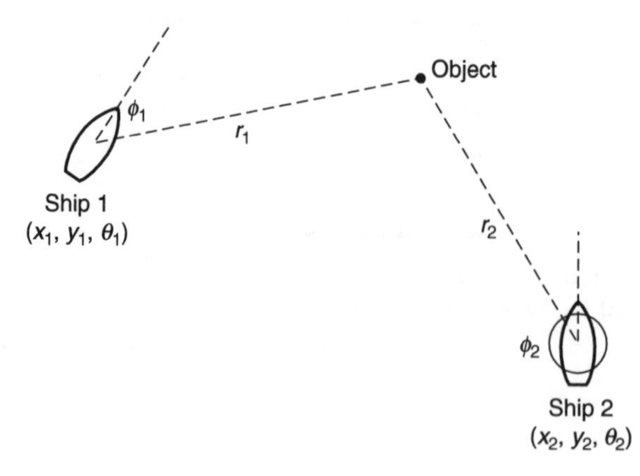

Figure 5.8 Two ships at positions (x_1, y_1) and (x_2, y_2) respectively. Ship 1 is traveling at heading θ_1, and Ship 2 is traveling at heading θ_2.

The function should have the following output arguments:

`xmin`—The value of x at which the minimum was found
`min_value`—The minimum value of $f(x)$ found
`xmax`—The value of x at which the maximum was found
`max_value`—The maximum value $f(x)$ found

Be sure to check that there are a valid number of input arguments, and that the MATLAB `help` and `lookfor` commands are properly supported.

5.18 Write a test program for the function generated in the previous exercise. The test program should pass to the function function the user-defined function $f(x) = x^3 - 5x^2 + 5x + 2$, and search for the minimum and maximum in 200 steps over the range $-1 \leq x \leq 3$. It should print out the resulting minimum and maximum values.

5.19 **Derivative of a Function** The *derivative* of a continuous function $f(x)$ is defined by the equation

$$\frac{d}{dx}f(x) = \lim_{\Delta x \to 0} \frac{f(x + \Delta x) - f(x)}{\Delta x} \tag{5-11}$$

In a sampled function, this definition becomes

$$f'(x_i) = \frac{f(x_{i+1}) - f(x_i)}{\Delta x} \tag{5-12}$$

where $\Delta x = x_{i+1} - x_i$. Assume that a vector `vect` contains `nsamp` samples of a function taken at a spacing of `dx` per sample. Write a function that will calculate the derivative of this vector from Equation 5-12. The function should check to make sure that `dx` is greater than zero to prevent divide-by-zero errors in the function.

To check you function, you should generate a data set whose derivative is known, and compare the result of the function with the known correct answer. A good choice for a test function is $\sin x$. From elementary calculus, we know that $\frac{d}{dx}(\sin x) = \cos x$. Generate an input vector containing 100 values of the function $\sin x$ starting at $x = 0$, and using a step size Δx of 0.05. Take the derivative of the vector with your function, and then compare the resulting answers to the known correct answer. How close did your function come to calculating the correct value for the derivative?

5.20 **Derivative in the Presence of Noise** We will now explore the effects of input noise on the quality of a numerical derivative. First, generate an input vector containing 100 values of the function $\sin x$ starting at $x = 0$, and using a step size Δx of 0.05, just as you did in the previous problem. Next, use function `random0` to generate a small amount of random noise with a maximum amplitude of ± 0.02, and add that random noise to the samples in your input vector. Note that the peak amplitude of the noise is only 2% of the peak amplitude of your signal, since the maximum value of $\sin x$ is 1. Now take the derivative of the function using the derivative

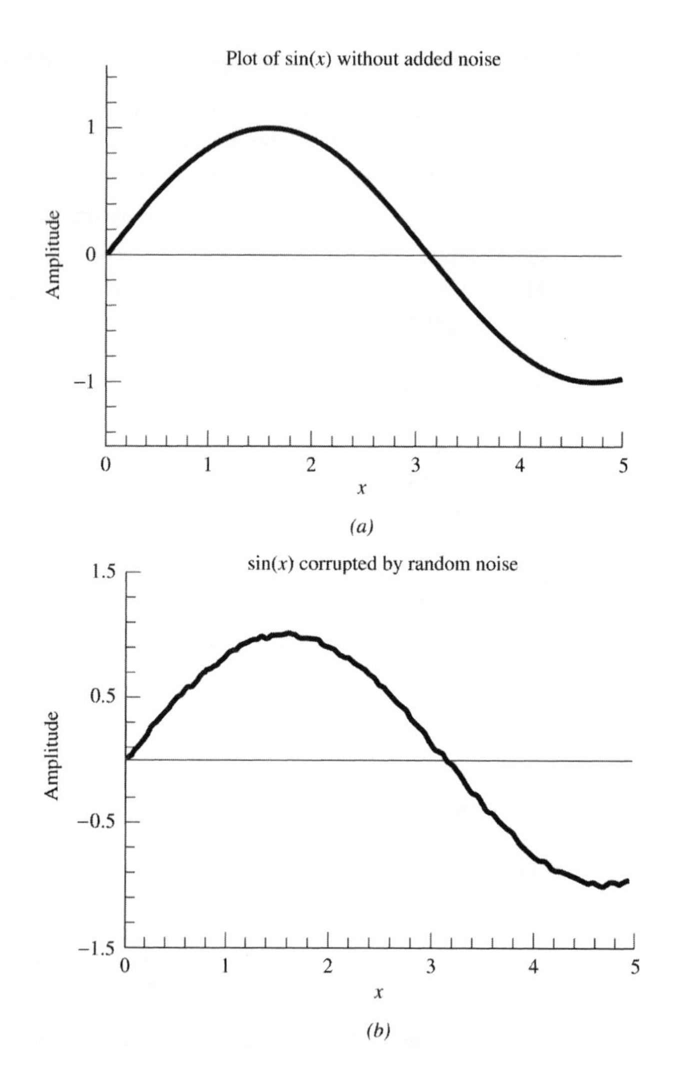

Figure 5.9 *(a)* A plot of sin x as a function of x with no noise added to the data. *(b)* A plot of sin x as a function of x with a 2% peak amplitude uniform random noise added to the data.

function that you developed in the last problem. How close to the theoretical value of the derivative did you come?

5.21 **Linear Least Squares Fit** Develop a function that will calculate slope m and intercept b of the least-squares line that best fits an input data set. The input data points (x, y) will be passed to the function in two input arrays, `x` and `y`. (The equations describing the slope and intercept of the least-squares line given in Example 4-6 in the previous chapter.)

Test your function using a test program and the following 20-point input data set:

Sample Data to Test Least Squares Fit Routine

No.	x	y	No.	x	y
1	−4.91	−8.18	11	−0.94	0.21
2	−3.84	−7.49	12	0.59	1.73
3	−2.41	−7.11	13	0.69	3.96
4	−2.62	−6.15	14	3.04	4.26
5	−3.78	−5.62	15	1.01	5.75
6	−0.52	−3.30	16	3.60	6.67
7	−1.83	−2.05	17	4.53	7.70
8	−2.01	−2.83	18	5.13	7.31
9	0.28	−1.16	19	4.43	9.05
10	1.08	0.52	20	4.12	10.95

5.22 Correlation Coefficient of Least Squares Fit Develop a function that will calculate both the slope m and intercept b of the least-squares line that best fits an input data set, and also the correlation coefficient of the fit. The input data points (x, y) will be passed to the function in two input arrays, `x` and `y`. The equations describing the slope and intercept of the least-squares line are given in Example 4-6, and the equation for the correlation coefficient is

$$r = \frac{n\left(\sum xy\right) - \left(\sum x\right)\left(\sum y\right)}{\sqrt{\left[\left(n\sum x^2\right) - \left(\sum x\right)^2\right]\left[\left(n\sum y^2\right) - \left(\sum y\right)^2\right]}} \tag{5-13}$$

where

$\sum x$ is the sum of the x values
$\sum y$ is the sum of the y values
$\sum x^2$ is the sum of the squares of the x values
$\sum y^2$ is the sum of the squares of the y values
$\sum xy$ is the sum of the products of the corresponding x and y values
n is the number of points included in the fit

Test your function using a test driver program and the 20-point input data set given in the previous problem.

5.23 Recursion. A function is said to be *recursive* if the function calls itself. MATLAB functions are designed to allow recursive operation. To test this feature, write a MATLAB function to evaluate the factorial function, which is defined as follows:

$$N! = \begin{cases} N(N-1)! & N \geq 1 \\ 1 & N = 0 \end{cases} \tag{5-14}$$

where N is a positive integer. The function should check to make sure that there is a single argument N, and that N is a non-negative integer. If it is not, generate an error using the `error` function. If the input argument is a non-negative integer, the function should evaluate $N!$ using Equation (5-14).

5.24 The Birthday Problem The Birthday Problem is: if there are a group of n people in a room, what is the probability that two or more of them have the same birthday? It is possible to determine the answer to this question by simulation. Write a function that calculates the probability that two or more of n people will have the same birthday, where n is a calling argument. (*Hint:* To do this, the function should create an array of size n and generate n birthdays in the range 1 to 365 randomly. It should then check to see if any of the n birthdays are identical. The function should perform this experiment at least 5000 times, and calculate the fraction of those times in which two or more people had the same birthday.) Write a test program that calculates and prints out the probability that two or more of n people will have the same birthday for $n = 2, 3, \ldots, 40$.

5.25 Use function `random0` to generate a set of three arrays of random numbers. The three arrays should be 100, 1000, and 2000 elements long. Then, use functions `tic` and `toc` to determine the time that it takes function `ssort` to sort each array. How does the elapsed time to sort increase as a function of the number of elements being sorted? (*Hint:* On a fast computer, you will need to sort each array many times and calculate the average sorting time in order to overcome the quantization error of the system clock.)

5.26 Gaussian (Normal) Distribution Function `random0` returns a uniformly-distributed random variable in the range [0, 1), which means that there is an equal probability of any given number in the range occurring on a given call to the function. Another type of random distribution is the Gaussian Distribution, in which the random value takes on the classic bell-shaped curve shown in Figure 5.10. A Gaussian Distribution with an average of 0.0 and a standard deviation of 1.0 is called a *standardized normal distribution,* and the probability of any given value occurring in the standardized normal distribution is given by the equation

$$p(x) = \frac{1}{\sqrt{2\pi}} e^{-x^2/2} \tag{5-15}$$

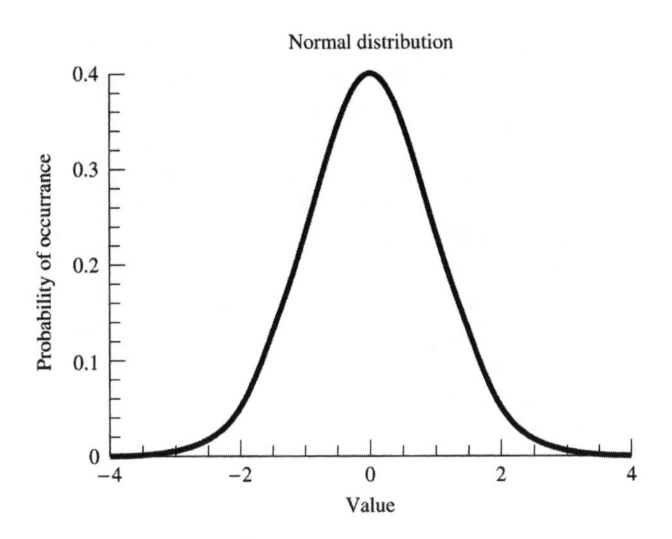

Figure 5.10 A Normal probability distribution.

It is possible to generate a random variable with a standardized normal distribution starting from a random variable with a uniform distribution in the range $[-1, 1)$ as follows:

1. Select two uniform random variables x_1 and x_2 from the range $[-1, 1)$ such that $x_1^2 + x_2^2 < 1$. To do this, generate two uniform random variables in the range $[-1, 1)$, and see if the sum of their squares happens to be less than 1. If so, use them. If not, try again.
2. Then each of the values y_1 and y_2 in the equations below will be a normally-distributed random variable.

$$y_1 = \sqrt{\frac{-2\ln r}{r}}\,x_1 \tag{5-16}$$

$$y_2 = \sqrt{\frac{-2\ln r}{r}}\,x_2 \tag{5-17}$$

where

$$r = x_1^2 + x_2^2 \tag{5-18}$$

Write a function that returns a normally-distributed random value each time that it is called. Test your function by getting 1000 random values, calculating the standard deviation, and plotting a histogram of the distribution. How close to 1.0 was the standard deviation?

5.27 **Gravitational Force** The gravitational force F between two bodies of masses m_1 and m_2 is given by the equation

$$F = \frac{Gm_1m_2}{r^2} \tag{5-19}$$

where G is the gravitation constant (6.672×10^{-11} N m^2/kg^2), m_1 and m_2 are the masses of the bodies in kilograms, and r is the distance between the two bodies. Write a function to calculate the gravitational force between two bodies given their masses and the distance between them. Test you function by determining the force on an 800 kg satellite in orbit 38,000 km above the Earth. (The mass of the Earth is 5.98×10^{24} kg.)

5.28 Rayleigh Distribution The Rayleigh distribution is another random number distribution that appears in many practical problems. A Rayleigh-distributed random value can be created by taking the square root of the sum of the squares of two normally-distributed random values. In other words, to generate a Rayleigh-distributed random value r, get two normally distributed random values (n_1 and n_2), and perform the following calculation:

$$r = \sqrt{n_1^2 + n_2^2} \tag{5-20}$$

(a) Create a function `rayleigh(n,m)` that returns an `n` × `m` array of Rayleigh-distributed random numbers. If only one argument is supplied [`rayleigh(n)`], the function should return an `n` × `n` array of Rayleigh-distributed random numbers. Be sure to design your function with input argument checking and with proper documentation for the MATLAB help system.

(b) Test your function by creating an array of 20,000 Rayleigh-distributed random values and plotting a histogram of the distribution. What does the distribution look like?

(c) Determine the mean and standard deviation of the Rayleigh distribution.

5.29 Constant False Alarm Rate (CFAR) A simplified radar receiver chain is shown in Figure 5.11*a*. When a signal is received in this receiver, it contains both the desired information (returns from targets) and thermal noise. After the detection step in the receiver, we would like to be able to pick out received target returns from the thermal noise background. We can do this by setting a threshold level, and then declaring that we see a target whenever the signal crosses that threshold. Unfortunately, it is occasionally

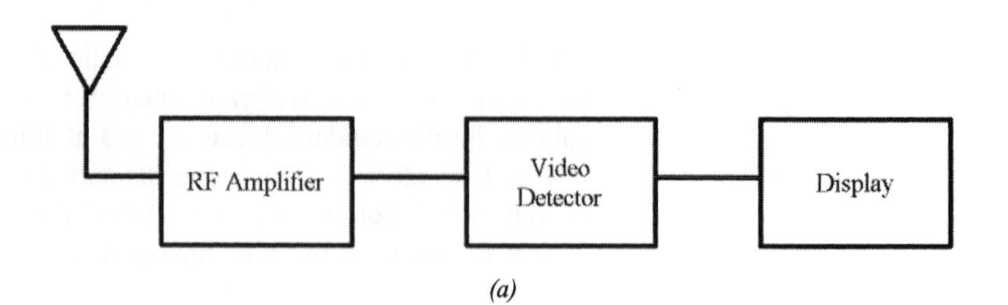

Figure 5.11 *(a)* A typical radar receiver.

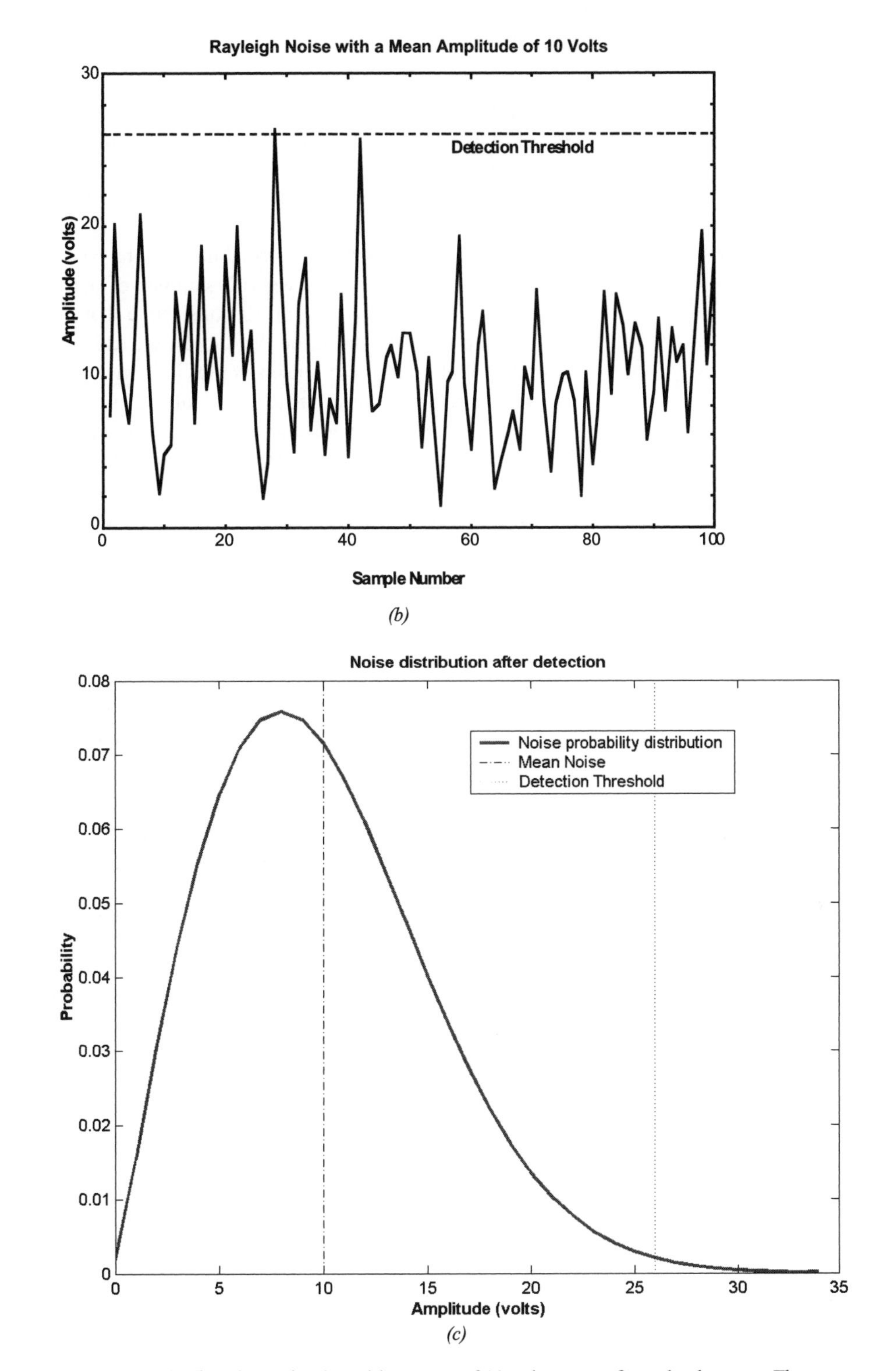

Figure 5.11 *(continued)* *(b)* Thermal noise with a mean of 10 volts output from the detector. The noise sometimes crosses the detection threshold. *(c)* Probability distribution of the noise out of the detector.

possible for the receiver noise to cross the detection threshold even if no target is present. If that happens, we will declare the noise spike to be a target, creating a *false alarm*. The detection threshold needs to be set as low as possible so that we can detect weak targets, but it must not be set too low, or we get many false alarms.

After video detection, the thermal noise in the receiver has a Rayleigh distribution. Figure 5.11*b* shows 100 samples of a Rayleigh-distributed noise with a mean amplitude of 10 volts. Note that there would be one false alarm even if the detection threshold were as high as 26! The probability distribution of these noise samples is shown in Figure 5.11*c*.

Detection thresholds are usually calculated as a multiple of the mean noise level, so that if the noise level changes, the detection threshold will change with it to keep false alarms under control. This is known as *constant false alarm rate* (CFAR) detection. A detection threshold is typical quoted in decibels. The relationship between the threshold in dB and the threshold in volts is

$$\text{Threshold (volts)} = \text{Mean Noise Level (volts)} \times 10^{\frac{\text{dB}}{20}} \tag{5-21}$$

or

$$\text{dB} = 20 \log_{10}\left(\frac{\text{Threshold (volts)}}{\text{Mean Noise Level (volts)}}\right) \tag{5-22}$$

The false alarm rate for a given detection threshold is calculated as:

$$P_{fa} = \frac{\text{Number of False Alarms}}{\text{Total Number of Samples}} \tag{5-23}$$

Write a program that generates 1,000,000 random noise samples with a mean amplitude of 10 volts and a Rayleigh noise distribution. Determine the false alarm rates when the detection threshold is set to 5, 6, 7, 8, 9, 10, 11, 12, and 13 dB above the mean noise level. At what level should the threshold be set to achieve a false alarm rate of 10^{-4}?

CHAPTER 6

Additional Data Types and Plot Types

In earlier chapters, we were introduced to three fundamental MATLAB data types: `double`, `logical`, and `char`. In this chapter, we will learn additional details about the `double` and `char` data types.

First, we will learn how to create, manipulate, and plot complex values in the `double` data type. Then, we will learn more about using the `char` data type, and how to extend MATLAB arrays of any type to more than two dimensions.

The chapter concludes with a discussion of additional types of plots available in MATLAB.

6.1 Complex Data

Complex numbers are numbers with both a real and an imaginary component. Complex numbers occur in many problems in science and engineering. For example, complex numbers are used in electrical engineering to represent alternating current voltages, currents, and impedances. The differential equations that describe the behavior of most electrical and mechanical systems also give rise to complex numbers. Because they are so ubiquitous, it is impossible to work as an engineer without a good understanding of the use and manipulation of complex numbers.

A complex number has the general form

$$c = a + bi \tag{6-1}$$

where c is a complex number, a and b are both real numbers, and i is $\sqrt{-1}$. The number a is called the *real part* and b is called the *imaginary part* of the complex

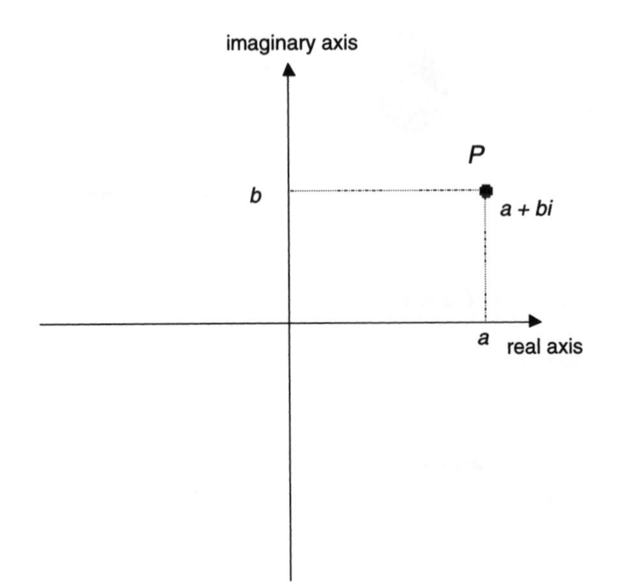

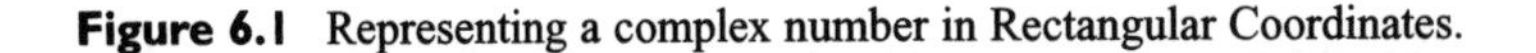
Figure 6.1 Representing a complex number in Rectangular Coordinates.

number c. Since a complex number has two components, it can be plotted as a point on a plane (see Figure 6.1). The horizontal axis of the plane is the real axis, and the vertical axis of the plane is the imaginary axis, so that any complex number $a + bi$ can be represented as a single point a units along the real axis and b units along the imaginary axis. A complex number represented this way is said to be in *rectangular coordinates*, since the real and imaginary axes define the sides of a rectangle.

A complex number can also be represented as a vector of length z and angle θ pointing from the origin of the plane to the point P (see Figure 6.2). A complex number represented this way is said to be in *polar coordinates*.

$$c = a + bi = z\angle\theta$$

The relationships among the rectangular and polar coordinate terms a, b, z, and θ are:

$$a = z \cos \theta \tag{6-2}$$

$$b = z \sin \theta \tag{6-3}$$

$$z = \sqrt{a^2 + b^2} \tag{6-4}$$

$$\theta = \tan^{-1} \frac{b}{a} \tag{6-5}$$

MATLAB uses rectangular coordinates to represent complex numbers. Each complex number consists of a pair of real numbers (a, b). The first number (a) is

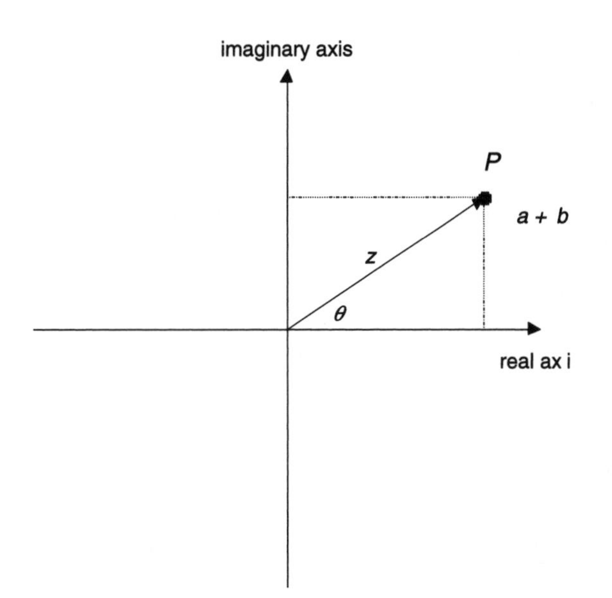

Figure 6.2 Representing a complex number in Polar Coordinates.

the real part of the complex number, and the second number (b) is the imaginary part of the complex number.

If complex numbers c_1 and c_2 are defined as $c_1 = a_1 + b_1 i$ and $c_2 = a_2 + b_2 i$, then the addition, subtraction, multiplication, and division of c_1 and c_2 are defined as:

$$c_1 + c_2 = (a_1 + a_2) + (b_1 + b_2)i \tag{6-6}$$

$$c_1 - c_2 = (a_1 - a_2) + (b_1 - b_2)i \tag{6-7}$$

$$c_1 \times c_2 = (a_1 a_2 - b_1 b_2) + (a_1 b_2 + b_1 a_2)i \tag{6-8}$$

$$\frac{c_1}{c_2} = \frac{a_1 a_2 + b_1 b_2}{a_2^2 + b_2^2} + \frac{b_1 a_2 - a_1 b_2}{a_2^2 + b_2^2} i \tag{6-9}$$

When two complex numbers appear in a binary operation, MATLAB performs the required additions, subtractions, multiplications, or divisions between the two complex numbers using versions of the preceding formulas.

6.1.1 Complex Variables

A complex variable is created automatically when a complex value is assigned to a variable name. This easiest way to create a complex value is to use the intrinsic values `i` or `j`, both of which are predefined to be $\sqrt{-1}$. For example, the following statement stores the complex value $4 + i3$ into variable `c1`.

```
» c1 = 4 + i*3
c1 =
     4.0000 + 3.0000i
```

Alternatively, the imaginary part can be specified by simply appending an `i` or `j` to the end of a number:

```
» c1 = 4 + 3i
c1 =
     4.0000 + 3.0000i
```

The function `isreal` can be used to determine whether a given array is real or complex. If any element of an array has an imaginary component, the array is complex, and `isreal(array)` returns a 0.

6.1.2 Using Complex Numbers with Relational Operators

It is possible to compare two complex numbers with the == relational operator to see if they are equal to each other, and to compare them with the ~= operator to see if they are not equal to each other. Both of these operators produce the expected results. For example, if $c_1 = 4 + i3$ and $c_2 = 4 - i3$, the relational operation $c_1 == c_2$ produces a 0, and the relational operation $c_1 \sim= c_2$ produces a 1.

However, *comparisons with the >, <, >=, or <= operators do not produce the expected results*. When complex numbers are compared with these relational operators, only the *real parts* of the numbers are compared. For example, if $c_1 = 4 + i3$ and $c_2 = 3 + i8$, the relational operation $c_1 > c_2$ produces a true (1) even though the magnitude of c_1 is really smaller than the magnitude of c_2.

If you ever need to compare two complex numbers with these operators, you will probably be more interested in the total magnitude of the number than we are in the magnitude of only its real part. The magnitude of a complex number can be calculated with the `abs` intrinsic function (see below), or directly from Equation 6-4.

$$|c| = \sqrt{a^2 + b^2} \tag{6-4}$$

If we compare the *magnitudes* of c_1 and c_2 above, the results are more reasonable: `abs(`c_1`) > abs(`c_2`)` produces a 0, since the magnitude of c_2 is greater than the magnitude of c_1.

Programming Pitfalls

Be careful when using the relational operators with complex numbers. The relational operators >, >=, <, and <= compare only the *real parts* of complex numbers, not their magnitudes. If you need these relational operators with complex numbers, it will probably be more sensible to compare the total magnitudes rather than only the real components.

6.1.3 Complex Functions

MATLAB includes many functions that support complex calculations. These functions fall into three general categories:

1. **Type conversion functions** These functions convert data from the complex data type to the real (`double`) data type. Function `real` converts the *real part* of a complex number into the double data type, and throws away the imaginary part of the complex number. Function `imag` converts the *imaginary part* of a complex number into a real number.
2. **Absolute value and angle functions** These functions convert a complex number to its polar representation. Function `abs(c)` calculates the absolute value of a complex number using the equation

$$abs(c) = \sqrt{a^2 + b^2}$$

   where $c = a + bi$. Function `angle(c)` calculates the angle of a complex number using the equation

```
angle(c) = atan2(imag(c),real(c))
```

   producing an answer in the range $-\pi \leq \theta \leq \pi$.
3. **Mathematical functions** Most elementary mathematical functions are defined for complex values. These functions include exponential functions, logarithms, trigonometric functions, and square roots. The functions `sin`, `cos`, `log`, `sqrt`, *etc.* will work as well with complex data as they will with real data.

Some of the intrinsic functions that support complex numbers are listed in Table 6.1.

Table 6.1 Some Functions that Support Complex Numbers

Function	Description
`conj(c)`	Computes the complex conjugate of a number `c`. If $c = a + bi$, then `conj(c)` $= a - bi$.
`real(c)`	Returns the real portion of the complex number `c`.
`imag(c)`	Returns the imaginary portion of the complex number `c`.
`isreal(c)`	Returns true (1) if no element of array `c` has an imaginary component. Therefore, `~isreal(c)` returns true (1) if an array is complex.
`abs(c)`	Returns the magnitude of the complex number `c`.
`angle(c)`	Returns the angle of the complex number `c`, computed from the expression `atan2(imag(c), real(c))`.

Example 6.1—The Quadratic Equation (Revisited)

The availability of complex numbers often simplifies the calculations required to solve problems. For example, when we solved the quadratic equation in Example 3-2, it was necessary to take three separate branches through the program depending on the sign of the discriminant. With complex numbers available, the square root of a negative number presents no difficulties, so we can greatly simplify these calculations.

Write a general program to solve for the roots of a quadratic equation, regardless of type. Use complex variables so that no branches will be required based on the value of the discriminant.

SOLUTION

1. **State the problem**
   Write a program that will solve for the roots of a quadratic equation, whether they are distinct real roots, repeated real roots, or complex roots, without requiring tests on the value of the discriminant.

2. **Define the inputs and outputs**
   The inputs required by this program are the coefficients *a*, *b*, and *c* of the quadratic equation

$$ax^2 + bx + c = 0 \tag{3-1}$$

   The output from the program will be the roots of the quadratic equation, whether they are real, repeated, or complex.

3. **Describe the algorithm**
   This task can be broken down into three major sections, whose functions are input, processing, and output:

```
Read the input data
Calculate the roots
Write out the roots
```

   We will now break each of the foregoing major sections into smaller, more detailed pieces. In this algorithm, the value of the discriminant is unimportant in determining how to proceed. The resulting pseudocode is:

```
Prompt the user for the coefficients a, b, and c.
Read a, b, and c
discriminant <- b^2 - 4 * a * c
x1 <- ( -b + sqrt(discriminant) ) / ( 2 * a )
x2 <- ( -b - sqrt(discriminant) ) / ( 2 * a )
```

```
Print 'The roots of this equation are: '
Print 'x1 = ', real(x1), ' +i ', imag(x1)
Print 'x2 = ', real(x2), ' +i ', imag(x2)
```

4. **Turn the algorithm into MATLAB statements**
   The final MATLAB code is as follows:

```
% Script file: calc_roots2.m
%
% Purpose:
%   This program solves for the roots of a quadratic equation
%   of the form a*x**2 + b*x + c = 0. It calculates the answers
%   regardless of the type of roots that the equation possesses.
%
% Record of revisions:
%     Date           Programmer          Description of change
%     ====           ==========          =====================
%   01/15/04       S. J. Chapman         Original code
%
% Define variables:
%   a            -- Coefficient of x^2 term of equation
%   b            -- Coefficient of x term of equation
%   c            -- Constant term of equation
%   discriminant -- Discriminant of the equation
%   x1           -- First solution of equation
%   x2           -- Second solution of equation

% Prompt the user for the coefficients of the equation
disp ('This program solves for the roots of a quadratic ');
disp ('equation of the form A*X^2 + B*X + C = 0. ');
a = input ('Enter the coefficient A: ');
b = input ('Enter the coefficient B: ');
c = input ('Enter the coefficient C: ');

% Calculate discriminant
discriminant = b^2 - 4 * a * c;

% Solve for the roots
x1 = ( -b + sqrt(discriminant) ) / ( 2 * a );
x2 = ( -b - sqrt(discriminant) ) / ( 2 * a );

% Display results
disp ('The roots of this equation are:');
fprintf ('x1 = (%f) +i (%f)\ n', real(x1), imag(x1));
fprintf ('x2 = (%f) +i (%f) \ n', real(x2), imag(x2));
```

5. **Test the program**

   Next, we must test the program using real input data. We will test cases in which the discriminant is greater than, less than, and equal to 0 to be certain that the program is working properly under all circumstances. From Equation (3-1), it is possible to verify the solutions to the equations given below:

$$x^2 + 5x + 6 = 0 \qquad x = -2, \text{ and } x = -3$$
$$x^2 + 4x + 4 = 0 \qquad x = -2$$
$$x^2 + 2x + 5 = 0 \qquad x = -1 \pm 2i$$

   When these coefficients are fed into the program, the results are

```
» calc_roots2
This program solves for the roots of a quadratic
equation of the form A*X^2 + B*X + C = 0.
Enter the coefficient A: 1
Enter the coefficient B: 5
Enter the coefficient C: 6
The roots of this equation are:
x1 = (-2.000000) +i (0.000000)
x2 = (-3.000000) +i (0.000000)
» calc_roots2
This program solves for the roots of a quadratic
equation of the form A*X^2 + B*X + C = 0.
Enter the coefficient A: 1
Enter the coefficient B: 4
Enter the coefficient C: 4
The roots of this equation are:
x1 = (-2.000000) +i (0.000000)
x2 = (-2.000000) +i (0.000000)
» calc_roots2
This program solves for the roots of a quadratic
equation of the form A*X^2 + B*X + C = 0.
Enter the coefficient A: 1
Enter the coefficient B: 2
Enter the coefficient C: 5
The roots of this equation are:
x1 = (-1.000000) +i (2.000000)
x2 = (-1.000000) +i (-2.000000)
```

The program gives the correct answers for our test data in all three possible cases. Note how much simpler this program is compared with the quadratic root solver found in Example 3-1. The complex data type has greatly simplified our program.

◄

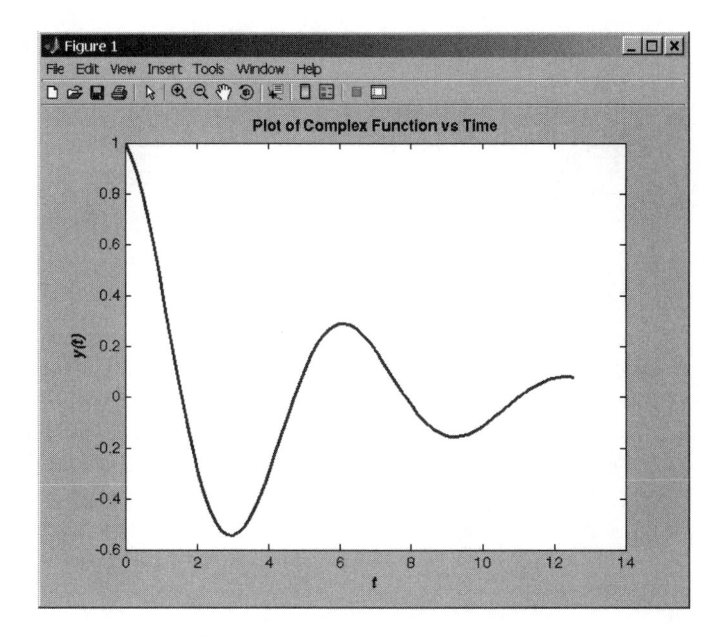

Figure 6.3 Plot of $y(t) = e^{-0.2t}(\cos t + i \sin t)$ using the command `plot(t,y)`.

6.1.4 Plotting Complex Data

Complex data has both real and imaginary components, and plotting complex data with MATLAB is a bit different from plotting real data. For example, consider the function

$$y(t) = e^{-0.2t}(\cos t + i \sin t) \tag{6-10}$$

If this function is plotted with the conventional `plot` command, only the real data will be plotted—the imaginary part will be ignored. The following statements produce the plot shown in Figure 6.3, together with a warning message that the imaginary part of the data is being ignored.

```
t = 0:pi/20:4*pi;
y = exp(-0.2*t).*(cos(t)+i*sin(t));
plot(t,y,'LineWidth',2);
title('\bfPlot of Complex Function vs Time');
xlabel('\bf\itt');
ylabel('\bf\ity(t)');
```

If both the real and imaginary parts of the function are of interest, the user has several choices. Both parts can be plotted as a function of time on the same axes using the statements shown below (see Figure 6.4).

```
t = 0:pi/20:4*pi;
y = exp(-0.2*t).*(cos(t)+i*sin(t));
plot(t,real(y),'b-','LineWidth',2);
hold on;
```

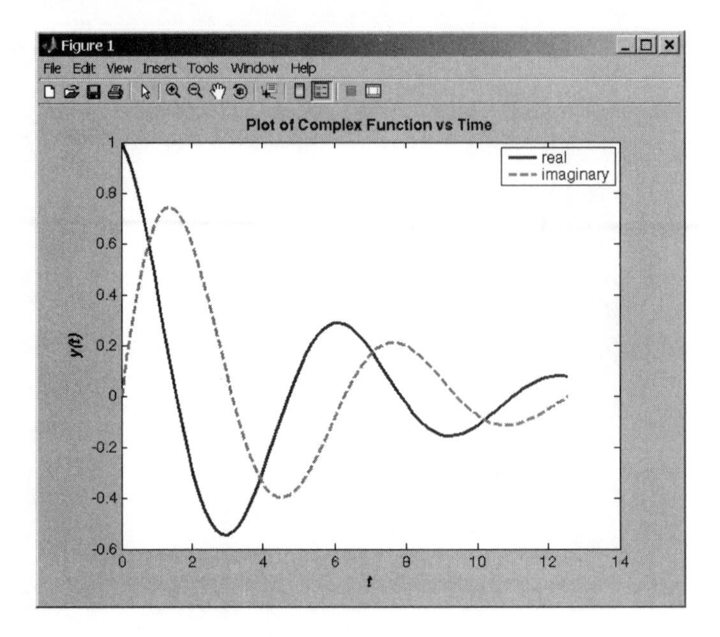

Figure 6.4 Plot of real and imaginary parts of $y(t)$ versus time.

```
plot(t,imag(y),'r--','LineWidth',2);
title('\bfPlot of Complex Function vs Time');
xlabel('\bf\itt');
ylabel('\bf\ity(t)');
legend ('real','imaginary');
hold off;
```

Alternatively, the real part of the function can be plotted versus the imaginary part. If a single complex argument is supplied to the `plot` function, it automatically generates a plot of the real part versus the imaginary part. The statements to generate this plot are shown below, and the result is shown in Figure 6.5.

```
t = 0:pi/20:4*pi;
y = exp(-0.2*t).*(cos(t)+i*sin(t));
plot(y,'b-','LineWidth',2);
title('\bfPlot of Complex Function');
xlabel('\bfReal Part');
ylabel('\bfImaginary Part');
```

Finally, the function can be plotted as a polar plot showing magnitude versus angle. The statements to generate this plot are shown below, and the result is shown in Figure 6.6.

```
t = 0:pi/20:4*pi;
y = exp(-0.2*t).*(cos(t)+i*sin(t));
polar(angle(y),abs(y));
title('\bfPlot of Complex Function');
```

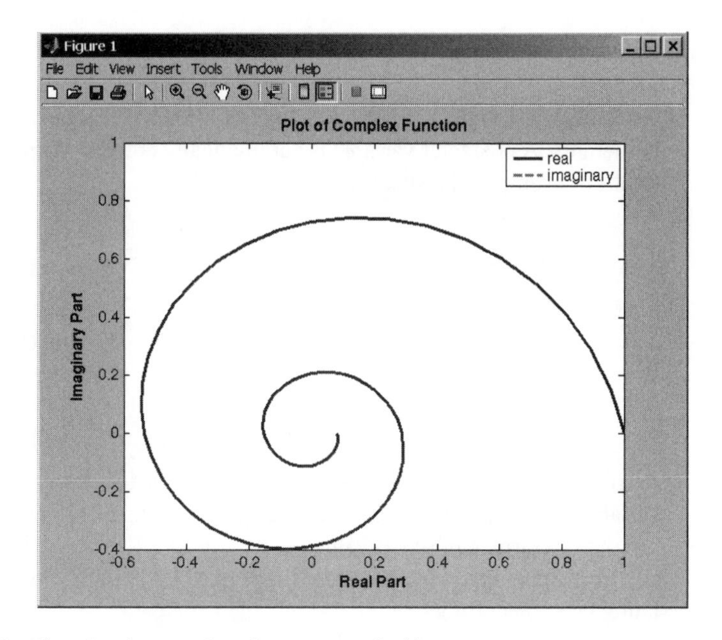

Figure 6.5 Plot of real versus imaginary parts of $y(t)$.

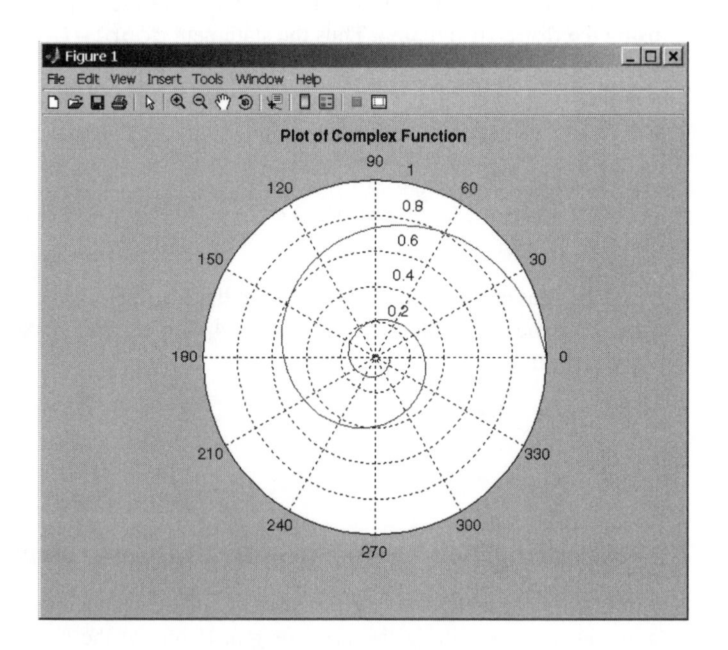

Figure 6.6 Polar plot of magnitude of $y(t)$ versus angle.

6.2 String Functions

A MATLAB string is an array of type `char`. Each character is stored in two bytes of memory. A character variable is automatically created when a string is assigned to it. For example, the statement

```
str = 'This is a test';
```

creates a 14-element character array. The output of **`whos`** for this array is

```
» whos str
  Name          Size            Bytes         Class

  str           1x14               28         char array

Grand total is 14 elements using 28 bytes
```

A special function `ischar` can be used to check for character arrays. If a given variable is of type character, then `ischar` returns a true (1) value. If it is not, `ischar` returns a false (0) value.

The following subsections describe MATLAB functions useful for manipulating character strings.

6.2.1 String Conversion Functions

Variables may be converted from the `char` data type to the `double` data type using the `double` function. Thus the statement `double(str)` yields the result:

```
» x = double(str)
x =
  Columns 1 through 12
   84 104 105 115  32 105 115  32  97  32 116 101
  Columns 13 through 14
  115 116
```

Variables can also be converted from the `double` data type to the `char` data type using the `char` function. If x is the 14-element array created above, then the statement `char(x)` yields the result:

```
» z = char(x)
z =
This is a test
```

6.2.2 Creating Two-Dimensional Character Arrays

It is possible to create two-dimensional character arrays, but *each row of such an array must have exactly the same length.* If one of the rows is shorter than the other rows, the character array is invalid and will produce an error. For example, the following statements are illegal because the two rows have different lengths.

```
name = ['Stephen J. Chapman';'Senior Engineer'];
```

The easiest way to produce two-dimensional character arrays is with the char function. This function will automatically pad all strings to the length of the largest input string.

```
» name = char('Stephen J. Chapman','Senior Engineer')
name =
Stephen J. Chapman
Senior Engineer
```

Two-dimensional character arrays can also be created with function strvcat, which is described in the next section.

Good Programming Practice

Use the char function to create two-dimensional character arrays without worrying about padding each row to the same length.

It is possible to remove any extra blanks from a string when it is extracted from an array using the deblank function. For example, the following statements remove the second line from array name and compare the results with and without blank trimming.

```
» line2 = name(2,:)
line2 =
Senior Engineer
» line2_trim = deblank(name(2,:))
line2_trim =
Senior Engineer
» size(line2)
ans =
     1   18
» size(line2_trim)
ans =
     1   15
```

6.2.3 Concatenating Strings

Function strcat concatenates two or more strings horizontally, ignoring any trailing blanks but preserving blanks within the strings. This function produces the result shown below

```
» result = strcat('String 1 ','String 2')
result =
String 1String 2
```

The result is 'String 1String 2'. Note that the trailing blanks in the first string were ignored.

Function `strvcat` concatenates two or more strings vertically, automatically padding the strings to make a valid two-dimensional array. This function produces the result shown below

```
» result = strvcat('Long String 1 ','String 2')
result =
Long String 1
String 2
```

6.2.4 Comparing Strings

Strings and substrings can be compared in several ways:

- Two strings, or parts of two strings, can be compared for equality.
- Two individual characters can be compared for equality.
- Strings can be examined to determine whether each character is a letter or whitespace.

Comparing Strings for Equality

You can use four MATLAB functions to compare two strings as a whole for equality. They are:

- `strcmp` determines whether two strings are identical.
- `strcmpi` determines whether two strings are identical, ignoring case.
- `strncmp` determines whether the first `n` characters of two strings are identical.
- `strncmpi` determines whether the first `n` characters of two strings are identical, ignoring case.

Function `strcmp` compares two strings, including any leading and trailing blanks, and returns a true (1) if the strings are identical.[1] Otherwise, it returns a false (0). Function `strcmpi` is the same as `strcmp`, except that it ignores the case of letters (that is, it treats `'a'` as equal to `'A'`.)

Function `strncmp` compares the first `n` characters of two strings, including any leading blanks, and returns a true (1) if the characters are identical. Otherwise, it returns a false (0). Function `strncmpi` is the same as `strncmp`, except that it ignores the case of letters.

To understand these functions, consider the three strings:

```
str1 = 'hello';
str2 = 'Hello';
str3 = 'help';
```

Strings `str1` and `str2` are not identical, but they differ only in the case of one letter. Therefore, `strcmp` returns false (0), while `strcmpi` returns true (1).

[1]**Caution:** The behavior of this function is different from that of the `strcmp` in C. C programmers can be tripped up by this difference.

```
» c = strcmp(str1,str2)
c =
      0
» c = strcmpi(str1,str2)
c =
      1
```

Strings `str1` and `str3` are also not identical, and both `strcmp` and `strcmpi` will return a false (0). However, the first three characters of `str1` and `str3` *are* identical, so invoking `strncmp` with any value up to 3 returns a true (1):

```
» c = strncmp(str1,str3,2)
c =
1
```

Comparing Individual Characters for Equality and Inequality

You can use MATLAB relational operators on character arrays to test for equality *one character at a time,* as long as the arrays you are comparing have equal dimensions or one is a scalar. For example, you can use the equality operator (==) to determine which characters in two strings match:

```
» a = 'fate';
» b = 'cake';
» result = a == b
result =
0 1 0 1
```

All of the relational operators (>, >=, <, <=, ==, ~=) compare the ASCII values of corresponding characters.

Unlike C, MATLAB does not have an intrinsic function to define a "greater than" or "less than" relationship between two strings taken as a whole. We will create such a function in an example at the end of this section.

Categorizing Characters Within a String

There are three functions for categorizing characters on a character-by-character basis inside a string:

- `isletter` determines whether a character is a letter.
- `isspace` determines whether a character is whitespace (blank, tab, or new line).
- `isstrprop('str', 'category')` is a more general function. It determines whether a character falls into a user-specified category, such as alphabetic, alphanumeric, upper case, lower case, numeric, control, etc.

To understand these functions, let's create a string named `mystring`:

```
mystring = 'Room 23a';
```

We will use this string to test the categorizing functions.

Function `isletter` examines each character in the string, producing a `logical` output vector of the same length as `mystring` that contains a true (1) in each location corresponding to a character, and a false (0) in the other locations. For example,

```
» a = isletter(mystring)
a =
1 1 1 1 0 0 0 1
```

The first four and the last elements in `a` are true (1) because the corresponding characters of `mystring` are letters.

Function `isspace` also examines each character in the string, producing a `logical` output vector of the same length as `mystring` that contains a true (1) in each location corresponding to whitespace, and a false (0) in the other locations. "Whitespace" is any character that separates tokens in MATLAB: a space, a tab, a linefeed, carriage return, etc. For example,

```
» a = isspace(mystring)
a =
0 0 0 0 1 0 0 0
```

The fifth element in `a` is true (1) because the corresponding character of `mystring` is a space.

Function `isstrprop` is new in MATLAB 7.0. It is a more flexible replacement for `isletter`, `isspace`, and several other functions. This function has two arguments, `'str'` and `'category'`. The first argument is the string to characterize, and the second argument is the type of category to check for. Some possible categories are given in Table 6.2.

This function examines each character in the string, producing a `logical` output vector of the same length as the input string that contains a true (1) in each location that matches the category, and a false (0) in the other locations. For example, the following function checks to see which characters in `mystring` are numbers:

```
» a = isstrprop(mystring,'digit')
a =
0 0 0 0 0 1 1 0
```

Also, the following function checks to see which characters in `mystring` are lower case letters:

```
» a = isstrprop(mystring,'lower')
a =
0 1 1 1 0 0 0 1
```

✱ Good Programming Practice

Use function `isstrprop` to determine the characteristics of each character in a string array. This function replaces the older functions `isletter` and `isspace`, which may be deleted in a future version of MATLAB.

Table 6.2 Selected Categories for Function `isstrprop`

Category	Description
`'alpha'`	Return true (1) for each character of the string that is alphabetic, and false (0) otherwise.
`'alphanum'`	Return true (1) for each character of the string that is alphanumeric, and false (0) otherwise. [**Note:** This category replaces function `isletter`.]
`'cntrl'`	Return true (1) for each character of the string that is a control character, and false (0) otherwise.
`'digit'`	Return true (1) for each character of the string that is a number, and false (0) otherwise.
`'lower'`	Return true (1) for each character of the string that is a lower-case letter, and false (0) otherwise.
`'wspace'`	Return true (1) for each character of the string that is whitespace, and false (0) otherwise. [**Note:** This category replaces function `isspace`.]
`'upper'`	Return true (1) for each character of the string that is an upper-case letter, and false (0) otherwise.
`'xdigit'`	Return true (1) for each character of the string that is a hexadecimal digit, and false (0) otherwise.

6.2.5 Searching and Replacing Characters Within a String

MATLAB provides several functions for searching and replacing characters in a string. Consider a string named `test`:

```
test = 'This is a test!';
```

Function `findstr` returns the starting position of all occurrences of the shorter of two strings within a longer string. For example, to find all occurrences of the string `'is'` inside `test`,

```
» position = findstr(test,'is')
position =
     3     6
```

The string `'is'` occurs twice within `test`, starting at positions 3 and 6.

Function `strmatch` is another matching function. This one looks at the beginning characters of the *rows* of a two-dimensional character array, and returns a list of those rows that start with the specified character sequence. The form of this function is

```
result = strmatch(str,array);
```

For example, suppose that we create a two-dimensional character array with the function `strvcat`:

```
array = strvcat('maxarray','min value','max value');
```

Then the following statement will return the row numbers of all rows beginning with the letters 'max':

```
» result = strmatch('max',array)
result =
          1
          3
```

Function strrep performs the standard search-and-replace operation. It finds all occurrences if one string within another one, and replaces them by a third string. The form of this function is

```
result = strrep(str,srch,repl)
```

where str is the string being checked, srch is the character string to search for, and repl is the replacement character string. For example,

```
» test = 'This is a test!'
» result = strrep(test,'test','pest')
result =
This is a pest!
```

The strtok function returns the characters before the first occurrence of a delimiting character in an input string. The default delimiting characters are the set of whitespace characters. The form of strtok is

```
[token,remainder] = strtok(string,delim)
```

where string is the input character string, delim is the (optional) set of delimiting characters, token is the first set of characters delimited by a character in delim, and remainder is the rest of the line. For example,

```
» [token,remainder] = strtok('This is a test!')
token =
This
remainder =
is a test!
```

You can use the strtok function to parse a sentence into words; for example:

```
function all_words = words(input_string)
remainder = input_string;
all_words = '';
while (any(remainder))
   [chopped,remainder] = strtok(remainder);
   all_words = strvcat(all_words,chopped);
end
```

6.2.6 Uppercase and Lowercase Conversion

Functions upper and lower convert all of the alphabetic characters within a string to uppercase and lowercase respectively. For example,

```
» result = upper('This is test 1!')
result =
THIS IS TEST 1!
» result = lower('This is test 2!')
result =
this is test 2!
```

Note that the alphabetic characters were converted to the proper case, while the numbers and punctuation were unaffected.

6.2.7 Trimming Whitespace from Strings

There are two functions that trim leading and/or trailing whitespace from a string. Whitespace characters consists of the spaces, newlines, carriage returns, tabs, vertical tabs, and formfeeds.

Function deblank removes any extra *trailing* whitespace from a string, and function strtrim removes any extra *leading and trailing* whitespace from a string.

For example, the following statements create a 21-character string with leading and trailing whitespace. Function deblank trims the trailing whitespace characters in the string only, while function strtrim trims both the leading and the trailing whitespace characters.

```
» test_string = '   This is a test.   '
test_string =
   This is a test.
» length(test_string)
ans =
    21
» test_string_trim1 = deblank(test_string)
test_string_trim1 =
   This is a test.
» length(test_string_trim1)
ans =
    18
» test_string_trim2 = strtrim(test_string)
test_string_trim2 =
This is a test.
» length(test_string_trim2)
ans =
    15
```

6.2.8 Numeric-to-String Conversions

MATLAB contains several functions to convert numeric values into character strings. We have already seen two such functions, num2str and int2str. Consider a scalar x:

```
x = 5317;
```

By default, MATLAB stores the number x as a 1 × 1 double array containing the value 5317. The int2str (integer to string) function converts this scalar into a 1-by-4 char array containing the string '5317':

```
» x = 5317;
» y = int2str(x);
» whos
  Name        Size         Bytes        Class

  x           1x1          8            double array
  y           1x4          8            char array

Grand total is 5 elements using 16 bytes
```

Function num2str converts a double value into a string, even if it does not contain an integer. It provides more control of the output string format than int2str. An optional second argument sets the number of digits in the output string, or specifies an actual format to use. The format specifications in the second argument as similar to those used by fprintf. For example,

```
» p = num2str(pi)
p =
3.1416
» p = num2str(pi,7)
p =
3.141593
» p = num2str(pi,'%10.5e')
p =
3.14159e+000
```

Both int2str and num2str are handy for labeling plots. For example, the following lines use num2str to prepare automated labels for the *x*-axis of a plot:

```
function plotlabel(x,y)
plot(x,y)
str1 = num2str(min(x));
str2 = num2str(max(x));
out = ['Value of f from ' str1 ' to ' str2];
xlabel(out);
```

There are also conversion functions designed to change numeric values into strings representing a decimal value in another base, such as a binary or hexadecimal representation. For example, the `dec2hex` function converts a decimal value into the corresponding hexadecimal string:

```
dec_num = 4035;
hex_num = dec2hex(dec_num)
hex_num =
FC3
```

Other functions of this type include `hex2num`, `hex2dec`, `bin2dec`, `dec2bin`, `base2dec`, and `dec2base`. MATLAB includes on-line help for all of these functions.

MATLAB function `mat2str` converts an array to a string that MATLAB can evaluate. This string is useful input for a function such as `eval`, which evaluates input strings just as if they were typed at the MATLAB command line. For example, if we define array `a` as

```
» a = [1 2 3; 4 5 6]
a =
        1        2        3
        4        5        6
```

then the function `mat2str` will return a string containing the result

```
» b = mat2str(a)
b =
[1 2 3; 4 5 6]
```

Finally, MATLAB includes a special function `sprintf` that is identical to function `fprintf`, except that the output goes into a character string instead of the Command Window. This function provides complete control over the formatting of the character string. For example,

```
» str = sprintf('The value of pi = %8.6f.',pi)
str =
The value of pi = 3.141593.
```

This function is extremely useful in creating complex titles and labels for plots.

6.2.9 String-to-Numeric Conversions

MATLAB also contains several functions to change character strings into numeric values. The most important of these function are `eval`, `str2double`, and `sscanf`.

Function `eval` evaluates a string containing a MATLAB expression, and returns the result. The expression can contain any combination of MATLAB functions, variables, constants, and operations. For example, the string `a` containing the characters `'2 * 3.141592'` can be converted to numeric form by the following statements:

```
» a = '2 * 3.141592';
» b = eval(a)
b =
        6.2832
» whos
  Name        Size        Bytes       Class

  a           1x8         16          char array
  b           1x1         8           double array

Grand total is 9 elements using 24 bytes
```

Function `str2double` converts character strings into an equivalent `double` value.[2] For example, the string a containing the characters `'3.141592'` can be converted to numeric form by the following statements:

```
» a = '3.141592';
» b = str2double(a)
b =
        3.1416
```

Strings can also be converted to numeric form using the function `sscanf`. This function converts a string into a number according to a format conversion character. The simplest form of this function is

```
value = sscanf(string,format)
```

where `string` is the string to scan, and `format` specifies the type of conversion to occur. The two most common conversion specifiers for sscanf are `'%d'` for decimals and `'%g'` for floating-point numbers. This function will be covered in much greater detail in Chapter 8.

The following examples illustrate the use of `sscanf`.

```
» a = '3.141592';
» value1 = sscanf(a,'%g')
value1 =
        3.1416
» value2 = sscanf(a,'%d')
value2 =
        3
```

[2]MATLAB also contains a function `str2num` that can convert a string into a number. For a variety of reasons mentioned in the MATLAB documentation, function `str2double` is better than function `str2num`. You should recognize function `str2num` when you see it, but always use function `str2double` in any new code that you write.

6.2.10 Summary

The common MATLAB string functions are summarized in Table 6.3.

Table 6.3 Common MATLAB String Functions

Category	Function	Description
General	`char`	(1) Convert numbers to the corresponding character values. (2) Create a two-dimensional character array from a series of strings.
	`double`	Convert characters to the corresponding numeric codes.
	`blanks`	Create a string of blanks.
	`deblank`	Remove trailing whitespace from a string.
	`strtrim`	Remove leading and trailing whitespace from a string.
String tests	`ischar`	Returns true (1) for a character array.
	`isletter`	Returns true (1) for letters of the alphabet.
	`isspace`	Returns true (1) for whitespace.
	`isstrprop`	Returns true (1) for characters matching the specified property.
String operations	`strcat`	Concatenate strings.
	`strvcat`	Concatenate strings vertically.
	`strcmp`	Returns true (1) if two strings are identical.
	`strcmpi`	Returns true (1) if two strings are identical, ignoring case.
	`strncmp`	Returns true (1) if first `n` characters of two strings are identical.
	`strncmpi`	Returns true (1) if first `n` characters of two strings are identical, ignoring case.
	`findstr`	Find one string within another one.
	`strjust`	Justify string.
	`strmatch`	Find matches for string.
	`strrep`	Replace one string with another.
	`strtok`	Find token in string.
	`upper`	Convert string to uppercase.
	`lower`	Convert string to lowercase.
Number to string conversion	`int2str`	Convert integer to string.
	`num2str`	Convert number to string.
	`mat2str`	Convert matrix to string.
	`sprintf`	Write formatted data to string.
String to number conversion	`eval`	Evaluate the result of a MATLAB expression.
	`str2double`	Convert string to a double value.
	`str2num`	Convert string to number.
	`sscanf`	Read formatted data from string.
Base Number Conversion	`hex2num`	Convert IEEE hexadecimal string to `double`.
	`hex2dec`	Convert hexadecimal string to decimal integer.
	`dec2hex`	Convert decimal to hexadecimal string.
	`bin2dec`	Convert binary string to decimal integer.
	`dec2bin`	Convert decimal integer to binary string.
	`base2dec`	Convert base B string to decimal integer.
	`dec2base`	Convert decimal integer to base B string.

Example 6.2—String Comparison Function

In C, function `strmcp` compares two strings according to the order of their characters in the ASCII table (called the **lexicographic order** of the characters), and returns a −1 if the first string is lexicographically less than the second string, a 0 if the strings are equal, and a +1 if the first string is lexicographically greater than the second string. This function is extremely useful for such purposes as sorting strings in alphabetic order.

Create a new MATLAB function `c_strcmp` that compares two strings in a similar fashion to the C function and returns similar results. The function should ignore trailing blanks in doing its comparisons. Note that the function must be able to handle the situation where the two strings are of different lengths.

SOLUTION

1. **State the problem**
   Write a function that will compare two strings `str1` and `str2`, and return the following results:

   - −1 if `str1` is lexicographically less than `str2`.
   - 0 if `str1` is lexicographically less than `str2`.
   - +1 if `str1` is lexicographically greater than `str2`.

   The function must work properly if `str1` and `str2` do not have the same length, and the function should ignore trailing blanks.

2. **Define the inputs and outputs**
   The inputs required by this function are two strings, `str1` and `str2`. The output from the function will be a −1, 0, or 1, as appropriate.

3. **Describe the algorithm**
   This task can be broken down into four major sections:

```
Verify input strings
Pad strings to be equal length
Compare characters from beginning to end, looking
   for the first difference
Return a value based on the first difference
```

   We will now break each of the above major sections into smaller, more detailed pieces. First, we must verify that the data passed to the function is correct. The function must have exactly two arguments, and the arguments must both be characters. The pseudocode for this step is:

```
% Check for a legal number of input arguments.
msg = nargchk(2,2,nargin)
error(msg)
```

```
% Check to see if the arguments are strings
if either argument is not a string
   error('str1 and str2 must both be strings')
else

   (add code here)

end
```

Next, we must pad the strings to equal lengths. The easiest way to do this is to combine both strings into a two-dimensional array using `strvcat`. Note that this step effectively results in the function ignoring trailing blanks, because both strings are padded out to the same length. The pseudocode for this step is:

```
% Pad strings
strings = strvcat(str1,str2)
```

Now we must compare each character until we find a difference, and return a value based on that difference. One way to do this is to use relational operators to compare the two strings, creating an array of 0's and 1's. We can then look for the first 1 in the array, which will correspond to the first difference between the two strings. The pseudocode for this step is:

```
% Compare strings
diff = strings(1,:) ~= strings(2,:)
if sum(diff) == 0
   % Strings match
   result = 0
else
   % Find first difference
   ival = find(diff)
   if strings(1,ival) > strings(2,ival)
      result = 1
   else
      result = -1
   end
end
```

4. **Turn the algorithm into MATLAB statements**
   The final MATLAB code is shown below.

```
function result = c_strcmp(str1,str2)
%C_STRCMP Compare strings like C function "strcmp"
% Function C_STRCMP compares two strings, and returns
% a -1 if str1 < str2, a 0 if str1 == str2, and a
% +1 if str1 > str2.
```

```
% Define variables:
%    diff      -- Logical array of string differences
%    msg       -- Error message
%    result    -- Result of function
%    str1      -- First string to compare
%    str2      -- Second string to compare
%    strings   -- Padded array of strings

%  Record of revisions:
%    Date        Programmer        Description of change
%    ====        ==========        =====================
%  01/16/04    S. J. Chapman     Original code

% Check for a legal number of input arguments.
msg = nargchk(2,2,nargin);
error(msg);

% Check to see if the arguments are strings
if  ~(isstr(str1) & isstr(str2))
    error('Both str1 and str2 must both be strings!')
else

   % Pad strings
   strings = strvcat(str1,str2);

   % Compare strings
   diff = strings(1,:) ~= strings(2,:);
   if sum(diff) == 0

      % Strings match, so return a zero!
      result = 0;
   else
      % Find first difference between strings
      ival = find(diff);
      if strings(1,ival(1)) > strings(2,ival(1))
         result = 1;
      else
         result = -1;
      end
   end
end
```

5. **Test the program**

   Next, we must test the function using various strings.

```
» result = c_strcmp('String 1','String 1')
result =
          0
```

```
» result = c_strcmp('String 1','String 1 ')
result =
            0
» result = c_strcmp('String 1','String 2')
result =
           -1
» result = c_strcmp('String 1','String 0')
result =
            1
» result = c_strcmp('String','str')
result =
           -1
```

The first test returns a zero, because the two strings are identical. The second test also returns a zero, because the two strings are identical *except for trailing blanks*, and trailing blanks are ignored. The third test returns a −1, because the two strings first differ in position 8, and `'1'` < `'2'` at that position. The fourth test returns a 1, because the two strings first differ in position 8, and `'1'` > `'0'` at that position. The fifth test returns a −1, because the two strings first differ in position 1, and `'S'` < `'s'` in the ASCII collating sequence.

This function appears to be working properly.

◄

Quiz 6.1

This quiz provides a quick check to see if you have understood the concepts introduced in Sections 6.1 through 6.2. If you have trouble with the quiz, reread the section, ask your instructor, or discuss the material with a fellow student. The answers to this quiz are found in the back of the book.

1. What is the value of `result` in the following statements?

(*a*)
```
x = 12 + i*5;
y = 5 - i*13;
result = x > y;
```

(*b*)
```
x = 12 + i*5;
y = 5 - i*13;
result = abs(x) > abs(y);
```

(*c*)
```
x = 12 + i*5;
y = 5 - i*13;
result = real(x) - imag(y);
```

2. If `array` is a complex array, what does the function `plot(array)` do?

3. How can you convert a vector of the `char` data type into a vector of the `double` data type?

For questions 4 through 11, determine whether these statements are correct. If they are, what is produced by each set of statements?

4.
```
str1 = 'This is a test!  ';
str2 = 'This line, too.';
res = strcat(str1,str2);
```

5.
```
str1 = 'Line 1';
str2 = 'line 2';
res = strcati(str1,str2);
```

6.
```
str1 = 'This is a test!  ';
str2 = 'This line, too.';
res = [str1; str2];
```

7.
```
str1 = 'This is another test!  ';
str2 = 'This line, too.';
res = strvcat(str1,str2);
```

8.
```
str1 = 'This is a test!  ';
str2 = 'This line, too.';
res = strncmp(str1,str2,5);
```

9.
```
str1 = 'This is a test!  ';
res = findstr(str1,'s');
```

10.
```
str1 = 'This is a test!  ';
str1(isspace(str1)) = x;
```

11.
```
str1 = 'aBcD 1234 !?';
res = isstrprop(str1,'alphanum');
```

12.
```
str1 = 'This is a test!  ';
str1(4:7) = upper(str1(4:7));
```

13.
```
str1 = '   456   '; % Note: Three blanks before & after
str2 = '   abc   '; % Note: Three blanks before & after
str3 = [str1 str2];
str4 = [strtrim(str1) strtrim(str2)];
str5 = [deblank(str1) deblank(str2)];
l1 = length(str1);
l3 = length(str3);
l4 = length(str4);
l5 = length(str4);
```

14.
```
str1 = 'This way to the egress.';
str2 = 'This way to the egret.'
res = strncmp(str1,str2);
```

6.3 Multidimensional Arrays

MATLAB also supports arrays with more than two dimensions. These **multidimensional arrays** are very useful for displaying data that intrinsically has more than two dimensions, or for displaying multiple versions of two-dimensional data sets. For example, measurements of pressure and velocity throughout a three-dimensional volume are very important in such studies as aerodynamics and fluid dynamics. These sorts of areas naturally use multidimensional arrays.

Multidimensional arrays are a natural extension of two-dimensional arrays. Each additional dimension is represented by one additional subscript used to address the data.

It is very easy to create a multidimensional array. They can be created either by assigning values directly in assignment statements or by using the same functions that are used to create one- and two-dimensional arrays. For example, suppose that you have a two-dimensional array created by the assignment statement

```
» a = [1 2 3 4; 5 6 7 8]
a =
         1         2         3         4
         5         6         7         8
```

This is a 2 × 4 array, with each element addressed by two subscripts. The array can be extended to be a three-dimensional 2 × 4 × 3 array with the following assignment statements.

```
» a(:,:,2) = [ 9 10 11 12; 13 14 15 16];
» a(:,:,3) = [ 17 18 19 20; 21 22 23 24]
a(:,:,1) =
         1         2         3         4
         5         6         7         8
a(:,:,2) =
         9        10        11        12
        13        14        15        16
a(:,:,3) =
        17        18        19        20
        21        22        23        24
```

Individual elements in this multidimensional array can be addressed by the array name followed by three subscripts, and subsets of the data can be created using the colon operators. For example, the value of a(2,2,2) is

```
» a(2,2,2)
ans =
     14
```

and the vector a(1,1,:) is

```
» a(1,1,:)
ans(:,:,1) =
     1
ans(:,:,2) =
     9
ans(:,:,3) =
     17
```

Multidimensional arrays can also be created using the same functions as other arrays, for example:

```
» b = ones(4,4,2)
b(:,:,1) =
     1     1     1     1
     1     1     1     1
     1     1     1     1
     1     1     1     1
b(:,:,2) =
     1     1     1     1
     1     1     1     1
     1     1     1     1
     1     1     1     1
» c = randn(2,2,3)
c(:,:,1) =
    -0.4326     0.1253
    -1.6656     0.2877
c(:,:,2) =
    -1.1465     1.1892
     1.1909    -0.0376
c(:,:,3) =
     0.3273    -0.1867
     0.1746     0.7258
```

The number of dimensions in a multidimensional array can be found using the ndims function, and the size of the array can be found using the size function.

```
» ndims(c)
ans =
     3
» size(c)
ans =
     2     2     3
```

If you are writing applications that need multidimensional arrays, see the MATLAB Users Guide for more details on the behavior of various MATLAB functions with multidimensional arrays.

✱ Good Programming Practice

Use multidimensional arrays to solve problems that are naturally multivariate in nature, such as aerodynamics and fluid flows.

Also, recall from Chapter 4 that the MATLAB just-in-time compiler cannot compile loops containing arrays with three or more dimensions. If you are working with such arrays, be sure to vectorize your code to increase its speed. Do not rely on the JIT compiler to do the job—it won't.

✱ Good Programming Practice

If you are working with multidimensional arrays, be sure to vectorize your code by hand. The MATLAB JIT compiler cannot handle loops containing multidimensional arrays.

6.4 Additional Two-Dimensional Plots

In previous chapters, we learned to create linear, log-log, semilog, and polar plots. MATLAB supports many additional types of plots that you can use to display your data. This section will introduce you to some of these additional plotting options.

6.4.1 Additional Types of Two-Dimensional Plots

In addition to the two-dimensional plots that we have already seen, MATLAB supports *many* other more specialized plots. In fact, the MATLAB help desk lists more than 20 types of two-dimensional plots! Examples include **stem plots, stair plots, bar plots, pie plots,** and **compass plots.** A *stem plot* is a plot in which each data value is represented by a marker and a line connecting the marker vertically to the x axis. A *stair plot* is a plot in which each data point is represented by a horizontal line, and successive points are connected by vertical lines, producing a stair-step effect. A *bar plot* is a plot in which each point is represented by a vertical bar or horizontal bar. A *pie plot* is a plot represented by "pie slices" of various sizes. Finally, a *compass plot* is a type of polar plot in which each value is represented by an arrow whose length is proportional to its value. These types of plots are summarized in Table 6.4, and examples of all of the plots are shown in Figure 6.7.

Table 6.4 Additional Two-Dimensional Plotting Functions

Function	Description
`bar(x,y)`	This function creates a *vertical* bar plot, with the values in `x` used to label each bar and the values in `y` used to determine the height of the bar.
`barh(x,y)`	This function creates a *horizontal* bar plot, with the values in `x` used to label each bar and the values in `y` used to determine the horizontal length of the bar.
`compass(x,y)`	This function creates a polar plot, with an arrow drawn from the origin to the location of each (x, y) point. Note that the locations of the points to plot are specified in Cartesian coordinates, not polar coordinates.
`pie(x)` `pie(x,explode)`	This function creates a pie plot. This function determines the percentage of the total pie corresponding to each value of `x` and plots pie slices of that size. The optional array `explode` controls whether or not individual pie slices are separated from the remainder of the pie.
`stairs(x,y)`	This function creates a stair plot, with each stair step centered on an (x, y) point.
`stem(x,y)`	This function creates a stem plot, with a marker at each (x, y) point and a stem drawn vertically from that point to the x axis.

Stair, stem, vertical bar, horizontal bar, and compass plots are all similar to `plot`, and they are used in the same manner. For example, the following code produces the stem plot shown in Figure 6.7*a*.

```
x = [1 2 3 4 5 6];
y = [2 6 8 7 8 5];
stem(x,y);
title('\bfExample of a Stem Plot');
xlabel('\bf\itx');
ylabel('\bf\ity');
axis([0 7 0 10]);
```

Stair, bar, and compass plots can be created by substituting `stairs`, `bar`, `barh`, or `compass` for `stem` in the above code. The details of all of these plots, including any optional parameters, can be found in the MATLAB on-line help system.

Function `pie` behaves differently from the other plots described previously. To create a pie plot, a programmer passes an array `x` containing the data to be plotted, and function `pie` determines the *percentage of the total pie* that each element of `x` represents. For example, if the array `x` is `[1 2 3 4]`, then `pie` will calculate that the first element `x(1)` is 1/10 or 10% of the pie, the second element `x(2)` is 2/10 or 20 percent of the pie, and so forth. The function then plots those percentages as pie slices.

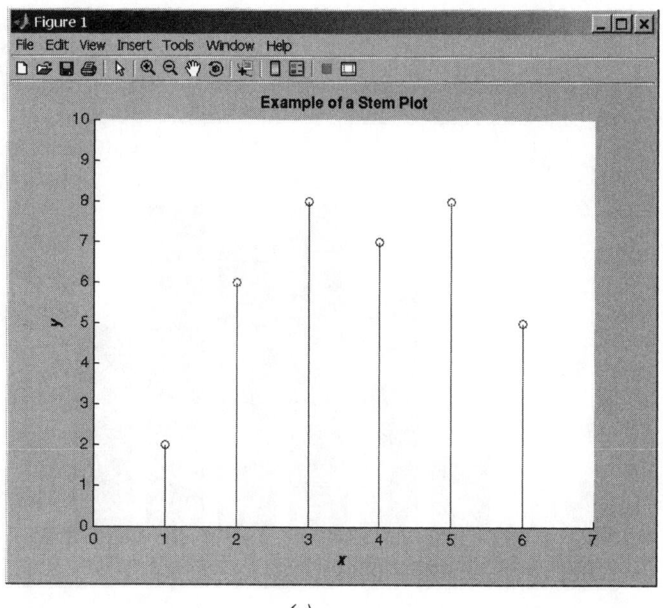

(a)

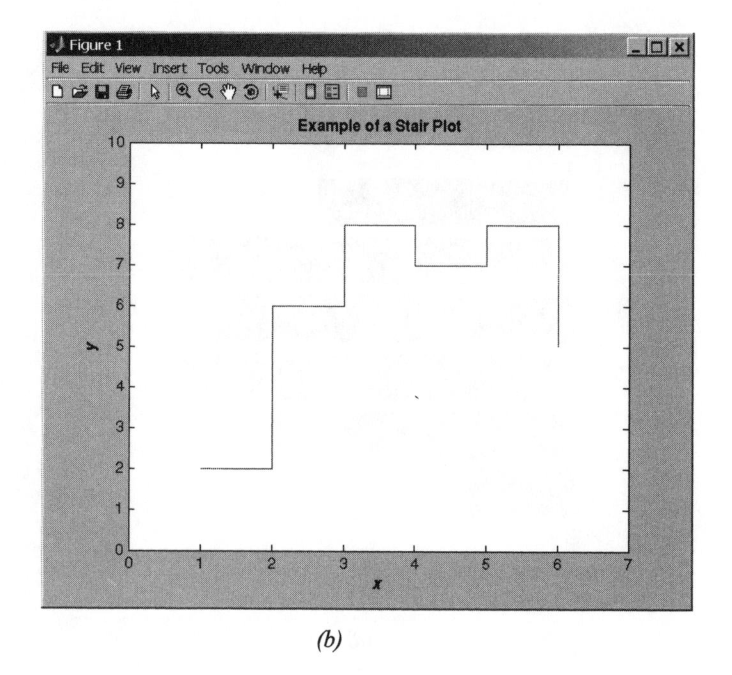

(b)

Figure 6.7 Additional types of two-dimensional plots: *(a)* stem plot; *(b)* stair plot.

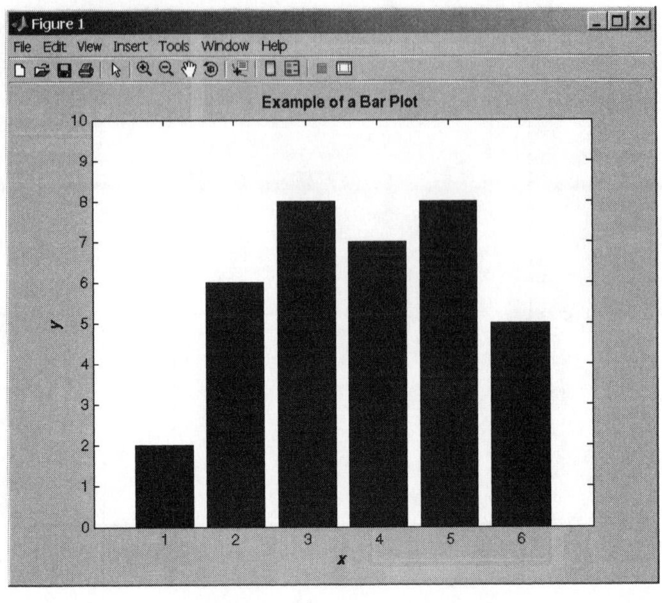

(c)

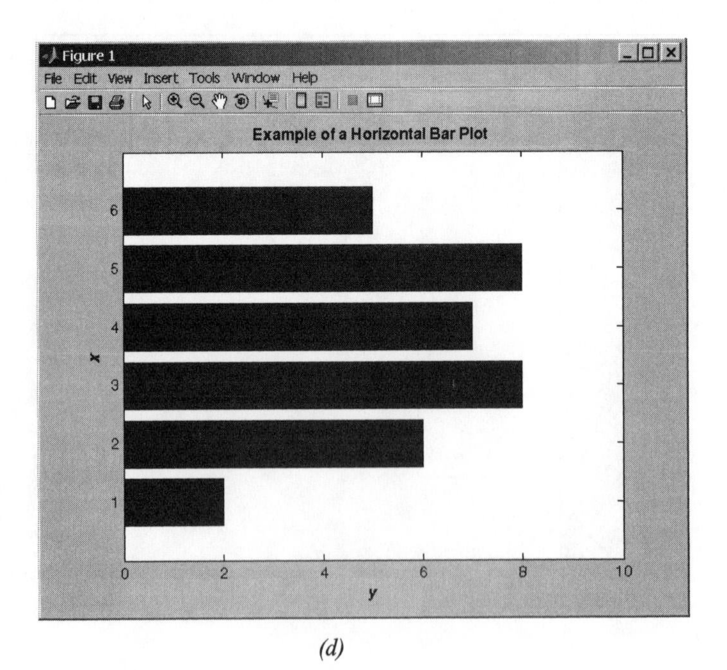

(d)

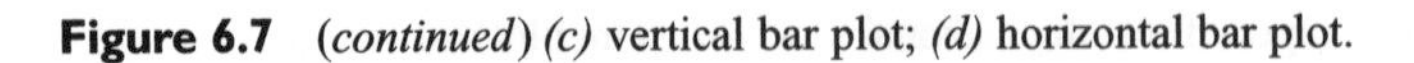

Figure 6.7 (*continued*) *(c)* vertical bar plot; *(d)* horizontal bar plot.

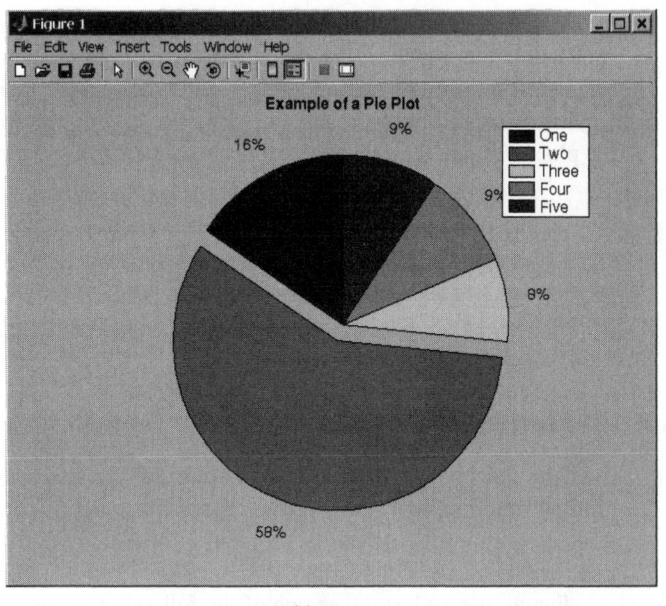

(e)

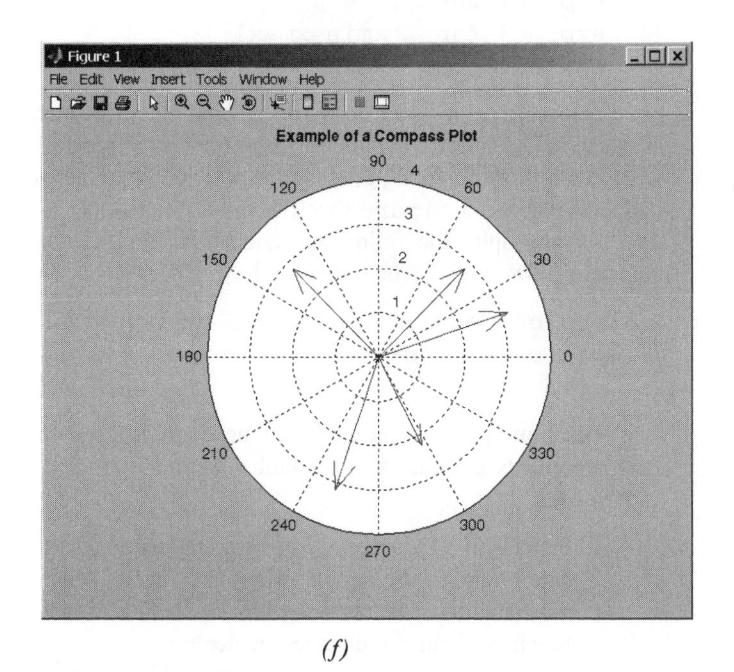

(f)

Figure 6.7 (*continued*) *(e)* pie plot; *(f)* compass plot.

Function `pie` also supports an optional parameter, `explode`. If present, `explode` is a logical array of 1s and 0s, with an element for each element in array `x`. If a value in `explode` is 1, then the corresponding pie slice is drawn slightly separated from the pie. For example, the code shown below produces the pie plot in Figure 6.6*e*. Note that the second slice of the pie is "exploded."

```
data = [ 10 37 5 6 6];
explode = [ 0 1 0 0 0];
pie(data,explode);
title('\bfExample of a Pie Plot');
legend('One','Two','Three','Four','Five');
```

6.4.2 Plotting Functions

In all previous plots, we have created arrays of data and passed those arrays to the plotting function. MATLAB also includes two functions that will plot a function directly, without the necessity of creating intermediate data arrays. These functions are `ezplot` and `fplot`.

Function `ezplot` takes one of the following forms.

```
ezplot( fun );
ezplot( fun, [xmin xmax] );
ezplot( fun, [xmin xmax], figure );
```

In each case, `fun` is a *character string* containing the functional expression to be evaluated. The optional parameter `[xmin xmax]` specifies the range of the function to plot. If it is absent, the function will be plotted between -2π and 2π. The optional parameter `figure` specifies the figure number to plot the function on.

For example, the following statements plot the function $f(x) = \sin x/x$ between -4π and 4π. The output of these statements is shown in Figure 6.8.

```
ezplot('sin(x)/x',[-4*pi 4*pi]);
title('Plot of sin x / x');
grid on;
```

Function `fplot` is similar to but more sophisticated than `ezplot`. The first two arguments are the same for both functions, but `fplot` has the following advantages:

1. Function `fplot` is *adaptive,* meaning that it calculates and displays more data points in the regions where the function being plotted is changing most rapidly. The resulting plot is more accurate at locations where a function's behavior changes suddenly.
2. Function `fplot` supports the use of $\TeX$ commands in titles and axis labels, while function `ezplot` does not.

In general, you should use `fplot` in preference to `ezplot` whenever you plot functions.

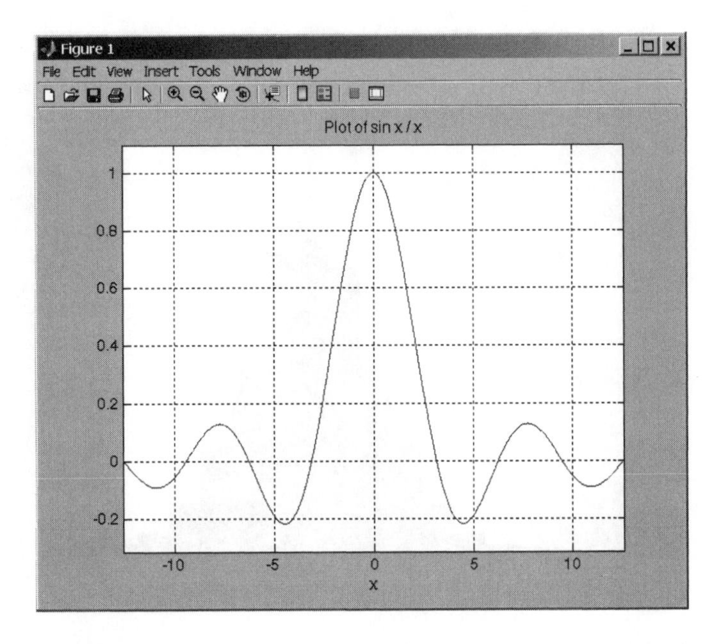

Figure 6.8 The function sin *x*/*x*, plotted with function `ezplot`.

Functions `ezplot` and `fplot` are examples of the "function functions" described in Chapter 5.

✷ Good Programming Practice

Use function `fplot` to plot functions directly without having to create intermediate data arrays.

6.4.3 Histograms

A *histogram* is a plot showing the distribution of values within a data set. To create a histogram, the range of values within the data set is divided into evenly spaced bins, and the number of data values falling into each bin is determined. The resulting count can then be plotted as a function of bin number.

The standard MATLAB histogram function is `hist`. The forms of this function are shown below:

```
hist(y)
hist(y,nbins)
hist(y,x);
[n,xout] = hist(y,...)
```

The first form of the function creates and plots a histogram with ten equally-spaced bins, while the second form creates and plots a histogram with `nbins`

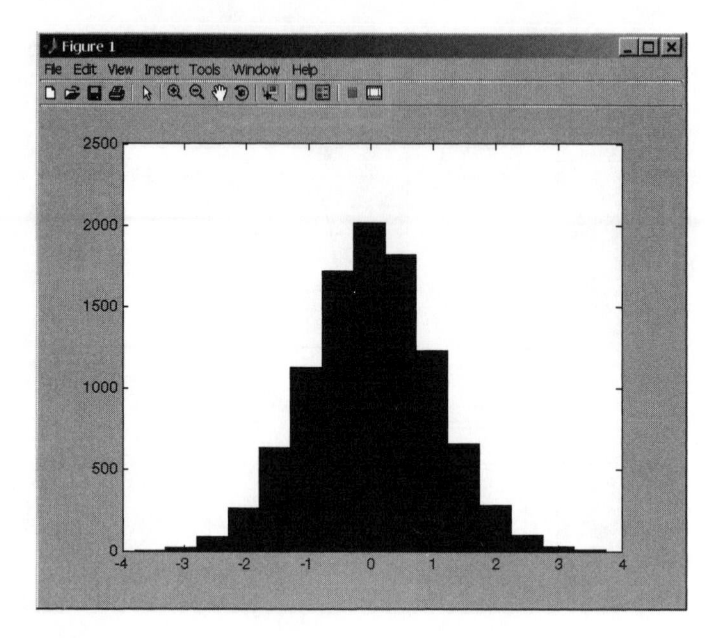

Figure 6.9 A histogram.

equally-spaced bins. The third form of the function allows the user to specify the bin centers to use in an array x; the function creates a bin centered around each element in the array. In all three of these cases, the function both creates and plots the histogram. The last form of the function creates a histogram and returns the bin centers in array xout and the count in each bin in array n, without actually creating a plot.

For example, the following statements create a data set containing 10,000 Gaussian random values, and generate a histogram of the data using 15 evenly spaced bins. The resulting histogram is shown in Figure 6.9.

```
y = randn(10000,1);
hist(y,15);.
```

MATLAB also includes a function rose to create and plot a histogram on radial axes. It is especially useful for distributions of angular data. You will be asked to use this function in an end-of-chapter exercise.

6.5 Three-Dimensional Plots

MATLAB also includes a rich variety of three-dimensional plots that can be useful for displaying certain types of data. In general, three-dimensional plots are useful for displaying two types of data:

1. Two variables that are functions of the same independent variable, when you wish to emphasize the importance of the independent variable
2. A single variable that is a function of two independent variables

6.5.1 Three-Dimensional Line Plots

A three-dimensional line plot can be created with the `plot3` function. This function is exactly like the two-dimensional `plot` function, except that each point is represented by x, y, and z values instead just of x and y values. The simplest form of this function is

```
plot(x,y,z);
```

where `x`, `y`, and `z` are equal-sized arrays containing the locations of data points to plot. Function `plot3` supports all the same line size, line style, and color options as `plot`, and you can use it immediately using the knowledge that we acquired in earlier chapters.

As an example of a three-dimensional line plot, consider the following functions

$$\begin{aligned} x(t) &= e^{-0.2t} \cos 2t \\ y(t) &= e^{-0.2t} \sin 2t \end{aligned} \tag{6-11}$$

These functions might represent the decaying oscillations of a mechanical system in two dimensions, so x and y together represent the location of the system at any given time. Note that x and y are both functions of the *same* independent variable t.

We could create a series of (x, y) points and plot them using the two-dimensional function `plot` (see Figure 6.10*a* on the next page), but if we do so, the importance of time to the behavior of the system will not be obvious in the graph. The following statements create the two-dimensional plot of the location of the object shown in Figure 6.10*a*. It is not possible from this plot to tell how rapidly the oscillations are dying out.

```
t = 0:0.1:10;
x = exp(-0.2*t) .* cos(2*t);
y = exp(-0.2*t) .* sin(2*t);
plot(x,y);
title('\bfTwo-Dimensional Line Plot');
xlabel('\bfx');
ylabel('\bfy');
grid on;
```

Instead, we could plot the variables with `plot3` to preserve the time information as well as the two-dimensional position of the object. The following statements will create a three-dimensional plot of Equations (6-11).

```
t = 0:0.1:10;
x = exp(-0.2*t) .* cos(2*t);
y = exp(-0.2*t) .* sin(2*t);
plot3(x,y,t);
title('\bfThree-Dimensional Line Plot');
xlabel('\bfx');
```

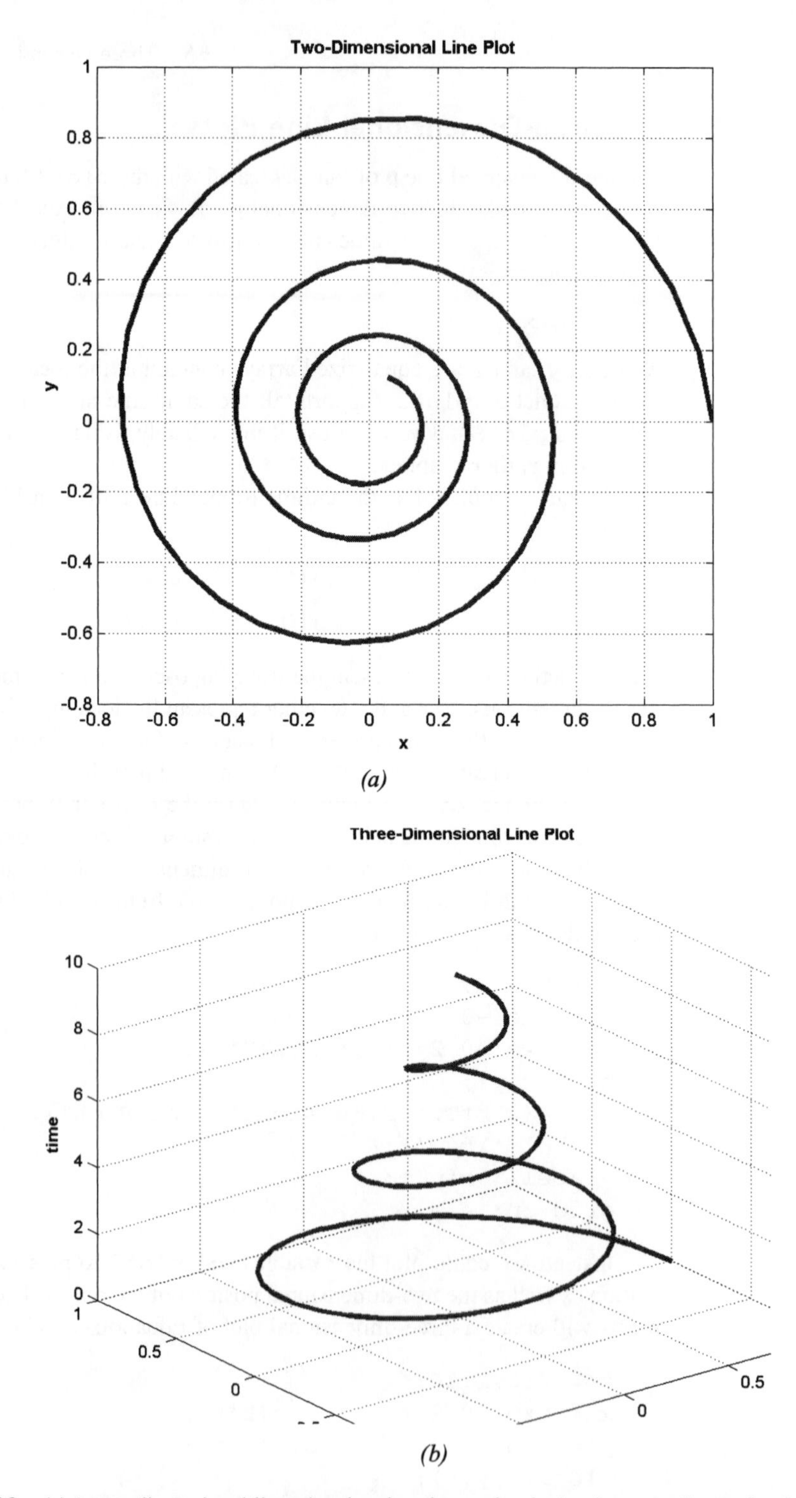

(a)

(b)

Figure 6.10 *(a)* A two-dimensional line plot showing the motion in (x, y) space of a mechanical system. This plot reveals nothing about the time behavior of the system. *(b)* A three-dimensional line plot showing the motion in (x, y) space versus time for the mechanical system. This plot clearly shows the time behavior of the system.

```
ylabel('\bfy');
zlabel('\bftime');
grid on;
```

The resulting plot is shown in Figure 6.10*b*. Note how this plot emphasizes time dependence of the two variables x and y.

6.5.2 Three-Dimensional Surface, Mesh, and Contour Plots

Surface, mesh, and contour plots are convenient ways to represent data that is a function of *two* independent variables. For example, the temperature at a point is a function of both the East-West location (x) and the North-South (y) location of the point. Any value that is a function of two independent variables can be displayed on a three-dimensional surface, mesh, or contour plot. The more common types of plots are summarized in Table 6.5, and examples of each plot are shown in Figure 6.11.[3]

To plot data using one of these functions, a user must create three equal-sized arrays. The three arrays must contain the x, y, and z values of every point to be plotted. As a simple example, suppose that we wanted to plot the four points $(-1, -1, 1)$, $(1, -1, 2)$, $(-1, 1, 1)$, and $(1, 1, 0)$. To plot these four points, we must create the arrays $\mathtt{x} = \begin{bmatrix} -1 & 1 \\ -1 & 1 \end{bmatrix}$, $\mathtt{y} = \begin{bmatrix} -1 & -1 \\ 1 & 1 \end{bmatrix}$, and $\mathtt{z} = \begin{bmatrix} 1 & 2 \\ 1 & 0 \end{bmatrix}$. Array `x` contains the x values associated with every point to plot, array `y` contains the y values associated with every point to plot, and array `z` contains the z values associated with every point to plot. These arrays are then passed to the plotting function.

Table 6.5 Selected Mesh, Surface, and Contour Plot Functions

Function	Description
`mesh(x,y,z)`	This function creates a mesh or wireframe plot, where `x` is a two-dimensional array containing the x values of every point to display, `y` is a two-dimensional array containing the y values of every point to display, and `z` is a two-dimensional array containing the z values of every point to display.
`surf(x,y,z)`	This function creates a surface plot. Arrays `x`, `y`, and `z` have the same meaning as for a mesh plot.
`contour(x,y,z)`	This function creates a contour plot. Arrays `x`, `y`, and `z` have the same meaning as for a mesh plot.

[3]There are many variations on these basic plot types. Consult the MATLAB Help Browser documentation for a complete description of these variations.

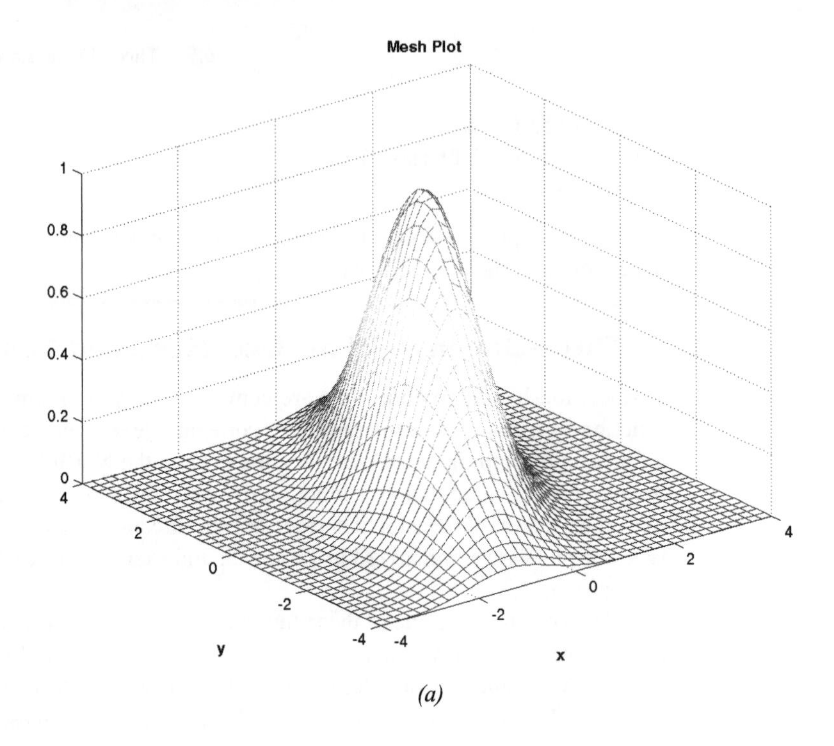

(a)

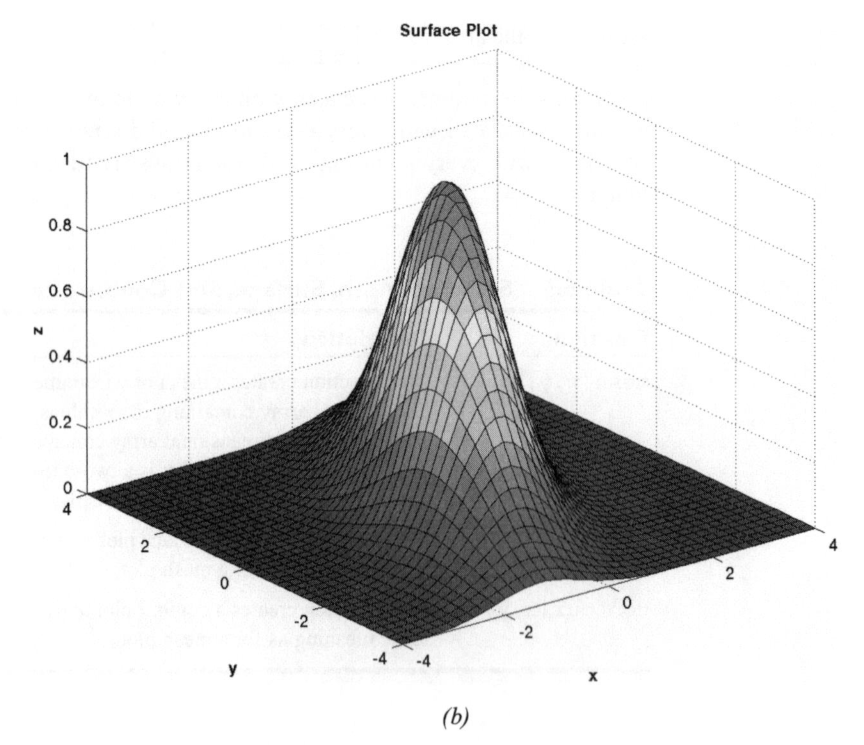

(b)

Figure 6.11 *(a)* A mesh plot of the function $z(x, y) = e^{-0.5[x^2+0.5(x-y)^2]}$. *(b)* A surface plot of the same function.

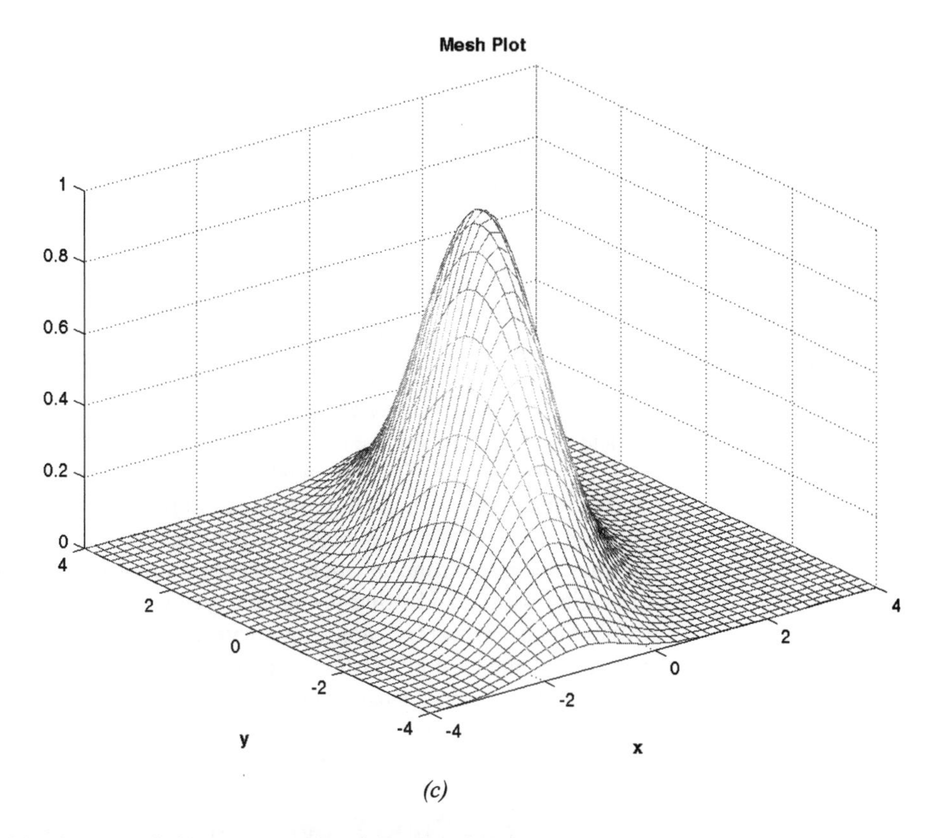

(c)

Figure 6.11 *(continued) (c)* A contour plot of the same function.

The MATLAB function meshgrid makes it easy to create the x and y arrays required for these plots. The form of this function is

```
[x, y] = meshgrid( xstart:xinc:xend, ystart:yinc:yend);
```

where xstart:xinc:xend specifies the x values to include in the grid, and ystart:yinc:yend specifies the y values to be included in the grid.

To create a plot, we use meshgrid to create the arrays of x and y values, and then evaluate the function to plot at each of those (x, y) locations. Finally, we call function mesh , surf, or contour to create the plot.

For example, suppose that we wish to create a mesh plot of the function

$$z(x, y) = e^{-0.5[x^2+0.5(x-y)^2]} \tag{6-12}$$

over the interval $-4 \leq x \leq 4$ and $-4 \leq y \leq 4$. The following statements will create the plot, which is shown in Figure 6.11*a*.

```
[x,y] = meshgrid(-4:0.2:4);
z = exp(-0.5*(x.^2+y.^2));
mesh(x,y,z);
```

```
xlabel('\bfx');
ylabel('\bfy');
zlabel('\bfz');
```

Surface and contour plots may be created by substituting the appropriate function for the mesh function.

6.6 Summary

MATLAB supports complex numbers as an extension of the double data type. They can be defined using the i or j, both of which are pre-defined as to be $\sqrt{-1}$. Using complex numbers is straightforward, except that the relational operators >, >=, <, and <= only compare the *real parts* of complex numbers, not their magnitudes. They must be used with caution when working with complex values.

String functions are functions designed to work with strings, which are arrays of type char. These functions allow a user to manipulate strings in a variety of useful ways, including concatenation, comparison, replacement, case conversion, and numeric-to-string and string-to-numeric type conversions.

Multidimensional arrays are arrays with more than two dimensions. They may be created and used in a fashion similar to one- and two-dimensional arrays. Multidimensional arrays appear naturally in certain classes of physical problems.

MATLAB includes a rich variety of two- and three-dimensional plots. In this chapter, we introduced stem, stair, bar, compass, mesh, surface, and contour plots.

6.6.1 Summary of Good Programming Practice

The following guidelines should be adhered to:

1. Use the char function to create two-dimensional character arrays without worrying about padding each row to the same length.
2. Use function isstrprop to determine the characteristics of each character in a string array. This function supercedes the older functions isletter and isspace, which may be deleted in a future version of MATLAB.
3. Use multidimensional arrays to solve problems that are naturally multivariate in nature, such as aerodynamics and fluid flows.
4. If you are working with multidimensional arrays, be sure to vectorize you code by hand. The MATLAB JIT compiler cannot handle loops containing multidimensional arrays.
5. Use function fplot to plot functions directly without having to create intermediate data arrays.

6.6.2 MATLAB Summary

The following summary lists all of the MATLAB commands and functions described in this chapter, along with a brief description of each one.

`abs`	Returns absolute value (magnitude) of a number.
`angle`	Returns the angle of a complex number, in radians.
`bar(x,y)`	Create a vertical bar plot.
`barh(x,y)`	Create a horizontal bar plot.
`base2dec`	Convert base B string to decimal integer.
`bin2dec`	Convert binary string to decimal integer.
`blanks`	Create a string of blanks.
`char`	(1) Convert numbers to the corresponding character values. (2) Create a two-dimensional character array from a series of strings.
`compass(x,y)`	Create a compass plot.
`conj`	Compute complex conjugate of a number.
`contour`	Create a contour plot.
`deblank`	Remove trailing whitespace from a string.
`dec2base`	Convert decimal integer to base B string.
`dec2bin`	Convert decimal integer to binary string.
`double`	Convert characters to the corresponding numeric codes.
`find`	Find indices and values of non-zero elements in a matrix.
`findstr`	Find one string within another one.
`hex2num`	Convert IEEE hexadecimal string to `double`.
`hex2dec`	Convert hexadecimal string to decimal integer.
`hist`	Create a histogram of a data set.
`full`	Convert a sparse matrix into a full matrix
`imag`	Returns the imaginary portion of the complex number.
`int2str`	Convert integer to string.
`ischar`	Returns true (1) for a character array.
`isletter`	Returns true (1) for letters of the alphabet.
`isreal`	Returns true (1) if no element of array has an imaginary component.
`isstrprop`	Returns true (1) a character has the specified property.
`isspace`	Returns true (1) for whitespace.
`lower`	Convert string to lowercase.
`mat2str`	Convert matrix to string.
`mesh`	Create a mesh plot.
`meshgrid`	Create the (x, y) grid required for mesh, surface, and contour plots.
`nnz`	Number of nonzero matrix elements.
`nonzeros`	Returns a column vector containing the nonzero elements in a matrix.

(Continued)

`num2str`	Convert number to string.
`nzmax`	Amount of storage allocated for nonzero matrix elements
`pie(x)`	Create a pie plot.
`plot(c)`	Plots the real versus the imaginary part of a complex array.
`real`	Returns the real portion of the complex number.
`rose`	Create a radial histogram of a data set.
`sscanf`	Read formatted data from string.
`stairs(x,y)`	Create a stair plot.
`stem(x,y)`	Create a stem plot.
`str2double`	Convert string to `double` value.
`str2num`	Convert string to number.
`strcat`	Concatenate strings.
`strcmp`	Returns true (1) if two strings are identical.
`strcmpi`	Returns true (1) if two strings are identical, ignoring case.
`strjust`	Justify string.
`strncmp`	Returns true (1) if first `n` characters of two strings are identical.
`strncmpi`	Returns true (1) if first `n` characters of two strings are identical, ignoring case.
`strmatch`	Find matches for string.
`strtrim`	Remove leading and trailing whitespace from a string.
`strrep`	Replace one string with another.
`strtok`	Find token in string.
`struct`	Pre-define a structure array.
`strvcat`	Concatenate strings vertically.
`surf`	Create a surface plot.
`upper`	Convert string to uppercase.

6.7 Exercises

6.1 Figure 6.12 shows a series *RLC* circuit driven by a sinusoidal AC voltage source whose value is $120\angle 0°$ volts. The impedance of the inductor in this circuit is $Z_L = j2\pi fL$, where j is $\sqrt{-1}$, f is the frequency of the voltage source in hertz, and L is the inductance in henrys. The impedance of the capacitor in this circuit is $Z_C = -j\frac{1}{2\pi fC}$, where C is the capacitance in farads. Assume that $R = 100\ \Omega$, $L = 0.1$ mH, and $C = 0.25$ nF.

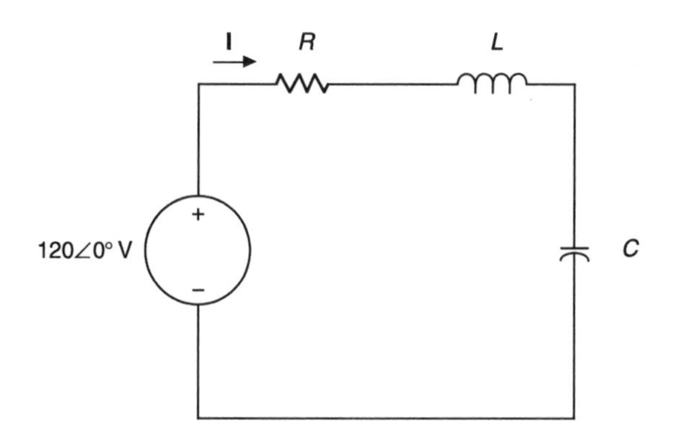

Figure 6.12 A series *RLC* circuit driven by a sinusoidal ac voltage source.

The current **I** flowing in this circuit is given by Kirchhoff's Voltage Law to be

$$\mathbf{I} = \frac{120\angle 0^\circ \text{ V}}{R + j2\pi fL - j\dfrac{1}{2\pi fC}} \tag{6-13}$$

(*a*) Calculate and plot the magnitude of this current as a function of frequency as the frequency changes from 100 kHz to 10 MHz. Plot this information on both a linear and a log-linear scale. Be sure to include a title and axis labels.

(*b*) Calculate and plot the phase angle in degrees of this current as a function of frequency as the frequency changes from 100 kHz to 10 MHz. Plot this information on both a linear and a log-linear scale. Be sure to include a title and axis labels.

(*c*) Plot both the magnitude and phase angle of the current as a function of frequency on two subplots of a single figure. Use log-linear scales.

6.2 Write a function `to_polar` that accepts a complex number `c` and returns two output arguments containing the magnitude `mag` and angle `theta` of the complex number. The output angle should be in degrees.

6.3 Write a function `to_complex` that accepts two input arguments containing the magnitude `mag` and angle `theta` of the complex number in degrees and returns the complex number `c`.

6.4 In a sinusoidal steady-state AC circuit, the voltage across a passive element is given by Ohm's Law:

$$\mathbf{V} = \mathbf{I}Z \tag{6-14}$$

where **V** is the voltage across the element, **I** is the current though the element, and Z is the impedance of the element. Note that all three of these values are complex, and that these complex numbers are usually specified

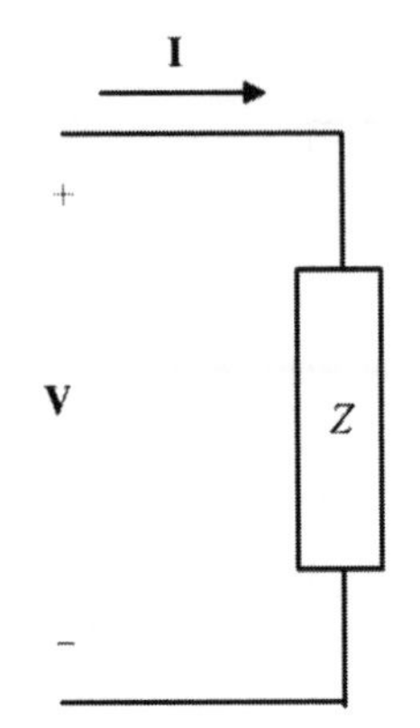

Figure 6.13 The voltage and current relationship on a passive ac circuit element.

in the form of a magnitude at a specific phase angle expressed in degrees. For example, the voltage might be $\mathbf{V} = 120\angle 30°$ V.

Write a program that reads the voltage across an element and the impedance of the element, and calculates the resulting current flow. The input values should be given as magnitudes and angles expressed in degrees, and the resulting answer should be in the same form. Use the function `to_complex` from Exercise 6.3 to convert the numbers to rectangular for the actual computation of the current, and the function `to_polar` from Exercise 6.2 to convert the answer into polar form for display.

6.5 Write a function that will accept a complex number `c`, and plot that point on a Cartesian coordinate system with a circular marker. The plot should include both the x and y axes, plus a vector drawn from the origin to the location of `c`.

6.6 Plot the function $v(t) = 10e^{(-0.2+j\pi)t}$ for $0 \le t \le 10$ using the function `plot(t,v)`. What is displayed on the plot?

6.7 Plot the function $v(t) = 10e^{(-0.2+j\pi)t}$ for $0 \le t \le 10$ using the function `plot(v)`. What is displayed on the plot this time?

6.8 Create a polar plot of the function $v(t) = 10e^{(-0.2+j\pi)t}$ for $0 \le t \le 10$.

6.9 Plot of the function $v(t) = 10e^{(-0.2+j\pi)t}$ for $0 \le t \le 10$ using function `plot3`, where the three dimensions to plot are the real part of the function, the imaginary part of the function, and time.

6.10 **Euler's Equation** Euler's equation defines e raised to an imaginary power in terms of sinusoidal functions as follows:

$$e^{i\theta} = \cos\theta + i\sin\theta \tag{6-15}$$

Create a two-dimensional plot of this function as θ varies from 0 to 2π. Create a three-dimensional line plot using function `plot3` as θ varies from 0 to 2π (the three dimensions are the real part of the expression, the imaginary part of the expression, and θ).

6.11 Create a mesh, surface plot, and contour plot of the function $z = e^{x+iy}$ for the interval $-1 \leq x \leq 1$ and $-2\pi \leq y \leq 2\pi$. In each case, plot the real part of z versus x and y.

6.12 Write a program that accepts an input string from the user, and determines the how many times a user-specified character appears within the string. (*Hint:* Look up the `'s'` option of the `input` function using the MATLAB Help Browser.)

6.13 Modify the previous program so that it determines how many times a user-specified character appears within the string without regard to the case of the character.

6.14 Write a program that accepts a string from a user with the `input` function, chops that string into a series of tokens, sorts the tokens into ascending order, and prints them out.

6.15 Write a program that accepts a series of strings from a user with the `input` function, sorts the strings into ascending order, and prints them out.

6.16 Write a program that accepts a series of strings from a user with the `input` function, sorts the strings into ascending order disregarding case, and prints them out.

6.17 MATLAB includes functions `upper` and `lower`, which shift a string to upper case and lower case respectively. Create a new function called `caps`, which capitalizes the first letter in each word and forces all other letters to be lowercase. (*Hint:* Take advantage of functions `upper`, `lower`, and `strtok`.)

6.18 Write a function that accepts a character string and returns a `logical` array with true values corresponding to each printable character that is *not* alphanumeric or whitespace (for example, $, %, #, etc.), and false values everywhere else.

6.19 Write a function that accepts a character string and returns a `logical` array with true values corresponding to each vowel, and false values everywhere else. Be sure that the function works properly for both lowercase and uppercase characters.

6.20 Plot the function $y = e^{-x} \sin x$ for x between 0 and 2 in steps of 0.1. Create the following plot types: *(a)* stem plot; *(b)* stair plot; *(c)* bar plot; *(d)* compass plot. Be sure to include titles and axis labels on all plots.

6.21 Suppose that George, Sam, Betty, Charlie, and Suzie contributed $5, $10, $7, $5, and $15 respectively to a colleague's going-away present. Create a pie chart of their contributions. What percentage of the cost was paid by Sam?

6.22 Plot the function $f(x) = 1/\sqrt{x}$ over the range $0.1 \leq x \leq 10.0$ using function `fplot`. Be sure to label your plot properly.

CHAPTER 7

Cell Arrays, Structures, and Handle Graphics

This chapter deals with three very useful features of MATLAB: cell arrays, structures, and handle graphics.

Cell arrays are very flexible type of array that can hold any sort of data. Each element of a cell array can hold any type of MATLAB data, and different elements within the same array can hold different types of data. They are used extensively in MATLAB Graphical User Interface (GUI) functions.

Structures are a special type of array with named subcomponents. Each structure can have any number of subcomponents, each with its own name and data type. Structures are the basis of MATLAB objects.

Handle graphics is the name of a set of low-level graphics functions that control the characteristics of graphics objects generated by MATLAB. These functions are normally hidden inside M-files, but they are very important to the programmer, since they allow him or her to have fine control of the appearance of the plots and graphs created by an executing program from within the program.

7.1 Cell Arrays

A **cell array** is a special MATLAB array whose elements are *cells,* containers that can hold other MATLAB arrays. For example, one cell of a cell array might contain an array of real numbers, another an array of strings, and yet another a vector of complex numbers (see Figure 7.1).

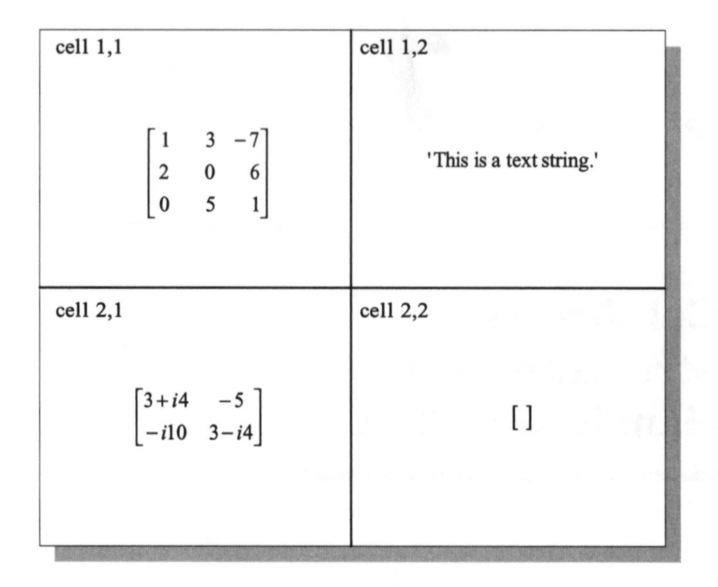

Figure 7.1 The individual elements of a cell array may point to real arrays, complex arrays, strings, other cell arrays, or even empty arrays.

In programming terms, each element of a cell array is a *pointer* to another data structure, and those data structures can be of different types. Figure 7.2 illustrates this concept. Cell arrays are great ways to collect information about a problem, since all of the information can be kept together and accessed by a single name.

Cell arrays use braces {} instead of parentheses () for selecting and displaying the contents of cells. This difference is due to the fact that *cell arrays contain data structures instead of data*. Suppose that the cell array `a` is defined as shown in Figure 7.2. Then the contents of element `a(1,1)` is a data structure containing a 3 × 3 array of numeric data, and a reference to `a(1,1)` displays the *contents* of the cell, which is the data structure.

```
» a(1,1)
ans =
    [3x3 double]
```

By contrast, a reference to `a{1,1}` displays *the contents of the contents of the cell.*

```
» a{1,1}
ans =
     1   3  -7
     2   0   6
     0   5   1
```

In summary, the notation `a(1,1)` refers to the contents of cell `a(1,1)` (which is a data structure), while the notation `a{1,1}` refers to the contents of the data structure within the cell.

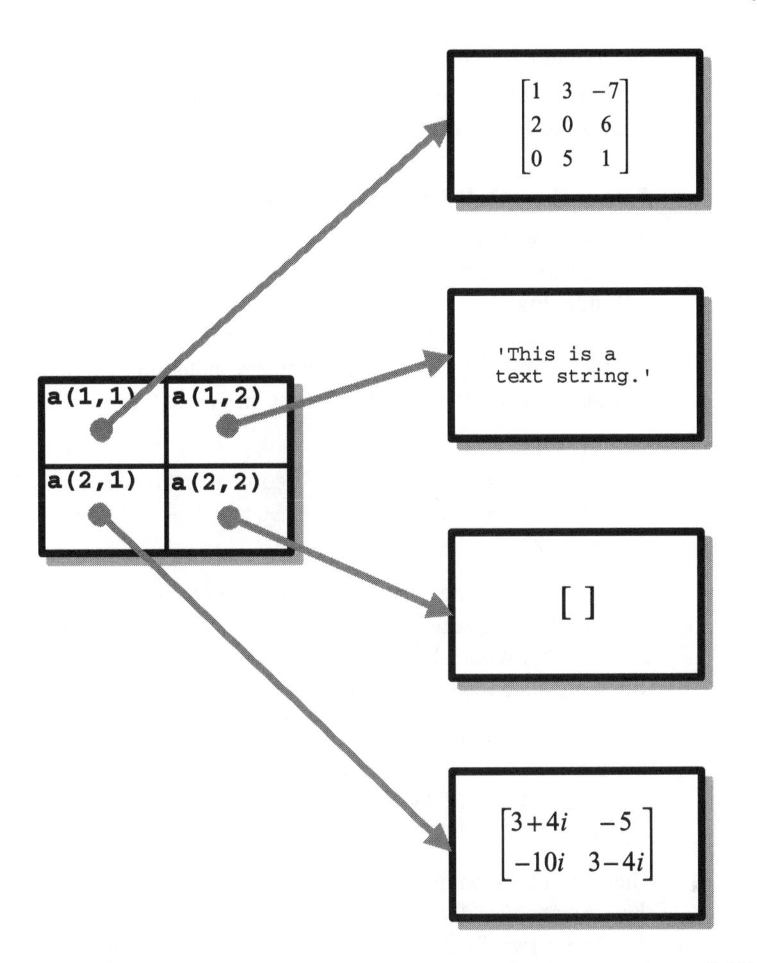

Figure 7.2 Each element of a cell array holds a *pointer* to another data structure, and different cells in the same cell array can point to different types of data structures.

Programming Pitfalls

Be careful not to confuse () with {} when addressing cell arrays. They are very different operations!

7.1.1 Creating Cell Arrays

Cell arrays can be created in two ways.

- By using assignment statements
- By preallocating a cell array using the `cell` function

The simplest way to create a cell array is to directly assign data to individual cells, one cell at a time. However, preallocating cell arrays is more efficient, so you should preallocate really large cell arrays.

Allocating Cell Arrays Using Assignment Statements

You can assign values to cell arrays one cell at a time using assignment statements. There are two ways to assign data to cells, known as **content indexing** and **cell indexing**.

Content indexing involves placing braces "{}" around the cell subscripts, together with cell contents in ordinary notation. For example, the following statement creates the 2 × 2 cell array in Figure 7.2:

```
a{1,1} = [1 3 -7; 2 0 6; 0 5 1];
a{1,2} = 'This is a text string.';
a{2,1} = [3+4*i -5; -10*i 3 - 4*i];
a{2,2} = [];
```

This type of indexing defines the *contents of the data structure contained in a cell.*

Cell indexing involves placing braces "{}" around the data to be stored in a cell, together with cell subscripts in ordinary subscript notation. For example, the following statement create the 2 × 2 cell array in Figure 7.2:

```
a(1,1) = {[1 3 -7; 2 0 6; 0 5 1]};
a(1,2) = {'This is a text string.'};
a(2,1) = {[3+4*i -5; -10*i 3 - 4*i]};
a(2,2) = {[]};
```

This type of indexing *creates a data structure containing the specified data, and then assigns that data structure to a cell.*

These two forms of indexing are completely equivalent, and they may be freely mixed in any program.

Programming Pitfalls

Do not attempt to create a cell array with the same name as an existing numeric array. If you do this, MATLAB will assume that you are trying to assign cell contents to an ordinary array, and it will generate an error message. Be sure to clear the numeric array before trying to create a cell array with the same name.

Preallocating Cell Arrays with the `cell` Function

The `cell` function allows you to preallocate empty cell arrays of the specified size. For example, the following statement creates an empty 2 × 2 cell array.

```
a = cell(2,2);
```

Once a cell array in created, you can use assignment statements to fill values in the cells.

7.1.2 Using Braces {} as Cell Constructors

It is possible to define many cells at once by placing all of the cell contents between a single set of braces. Individual cells on a row are separated by commas, and rows are separated by semicolons. For example, the following statement creates a 2×3 cell array:

```
b = {[1 2], 17, [2;4]; 3-4*i, 'Hello', eye(3)}
```

7.1.3 Viewing the Contents of Cell Arrays

MATLAB displays the data structures in each element of a cell array in a condensed form that limits each data structure to a single line. If the entire data structure can be displayed on the single line, it is. Otherwise, a summary is displayed. For example, cell arrays a and b would be displayed as:

```
» a
a =
       [3x3 double]      [1x22 char]
       [2x2 double]               []
» b
b =
            [1x2 double]    [    17] [2x1 double]
       [3.0000- 4.0000i]    'Hello'  3x3 double]
```

Note that MATLAB *is displaying the data structures,* complete with brackets or apostrophes, not the entire contents of the data structures.

If you would like to see the full contents of a cell array, use the celldisp function. This function displays *the contents of the data structures in each cell.*

```
» celldisp(a)
a{1,1} =
     1    3    -7
     2    0     6
     0    5     1
a{2,1} =
   3.0000 + 4.0000i    -5.0000
        0 -10.0000i     3.0000 - 4.0000i
a{1,2} =
This is a text string.
a{2,2} =
     []
```

For a high-level graphical display of the structure of a cell array, use the function cellplot. For example, the function cellplot(b) produces the plot shown in Figure 7.3.

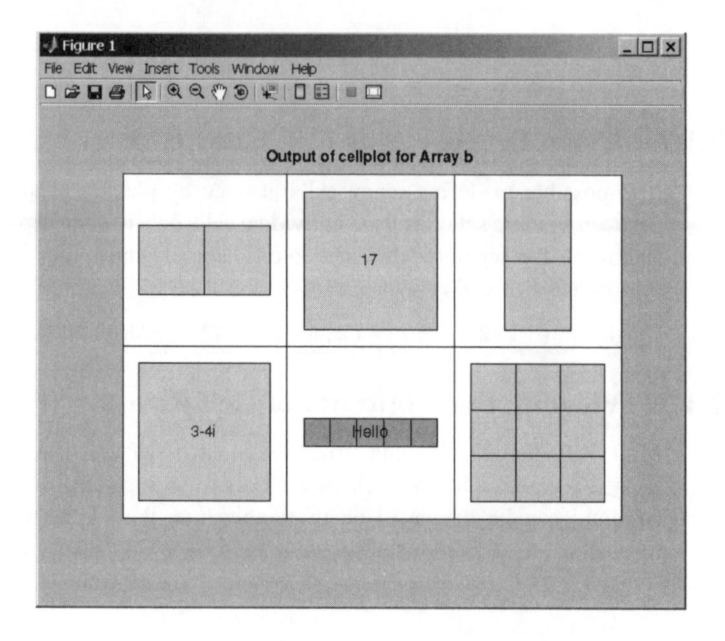

Figure 7.3 The structure of cell array b is displayed as a nested series of boxes by function `cellplot`.

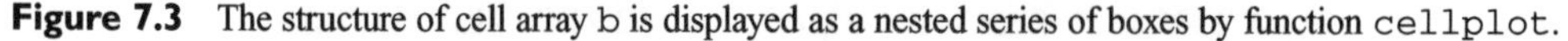

7.1.4 Extending Cell Arrays

If a value is assigned to a cell array element that does not currently exist, the element will be automatically created, and any additional cells necessary to preserve the shape of the array will be automatically created. For example, suppose that array a has been defined to be a 2 × 2 cell array as shown in Figure 7.1. If the following statement is executed

```
a{3,3} = 5
```

the cell array will be automatically extended to 3 × 3, as shown in Figure 7.4.

Preallocating cell arrays with the `cell` function is much more efficient than extending them the arrays element at a time using assignment statements. When a new element is added to an existing array as we did above, MATLAB must create a new array large enough to include this new element, copy the old data into the new array, add the new value to the array, and then delete the old array. This is a very time-consuming process. Instead, you should always allocate the cell array to be the largest size that you will need, and then add values to it an element at a time. If you do that, only the new element needs to be added—the rest of the array can remain undisturbed.

The program shown below illustrates the advantages of preallocation. It creates a cell array containing 50,000 strings added one at a time, with and without preallocation.

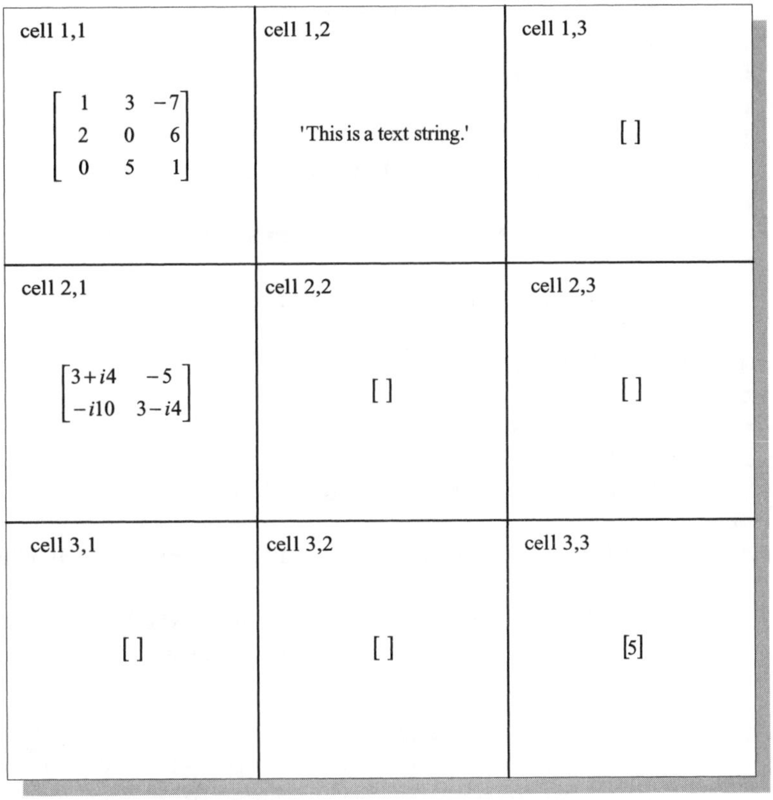

cell 1,1	cell 1,2	cell 1,3
$\begin{bmatrix} 1 & 3 & -7 \\ 2 & 0 & 6 \\ 0 & 5 & 1 \end{bmatrix}$	'This is a text string.'	[ ]
cell 2,1	cell 2,2	cell 2,3
$\begin{bmatrix} 3+i4 & -5 \\ -i10 & 3-i4 \end{bmatrix}$	[ ]	[ ]
cell 3,1	cell 3,2	cell 3,3
[ ]	[ ]	[5]

Figure 7.4 The result of assigning a value to `a{3,3}`. Note that four other empty cells were created to preserve the shape of the cell array.

```
%  Script file: test_preallocate.m
%
%  Purpose:
%    This program tests the creation of cell arrays with
%    and without preallocation.
%
%  Record of revisions:
%      Date         Programmer          Description of change
%      ====         ==========          =====================
%    01/18/05     S. J. Chapman         Original code
%
% Define variables:
%   a              -- Cell array
%   maxvals        -- Maximum values in cell array
```

```
% Create array without preallocation
clear all
maxvals = 50000;
tic
for ii = 1:maxvals
   a{ii} = ['Element ' int2str(ii)];
end
disp( ['Elapsed time without preallocation = ' num2str(toc)] );

% Create array with preallocation
clear all
maxvals = 50000;
tic
a = cell(1,maxvals);
for ii = 1:maxvals
   a{ii} = ['Element ' int2str(ii)];
end
disp( ['Elapsed time with preallocation = ' num2str(toc)] );
```

When this program is executed using MATLAB 7.0 on a 2.4 GHz Pentium IV computer, it produces the following results. The advantages of preallocation are obvious.

```
» test_preallocate
Elapsed time without preallocation = 13.079
Elapsed time with preallocation    = 4.813
```

Good Programming Practice

Always preallocate all cell arrays before assigning values to the elements of the array. This practice greatly increases the execution speed of a program.

7.1.5 Deleting Cells in Arrays

To delete an entire cell array, use the `clear` command. Subsets of cells may be deleted by assigning an empty array to them. For example, assume that `a` is the 3 × 3 cell array defined above.

```
» a
a =
    [3x3 double]    [1x22 char]    []
    [2x2 double]             []    []
              []             []   [5]
```

It is possible to delete the entire third row with the statement

```
» a(3,:) = []
a =
    [3x3 double]    [1x22 char]    []
    [2x2 double]             []    []
```

7.1.6 Using Data in Cell Arrays

The data stored inside the data structures within a cell array may be used at any time, with either content indexing or cell indexing. For example, suppose that a cell array `c` is defined as

```
c = {[1 2;3 4], 'dogs'; 'cats', i}
```

The contents of the array stored in cell `c(1,1)` can be accessed as follows

```
» c{1,1}
ans =
     1    2
     3    4
```

and the contents of the array in cell `c(2,1)` can be accessed as follows

```
» c{2,1}
ans =
cats
```

Subsets of a cell's contents can be obtained by concatenating the two sets of subscripts. For example, suppose that we would like to get the element (1, 2) from the array stored in cell `c(1,1)` of cell array `c`. To do this, we would use the expression `c{1,1}(1,2)`, which says: select element (1, 2) from the contents of the data structure contained in cell `c(1,1)`.

```
» c{1,1}(1,2)
ans =
     2
```

7.1.7 Cell Arrays of Strings

It is often convenient to store groups of strings in a cell array instead of storing them in rows of a standard character array, because each string in a cell array can have a different length, while every row of a standard character array must have an identical length. This fact means that *strings in cell arrays do not have to be padded with blanks*. Many MATLAB Graphical User Interface functions use cell arrays for precisely this reason.

Cell arrays of strings can be created in one of two ways. Either the individual strings can be inserted into the array with brackets, or else function `cellstr` can be used to convert a two-dimensional string array into a cell array of strings.

The following example creates a cell array of strings by inserting the strings into the cell array one at a time and displays the resulting cell array. Note that the individual strings can be of different lengths.

```
» cellstring{1} = 'Stephen J. Chapman';
» cellstring{2} = 'Male';
» cellstring{3} = 'SSN 999-99-9999';
» cellstring
   'Stephen J. Chapman'    'Male'    'SSN 999-99-9999'
```

Function `cellstr` creates a cell array of strings from a two-dimensional string array. Consider the character array

```
» data = ['Line 1         ';'Additional Line']
data =
Line 1
Additional Line
```

This 2 × 15 character array can be converted into an cell array of strings with the function `cellstr` as follow:

```
» c = cellstr(data)
c =
   'Line 1'
   'Additional Line'
```

and it can be converted back to a standard character array using function `char`

```
» newdata = char(c)
newdata =
Line 1
Additional Line
```

7.1.8 The Significance of Cell Arrays

Cell arrays are extremely flexible, since any amount of any type of data can be stored in each cell. As a result, cell arrays are used in many internal MATLAB data structures. We must understand them in order to use many features of Handle Graphics and Graphical User Interfaces.[1]

In addition, the flexibility of cell arrays makes them regular features of functions with variable numbers of input arguments and output arguments. A special input argument, `varargin`, is available within user-defined MATLAB functions to support variable numbers of input arguments. This argument appears as the last item in an input argument list, and it returns a cell array, so *a single dummy input*

---

[1]Graphical User Interfaces are beyond the scope of this book.

argument can support any number of actual arguments. Each actual argument becomes one element of the cell array returned by `varargin`. If it is used, `varargin` must be the *last* input argument in a function, after all of the required input arguments.

For example, suppose that we are writing a function that may have any number of input arguments. This function could be implemented as shown:

```
function test1(varargin)
disp(['There are ' int2str(nargin) ' arguments.']);
disp('The input arguments are:');
disp(varargin);

end % function test1
```

When this function is executed with varying numbers of arguments, the results are:

```
» test1
There are 0 arguments.
The input arguments are:
» test1(6)
There are 1 arguments.
The input arguments are:
    [6]
» test1(1,'test 1',[1 2;3 4])
There are 3 arguments.
The input arguments are:
    [1]    'test 1'    [2x2 double]
```

As you can see, the arguments become a cell array within the function.

A sample function making use of variable numbers of arguments is shown below. Function `plotline` accepts an arbitrary number of 1×2 row vectors, with each vector containing the (x, y) position of one point to plot. The function plots a line connecting all of the (x, y) values together. Note that this function also accepts an optional line specification string, and passes that specification on to the `plot` function.

```
function plotline(varargin)
%PLOTLINE Plot points specified by [x,y] pairs.
% Function PLOTLINE accepts an arbitrary number of
% [x,y] points and plots a line connecting them.
% In addition, it can accept a line specification
% string, and pass that string on to function plot.

% Define variables:
%  ii         -- Index variable
%  jj         -- Index variable
%  linespec   -- String defining plot characteristics
%  msg        -- Error message
```

```
%  varargin  -- Cell array containing input arguments
%  x         -- x values to plot
%  y         -- y values to plot

%  Record of revisions:
%     Date       Programmer       Description of change
%     ====       ==========       =====================
%   01/18/05    S. J. Chapman    Original code

% Check for a legal number of input arguments.
% We need at least 2 points to plot a line...
msg = nargchk(2,Inf,nargin);
error(msg);

% Initialize values
jj = 0;
linespec = '';

% Get the x and y values, making sure to save the line
% specification string, if one exists.
for ii = 1:nargin

  % Is this argument an [x,y] pair or the line
  % specification?
  if ischar(varargin{ii})

     % Save line specification
     linespec = varargin{ii};

  else

     % This is an [x,y] pair. Recover the values.
     jj = jj + 1;
      x(jj) = varargin{ii}(1);
      y(jj) = varargin{ii}(2);

  end
end

% Plot function.
if isempty(linespec)
   plot(x,y);
else
   plot(x,y,linespec);
end
```

When this function is called with the arguments shown below, the resulting plot is shown in Figure 7.5. Try the function with different numbers of arguments and see for yourself how it behaves.

```
plotline([0 0],[1 1],[2 4],[3 9],'k--');
```

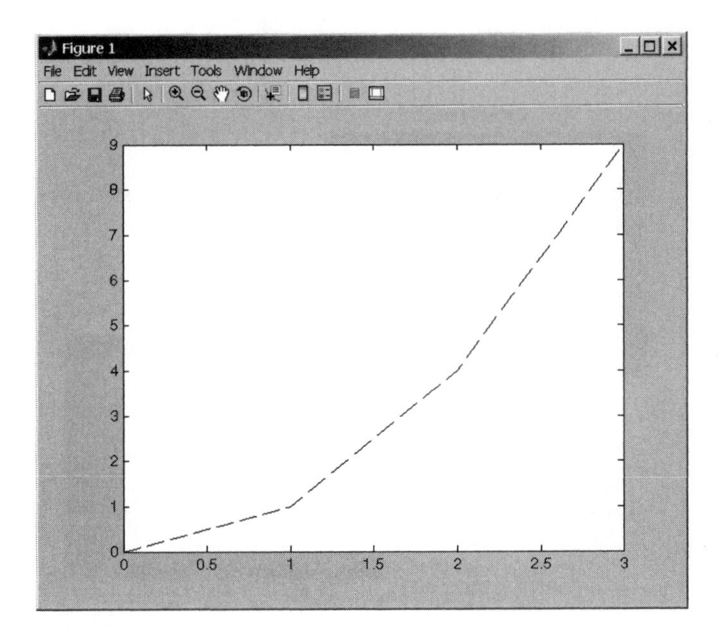

Figure 7.5 The plot produced by function `plotline`.

There is also a special output argument, `varargout`, to support variable numbers of output arguments. This argument appears as the last item in an output argument list, and it returns a cell array, so *a single dummy output argument can support any number of actual arguments*. Each actual argument becomes one element of the cell array stored in `varargout`.

If it is used, `varargout` must be the *last* output argument in a function, after all of the required input arguments. The number of values to be stored in `varargout` can be determined from function `nargout`, which specifies the number of actual output arguments for any given function call.

A sample function `test2` is shown below. This function detects the number of output arguments expected by the calling program, using the function `nargout`. It returns the number of random values in the first output argument, and then fills the remaining output arguments with random numbers taken from a Gaussian distribution. Note that the function uses `varargout` to hold the random numbers, so that there can be an arbitrary number of output values.

```
function [nvals,varargout] = test2(mult)
% nvals is the number of random values returned
% varargout contains the random values returned
nvals = nargout - 1;
for ii = 1:nargout-1
```

```
    varargout{ii} = randn * mult;
end

end % function test2
```

When this function is executed, the results are as shown below.

```
» test2(4)
ans =
    -1
» [a b c d] = test2(4)
a =
    3
b =
   -1.7303
c =
   -6.6623
d =
    0.5013
```

✱ Good Programming Practice

Use cell array arguments `varargin` and `varargout` to create functions that support varying numbers of input and output arguments.

7.1.9 Summary of `cell` Functions

The common MATLAB cell functions are summarized in Table 7.1.

Table 7.1 Common MATLAB Cell Functions

Function	Description
`cell`	Predefine a cell array structure.
`celldisp`	Display contents of a cell array.
`cellplot`	Plot structure of a cell array.
`cellstr`	Convert a two-dimensional character array to a cell array of strings.
`char`	Convert a cell array of strings into a two-dimensional character array.

7.2 Structure Arrays

An *array* is a data type in which there is a name for the whole data structure, but individual elements within the array are known only by number. Thus the fifth element in the array named `arr` would be accessed as `arr(5)`. All of the individual elements in an array must be of the *same* type.

A *cell array* is a data type in which there is a name for the whole data structure, but individual elements within the array are known only by number. However, the individual elements in the cell array may be of *different* types.

In contrast, a **structure** is a data type in which each individual element is has a name. The individual elements of a structure are known as **fields**, and each field in a structure may have a different type. The individual fields are addressed by combining the name of the structure with the name of the field, separated by a period.

Figure 7.6 shows a sample structure named `student`. This structure has five fields, called `name`, `addr1`, `city`, `state`, and `zip`. The field called "`name`" would be addressed as `student.name`.

A **structure array** is an array of structures. Each structure in the array will have identically the same fields, but the data stored in each field can differ. For example, a class could be described by an array of the structure `student`. The first student's name would be addressed as `student(1).name`, the second student's city would be addressed as `student(2).city`, and so forth.

7.2.1 Creating Structure Arrays

Structure arrays can be created in two ways.

- A field at a time using assignment statements
- All at once using the `struct` function

Building a Structure with Assignment Statements

You can build a structure a field at a time using assignment statements. Each time that data is assigned to a field, that field is automatically created. For example, the structure shown in Figure 7.6 can be created with the following statements.

```
» student.name='John Doe';
» student.addr1='123 Main Street';
» student.city ='Anytown';
» student.zip='71211'
student =
     name:   'John Doe'
    addr1:   '123 Main Street'
     city:   'Anytown'
    state:   'LA'
      zip:   '71211'
```

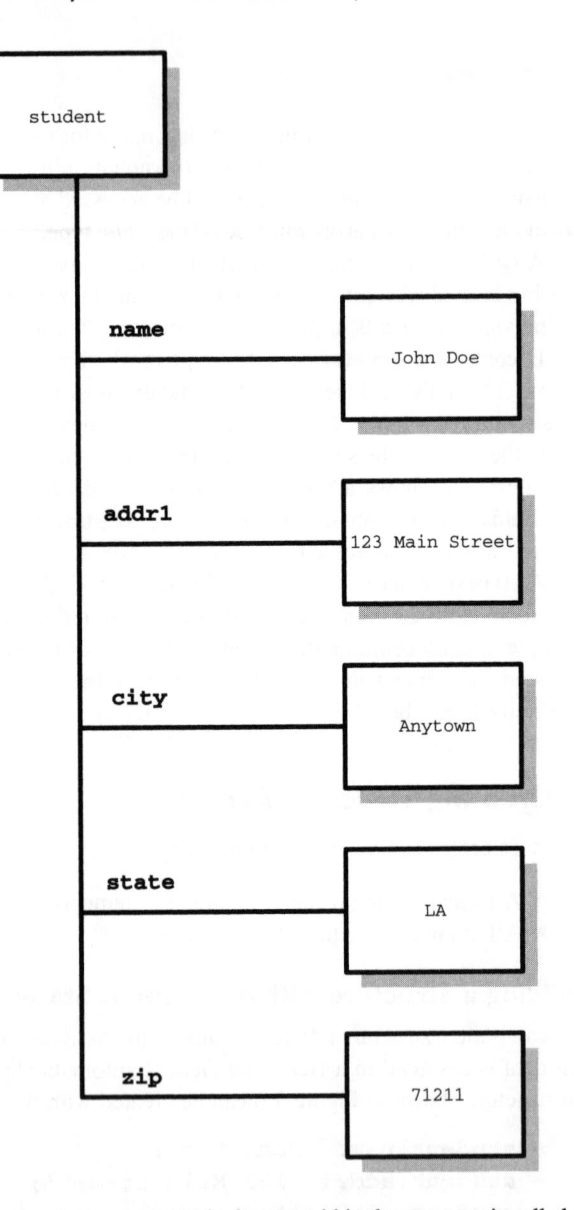

Figure 7.6 A sample structure. Each element within the structure is called a field, and each field is addressed by name.

A second student can be added to the structure by adding a subscript to the structure name (*before* the period).

```
» student(2).name = 'Jane Q. Public'
student =
```

```
1x2 struct array with fields:
    name
    addr1
    city
    state
    zip
```

student is now a 1 × 2 array. Note that when a structure array has more than one element, only the field names are listed, not their contents. The contents of each element can be listed by typing the element separately in the Command window:

```
» student(1)
ans =
      name: 'John Doe'
     addr1: '123 Main Street'
      city: 'Anytown'
     state: 'LA'
       zip: '71211'
» student(2)
ans =
      name:  'Jane Q. Public'
     addr1:  []
      city:  []
     state:  []
       zip:  []
```

Note that *all of the fields of a structure are created for each array element whenever that element is defined,* even if they are not initialized. The uninitialized fields will contain empty arrays, which can be initialized with assignment statements at a later time.

The field names used in a structure can be recovered at any time using the `fieldnames` function. This function returns a list of the field names in a cell array of strings and is very useful for working with structure arrays within a program.

Creating Structures with the `struct` Function

The `struct` function allows you to preallocate a structure or an array of structures. The basic form of this function is

```
str_array = struct('field1',val1,'field2',val2, ...)
```

where the arguments are field names and their initial values. With this syntax, function `struct` initializes every field to the specified value.

To preallocate an entire array with the `struct` function, simply assign the output of the `struct` function to the *last value* in the array. All of the values before that will be automatically created at the same time. For example, the following statements create an array containing 1,000 structures of type `student`.

```
student(1000) = struct('name',[].'addr1',[], ...
                       'city',[],'state',[],'zip',[])
student =
1x1000 struct array with fields:
    name
    addr1
    city
    state
    zip
```

All of the elements of the structure are preallocated, which will speed up any program using the structure.

There is another version of the `struct` function that will preallocate an array and at the same time assign initial values to all of its fields. You will be asked to do this in an end-of-chapter exercise.

7.2.2 Adding Fields to Structures

If a new field name is defined for any element in a structure array, the field is automatically added to all of the elements in the array. For example, suppose that we add some exam scores to Jane Public's record:

```
» student(2).exams = [90 82 88]
student =
1x2 struct array with fields:
    name
    addr1
    city
    state
    zip
    exams
```

There is now a field called exams in every record of the array, as shown below. This field will be initialized for `student(2)`, and will be an empty array for all other students until appropriate assignment statements are issued.

```
» student(1)
ans =
      name: 'John Doe'
     addr1: '123 Main Street'
      city: 'Anytown'
     state: 'LA'
       zip: '71211'
     exams: []
» student(2)
ans =
      name: 'Jane Q. Public'
```

```
 addr1: []
  city: []
 state: []
   zip: []
 exams: [90 82 88]
```

7.2.3 Removing Fields from Structures

A field may be removed from a structure array using the `rmfield` function. The form of this function is:

```
struct2 = rmfield(str_array,'field')
```

where `str_array` is a structure array, `'field'` is the field to remove, and `struct2` is the name of a new structure with that field removed. For example, we can remove the field `'zip'` from structure array `student` with the following statement:

```
» stu2 = rmfield(student,'zip')
stu2 =
1x2 struct array with fields:
    name
    addr1
    city
    state
    exams
```

7.2.4 Using Data in Structure Arrays

Now let's assume that structure array student has been extended to include three students and all data has been filled in as shown in Figure 7.7. How do we use the data in this structure array?

To access the information in any field of any array element, just name the array element followed by a period and the field name:

```
» student(2).addr1
ans =
P. O. Box 17
» student(3).exams
ans =
    65    84    81
```

To access an individual item within a field, add a subscript after the field name. For example, the second exam of the third student is

```
» student(3).exams(2)
ans =
    84
```

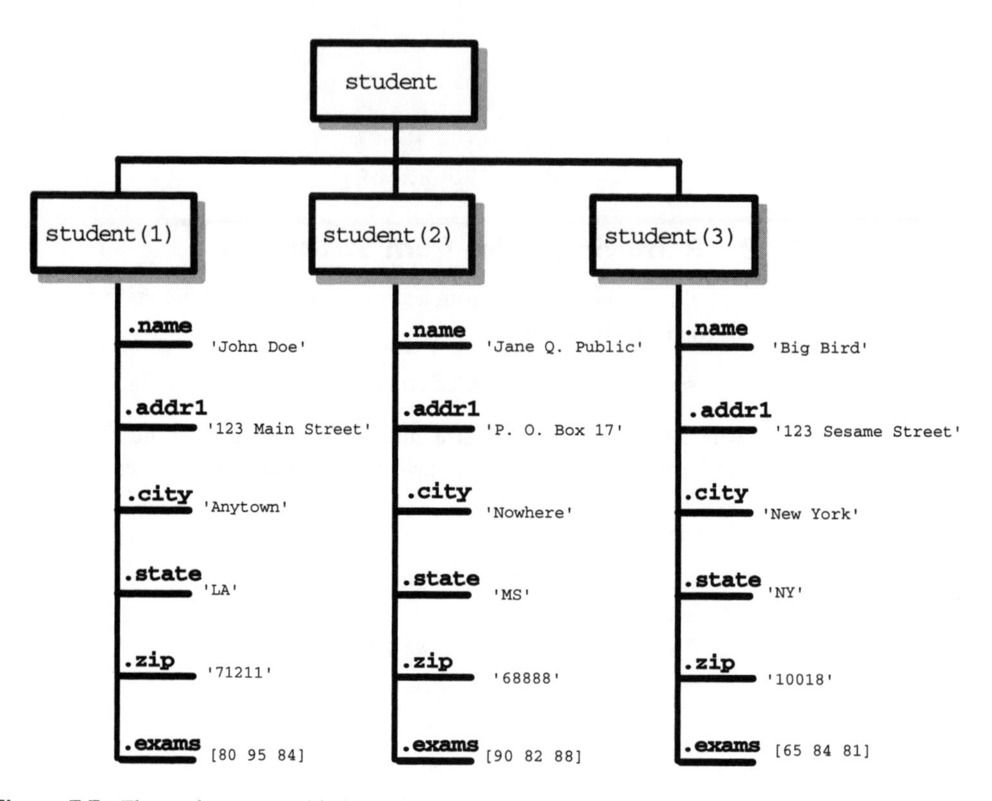

Figure 7.7 The student array with three elements and all fields filled in.

The fields in a structure array can be used as arguments in any function that supports that type of data. For example, to calculate `student(2)`'s exam average, we could use the function

```
» mean(student(2).exams)
ans =
    86.6667
```

Unfortunately, we can *not* extract the values from a given field across multiple array elements at the same time. For example, we cannot get access to an array of zip codes with the expression `student.zip`. That expression returns the three zip codes of the three students in three separate arrays. If we want to get the zip codes of all of the students in a single array, we must use a `for` loop:

```
for ii = 1:length(student)
   zip(ii) = student(ii).zip;
end
```

Similarly, if we wanted to get the average of *all* exams from *all* students, we cannot use the function `mean(student.exams)`. Instead, we must build up an array containing all the exam scores by accessing each student's exams separately, and then call `mean` with that array.

```
exam_list = [];
for ii = 1:length(student)
   exam_list = [exam_list student(ii).exams];
end
mean(exam_list)
```

7.2.5 The getfield and setfield Functions

Two MATLAB functions are available to make structure arrays easier to use in programs. Function `getfield` gets the current value stored in a field, and function `setfield` inserts a new value into a field. The structure of function `getfield` is

```
f = getfield(array,{array_index},'field',{field_index})
```

where the `field_index` is optional, and `array_index` is optional for a 1-by-1 structure array. The function call corresponds to the statement

```
f = array(array_index).field(field_index);
```

but it can be used even if the programmer doesn't know the names of the fields in the structure array at the time the program is written.

For example, suppose that we needed to write a function to read and manipulate the data in an unknown structure array. This function could determine the field names in the structure using a call to `fieldnames`, and then could read the data using function `getfield`. To read the zip code of the second student, the function would be

```
» zip = getfield(student,{2},'zip')
zip =
      68888
```

Similarly, a program could modify values in the structure using function `setfield`. The structure of function `setfield` is

```
f = setfield(array,{array_index},'field',{field_index},value)
```

where `f` is the output structure array, the `field_index` is optional, and `array_index` is optional for a 1-by-1 structure array. The function call corresponds to the statement

```
array(array_index).field(field_index) = value;
```

7.2.6 Dynamic Field Names

Beginning with MATLAB 7.0, there is an alternative way to access the elements of a structure: **dynamic field names**. A dynamic field name is a string enclosed in parentheses at a location where a field name is expected. For example, the name of student 1 can be retrieved with either static or dynamic field names as shown below:

```
» student(1).name                   % Static field name
ans =
John Doe
» student(1).('name')               % Dynamic field name
ans =
John Doe
```

Dynamic field names perform the same function as static field names, but *dynamic field names can be changed during program execution*. This allows a user to access different information in the same function within a program.

For example, the following function accepts a structure array and a field name, and calculates the average of the values in the specified field for all elements in the structure array. It returns that average (and optionally the number of values averaged) to the calling program.

```
function [ave, nvals] = calc_average(structure,field)
%CALC_AVERAGE Calculate the average of values in a field.
% Function CALC_AVERAGE calculates the average value
% of the elements in a particular field of a structure
% array. It returns the average value and (optionally)
% the number of items averaged.

% Define variables:
%   arr        -- Array of values to average
%   ave        -- Average of arr
%   ii         -- Index variable
%
%  Record of revisions:
%      Date          Programmer          Description of change
%      ====          ==========          =====================
%    01/18/05     S. J. Chapman       Original code
%
% Check for a legal number of input arguments.
msg = nargchk(2,2,nargin);
error(msg);

% Create an array of values from the field
arr = [];
for ii = 1:length(structure)
   arr = [arr structure(ii).(field)];
end
```

```
% Calculate average
ave = mean(arr);

% Return number of values averaged
if nargout == 2
   nvals = length(arr);
end
```

A program can average the values in different fields by simply calling this function multiple times with different structure names and different field names. For example, we can calculate the average values in fields `exams` and `zip` as follows:

```
» [ave,nvals] = calc_average(student,'exams')
ave =
   83.2222
nvals =
     9
» ave = calc_average(student,'zip')
ave =
      50039
```

7.2.7 Using the `size` Function with Structure Arrays

When the `size` function is used with a structure array, it returns the size of the structure array itself. When the `size` function is used with a *field* from a particular element in a structure array, it returns the size of that field instead of the size of the whole array. For example,

```
» size(student)
ans =
      1     3
» size(student(1).name)
ans =
      1     8
```

7.2.8 Nesting Structure Arrays

Each field of a structure array can be of any data type, including a cell array or a structure array. For example, the following statements define a new structure array as a field under array `student` to carry information about each class that the student in enrolled in.

```
student(1).class(1).name = 'COSC 2021'
student(1).class(2).name = 'PHYS 1001'
student(1).class(1).instructor = 'Mr. Jones'
student(1).class(2).instructor = 'Mrs. Smith'
```

After these statements are issued, `student(1)` contains the following data. Note the technique used to access the data in the nested structures.

```
» student(1)
ans =
        name:   'John Doe'
       addr1:   '123 Main Street'
        city:   'Anytown'
       state:   'LA'
         zip:   '71211'
       exams:   [80 95 84]
       class:   [1x2 struct]
» student(1).class
ans =
1x2 struct array with fields:
    name
    instructor
» student(1).class(1)
ans =
          name:   'COSC 2021'
    instructor:   'Mr. Jones'
» student(1).class(2)
ans =
          name:   'PHYS 1001'
    instructor:   'Mrs. Smith'
» student(1).class(2).name
ans =
PHYS 1001
```

7.2.9 Summary of structure Functions

The common MATLAB structure functions are summarized in Table 7.2.

Table 7.2 Common MATLAB Structure Functions

Function	Description
`fieldnames`	Return a list of field names in a cell array of strings.
`getfield`	Get current value from a field.
`rmfield`	Remove a field from a structure array.
`setfield`	Set new value into a field.
`struct`	Predefine a structure array.

Quiz 7.1

This quiz provides a quick check to see if you have understood the concepts introduced in Sections 7.1 through 7.2. If you have trouble with the quiz, reread the section, ask your instructor, or discuss the material with a fellow student. The answers to this quiz are found in the back of the book.

1. What is a cell array? How does it differ from an ordinary array?
2. What is the difference between content indexing and cell indexing?
3. What is a structure? How does it differ from ordinary arrays and cell arrays?
4. What is the purpose of `varargin`? How does it work?
5. Given the definition of array `a` shown below, what will be produced by each of the following sets of statements? (*Note:* some of these statements may be illegal. If a statement is illegal, explain why.)

```
a{1,1} = [1 2 3; 4 5 6; 7 8 9];
a(1,2) = {'Comment line'};
a{2,1} = j;
a{2,2} = a{1,1} - a{1,1}(2,2);
```

(*a*) `a(1,1)`
(*b*) `a{1,1}`
(*c*) `2*a(1,1)`
(*d*) `2*a{1,1}`
(*e*) `a{2,2}`
(*f*) `a(2,3) = {[-17; 17]}`
(*g*) `a{2,2}(2,2)`

6. Given the definition of structure array `b` shown below, what will be produced by each of the following sets of statements? (*Note:* some of these statements may be illegal. If a statement is illegal, explain why.)

```
b(1).a = -2*eye(3);
b(1).b = 'Element 1';
b(1).c = [1 2 3];
b(2).a = [b(1).c' [-1; -2; -3] b(1).c'];
b(2).b = 'Element 2';
b(2).c = [1 0 -1];
```

(*a*) `b(1).a - b(2).a`
(*b*) `strncmp(b(1).b,b(2).b,6)`
(*c*) `mean(b(1).c)`
(*d*) `mean(b.c)`
(*e*) `b`
(*f*) `b(1).('b')`
(*g*) `b(1)`

7.3 Handle Graphics

Handle graphics is the name of a set of low-level graphics functions that control the characteristics of graphics objects generated by MATLAB. These functions are normally hidden inside M-files, but they are very important to the programmer, since they allow him/her to have fine control of the appearance of the plots and graphs that they generate. For example, it is possible to use handle graphics to turn on a grid on the *x*-axis only, or to choose a line color like orange, which is not supported by the standard `LineSpec` option of the `plot` command.

This section introduces the structure of the MATLAB graphics system, and explains how to control the properties of graphical objects to create a desired display.

7.3.1 The MATLAB Graphics System

The MATLAB graphics system is based on a hierarchical system of **graphics objects**, each of which is known by a unique number called a **handle**. Each graphics object has special data called **properties** associated with it, and modifying those properties will modify the behavior of the object. For example, a **`line`** is one type of graphics object. The properties associated with a line object include: *x*-data, *y*-data, color, line style, line width, marker type, *etc*. Modifying any of these properties will change the way that the line is displayed in a Figure Window.

Every component of a MATLAB graph is a graphical object. For example, each line, axes and text string is a separate object with its own unique identifying number (handle) and characteristics. All graphical objects are arranged in a hierarchy with **parent objects** and **child objects**, as shown in Figure 7.8. When a child object is created, it inherits many of its properties from its parent.

The highest-level graphics object in MATLAB is the `root`, which can be thought of as the entire computer screen. The handle of the `root` object is always 0. It is created automatically when MATLAB starts up, and it is always present until the program is shut down. The properties associated with the root object are the defaults that apply to all MATLAB windows.

Under the root there can be one or more Figure Windows, or just **`figures`**. Each `figure` is a separate window on the computer screen that can display graphical data, and each figure has its own properties. The properties associated with a `figure` include color, color map, paper size, paper orientation, pointer type, etc.

Each `figure` can contain seven types of objects: `uimenus`, `uicontextmenus`, `uicontrols`, `uitoolbars`, `uipanels`, `uibuttongroups`, and `axes`. `Uimenus`, `uicontextmenus`, `uicontrols`, `uitoolbars`, `uipanels`, and `uibuttongroups` are special graphics objects used to create graphical user interfaces—they are not discussed in this book. `Axes` are regions within a figure where data is actually plotted. There can be more than one set of axes in a single figure.

Each set of axes can contain as many `lines`, `text` strings, `patches`, etc. as necessary to create the plot of interest.

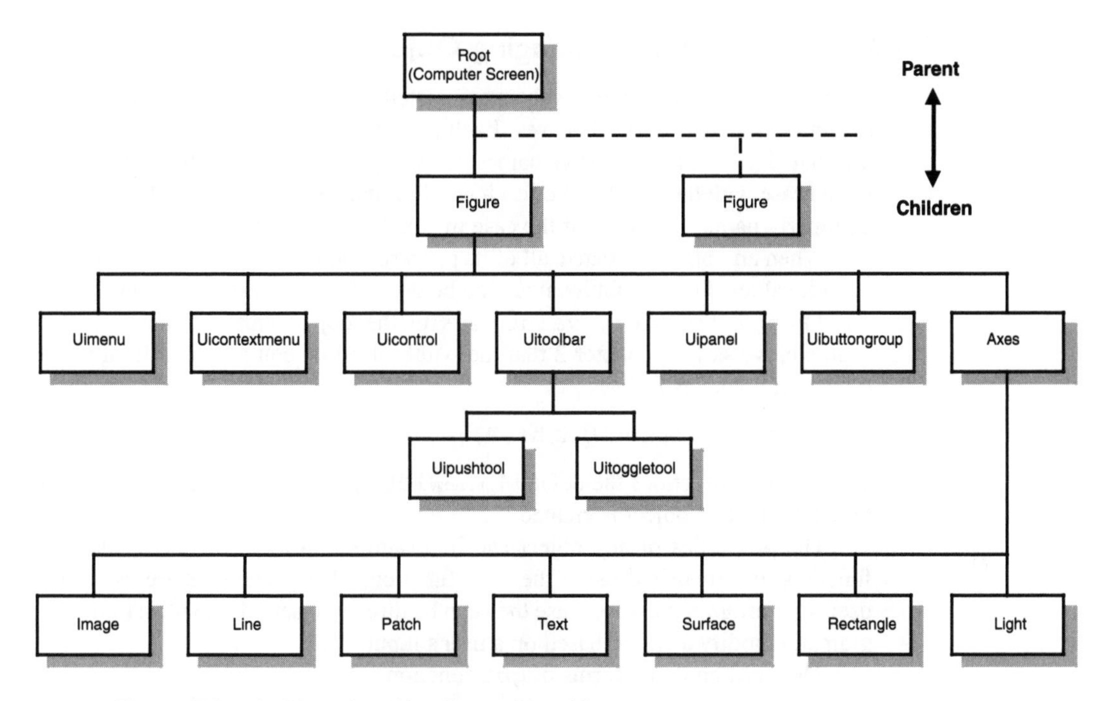

Figure 7.8 The hierarchy of handle graphics objects.

7.3.2 Object Handles

Each graphics object has a unique name called a **handle**. The handle is a unique integer or real number that is used by MATLAB to identify the object. A handle is automatically returned by any function that creates a graphics object. For example, the function call

```
» hndl = figure;
```

creates a new figure and returns the handle of that figure in variable `hndl`. Another example is the `plot` function. The statement

```
» hndl = plot(x,y);
```

plots a line on the current axes (first creating a figure and axes, if one does not exist), and returns the handle of the line in variable `hndl`.

The handle of the root object is always 0, and the handle of each figure is normally a small positive integer, such as 1, 2, 3, . . . The handles of all other graphics objects are arbitrary floating-point numbers.

There are MATLAB functions available to get the handles of figures, axes, and other objects. For example, the function `gcf` returns the handle of the currently selected figure, `gca` returns the handle of the currently selected axes within the currently selected figure, and `gco` returns the handle of the currently selected object. These functions are discussed in more detail subsequently.

By convention, handles are usually stored in variables that begin with the letter `h`. This practice helps us to recognize handles in MATLAB programs.

7.3.3 Examining and Changing Object Properties

Object properties are special values associated with an object that control some aspect of how that object behaves. Each property has a **property name** and an associated value. The property names are strings that are typically displayed in mixed case with the first letter of each word capitalized, but MATLAB recognizes a property name regardless of the case in which it is written.

When an object is created, all of its properties are automatically initialized to default values. These default values can be overridden at creation time by including `'PropertyName', value` pairs in the object creation function.[2] For example, we saw in Chapter 2 that the width of a line could be modified in the `plot` command as follows.

```
plot(x,y,'LineWidth',2);
```

This function overrides the default `LineWidth` property with the value 2 at the time that the line object is created.

The properties of any object can be examined at any time using the `get` function, and modified using the `set` function. These functions are especially useful for programmers, because they can be directly inserted into MATLAB programs to modify a figure based on a user's input.

The most common forms of `get` function are

```
value = get(handle,'PropertyName');
value = get(handle);
```

where `value` is the value contained in the specified property of the object whose handle is supplied. If only the handle is included in the function call, then the function returns a structure array in which the field names are all of the properties of the object, and the field values are the property values.

The most common form of the `set` function is

```
set(handle,'PropertyName1',value1,...);
```

where there can be any number of `'PropertyName',value` pairs in a single function.

For example, suppose that we plotted the function $y(x) = x^2$ from 0 to 2 with the following statements:

```
x = 0:0.1:2;
y = x.^2;
hndl = plot(x,y);
```

The resulting plot is shown in Figure 7.9*a*. The handle of the plotted line is stored in `hndl`, and we can use it to examine or modify the properties of the line. The

---

[2]Examples of object creation functions include `figure`, which creates a new figure, `axes`, which creates a new set of axes within a figure, and `line`, which creates a line within a set of axes. High-level functions such as `plot` are also object creation functions.

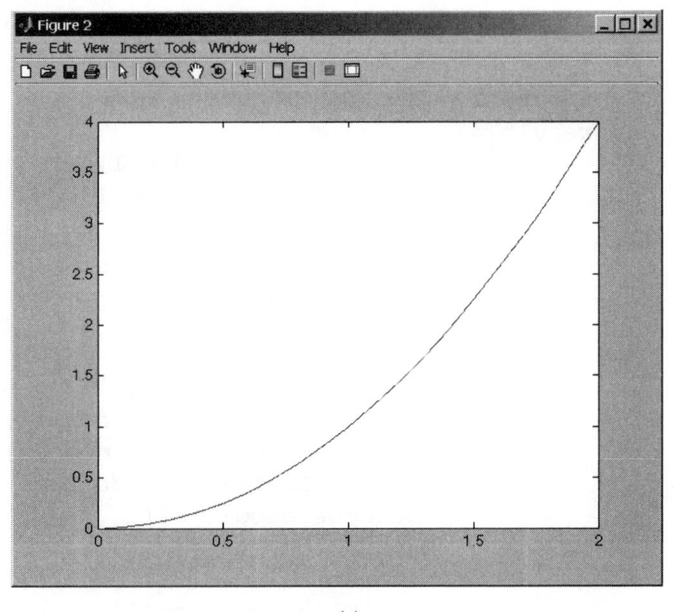

(a)

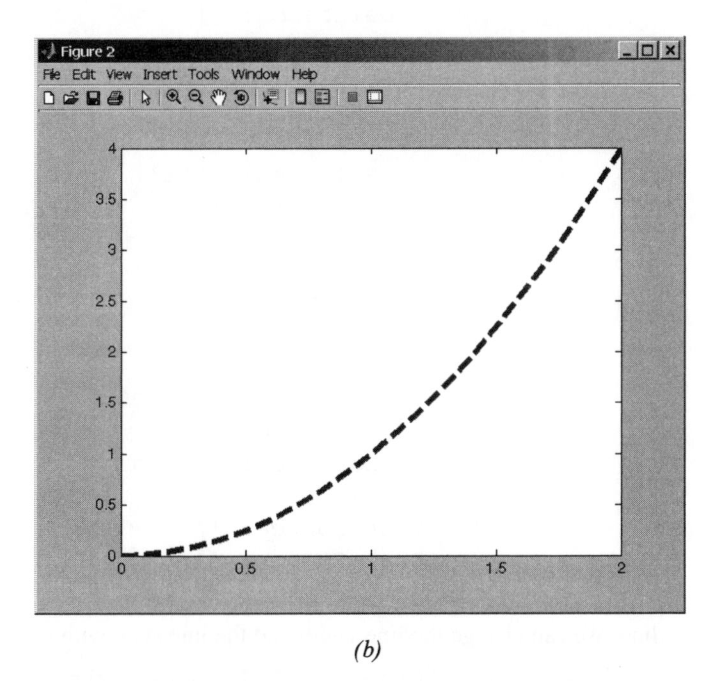

(b)

Figure 7.9 *(a)* Plot of the function $y = x^2$ using the default linewidth. *(b)* Plot of the function after modifying the `LineWidth` and `LineStyle` properties.

function get(hndl) will return all of the properties of this line in a structure, with each property name being an element of the structure.

```
» result = get(hndl)
result =
                Color: [0 0 1]
            EraseMode: 'normal'
            LineStyle: '-'
            LineWidth: 0.5000
               Marker: 'none'
           MarkerSize: 6
      MarkerEdgeColor: 'auto'
      MarkerFaceColor: 'none'
                XData: [1x21 double]
                YData: [1x21 double]
                ZData: [1x0 double]
         BeingDeleted: 'off'
        ButtonDownFcn: []
             Children: [0x1 double]
             Clipping: 'on'
            CreateFcn: []
            DeleteFcn: []
           BusyAction: 'queue'
     HandleVisibility: 'on'
              HitTest: 'on'
        Interruptible: 'on'
             Selected: 'off'
   SelectionHighlight: 'on'
                  Tag: ''
                 Type: 'line'
        UIContextMenu: []
             UserData: []
              Visible: 'on'
               Parent: 303.0004
          DisplayName: ''
            XDataMode: 'manual'
          XDataSource: ''
          YDataSource: ''
          ZDataSource: ''
```

Note that the current line width is 0.5 pixels and the current line style is a solid line. We can change the line width and the line style with the commands

```
» set(hndl,'LineWidth',4,'LineStyle','--')
```

The plot after this command is issued is shown in Figure 7.9*b*.

For the end user, however, it is often easier to change the properties of a MATLAB object interactively. The Property Editor is a GUI-based tool designed for this purpose. The Property Editor is started by first selecting the Edit Button ( ) on the figure toolbar, and then clicking on the object that you want to modify with the mouse. Alternatively, the property editor can be started from the command line.

```
propedit(HandleList);
propedit;
```

For example, the following statements will create a plot containing the line $y = x^2$ over the range 0 to 2, and open the Property Editor to allow the user to interactively change the properties of the line.

```
figure(2);
x = 0:0.1:2;
y = x.^2;
hndl = plot(x,y);
propedit(hndl);
```

The Property Editor invoked by these statements is shown in Figure 7.10. The Property Editor contains a series of panes that vary depending on the type of object being modified.

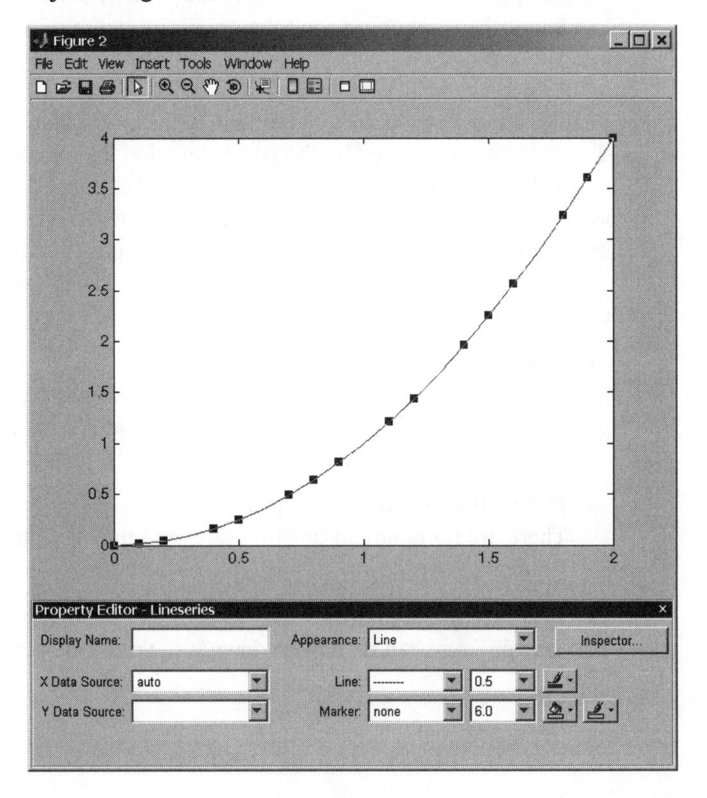

Figure 7.10 The Property Editor when editing a line object. Changes in style are immediately displayed on the figure as the object is edited.

Example 7.1—Using Low-Level Graphics Commands

The function sinc(x) is defined by the equation

$$\operatorname{sinc} x = \begin{cases} \dfrac{\sin x}{x} & x \neq 0 \\ 1 & x = 0 \end{cases} \tag{7-1}$$

Plot this function from $x = -3\pi$ to $x = 3\pi$. Use handle graphics functions to customize the plot as follows:

1. Make the figure background pink.
2. Use y-axis grid lines only (no x-axis grid lines).
3. Plot the function as a 2-point-wide solid orange line.

SOLUTION To create this graph, we need to plot the function sinc$x$ from $x = -3\pi$ to $x = 3\pi$ using the `plot` function. The plot function will return a handle for the line that we can save and use later.

After plotting the line, we need to modify the color of the *figure* object, the grid status of the *axes* object, and the color and width of the *line* object. These modifications require us to have access to the handles of the `figure`, `axes`, and `line` objects. The handle of the `figure` object is returned by the `gcf` function, the handle of the `axes` object is returned by the `gca` function, and the handle of the `line` object is returned by the `plot` function that created it.

The low-level graphics properties that need to be modified can be found by referring to the on-line MATLAB Help Browser documentation, under the topic "Handle Graphics." They are the `'Color'` property of the current figure, the `'YGrid'` property of the current axes, and the `'LineWidth'` and `'Color'` properties of the line.

1. **State the problem**
   Plot the function sinc x from $x = -3\pi$ to $x = 3\pi$ using a figure with a pink background, y-axis grid lines only, and a 2-point-wide solid orange line.

2. **Define the inputs and outputs**
   There are no inputs to this program, and the only output is the specified figure.

3. **Describe the algorithm**
   This program can be broken down into three major steps

   ```
   Calculate sinc(x)
   Plot sinc(x)
   Modify the required graphics object properties
   ```

   The first major step is to calculate sinc x from $x = -3\pi$ to $x = 3\pi$. This can be done with vectorized statements, but the vectorized statements will produce a NaN at $x = 0$, since the division of 0/0 is undefined. We must replace the NaN with a 1.0 before plotting the function. The detailed pseudocode for this step is:

```
% Calculate sinc(x)
x = -3*pi:pi/10:3*pi
y = sin(x) ./ x

% Find the zero value and fix it up. The zero is
% located in the middle of the x array.
index = fix(length(y)/2) + 1
y(index) = 1
```

Next, we must plot the function, saving the handle of the resulting line for further modifications. The detailed pseudocode for this step is:

```
hndl = plot(x,y);
```

Now we must use handle graphics commands to modify the figure background, y-axis grid, and line width and color. Remember that the figure handle can be recovered with the function `gcf`, and the axis handle can be recovered with the function `gca`. The color pink can be created with the RGB vector `[1 0.8 0.8]`, and the color orange can be created with the RGB vector `[1 0.5 0]`. The detailed pseudocode for this step is:

```
set(gcf,'Color',[1 0.8 0.8])
set(gca,'YGrid','on')
set(hndl,'Color',[1 0.5 0],'LineWidth',2)
```

4. **Turn the algorithm into MATLAB statements.**
   The final MATLAB program is shown below.

```
% Script file: plotsinc.m
%
% Purpose:
%   This program illustrates the use of handle graphics
%   commands by creating a plot of sinc(x) from -3*pi to
%   3*pi, and modifying the characteristics of the figure,
%   axes, and line using the "set" function.
%
% Record of revisions:
%    Date        Programmer       Description of change
%    ====        ==========       =====================
%   01/24/05    S. J. Chapman     Original code
%
% Define variables:
%   hndl          -- Handle of line
%   x             -- Independent variable
%   y             -- sinc(x)
```

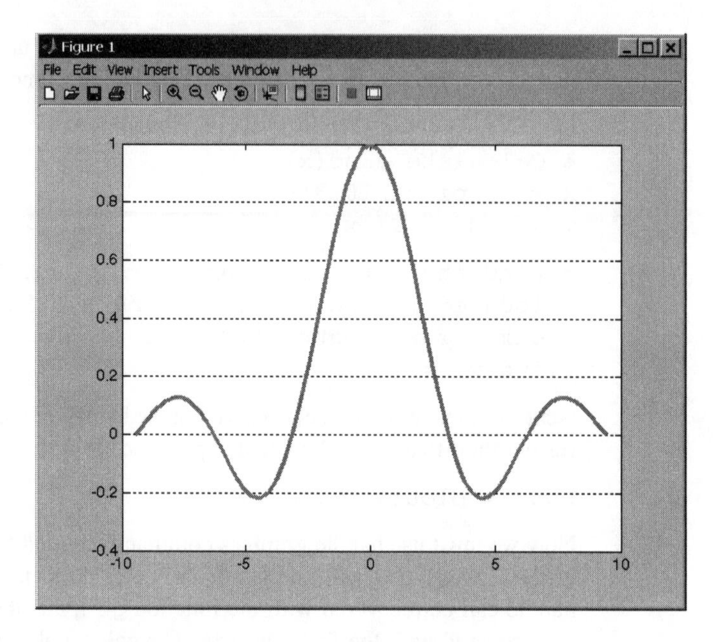

Figure 7.11 Plot of sinc *x* versus *x*.

```
% Calculate sinc(x)
x = -3*pi:pi/10:3*pi;
y = sin(x) ./ x;

% Find the zero value and fix it up. The zero is
% located in the middle of the x array.
index = fix(length(y)/2) + 1;
y(index) = 1;

% Plot the function.
hndl = plot(x,y);

% Now modify the figure to create a pink background,
% modify the axis to torn on y-axis grid lines, and
% modify the line to be a 2-point wide orange line.
set(gcf,'Color',[1 0.8 0.8]);
set(gca,'YGrid','on');
set(hndl,'Color',[1 0.5 0],'LineWidth',2);
```

5. **Test the program.**
   Testing this program is very simple—we just execute it and examine the resulting plot. The plot created is shown in Figure 7.11, and it does have the characteristics that we wanted.

7.3.4 Using set to List Possible Property Values

The set function can be used to provide lists of possible property values. If a set function call contains a property name but not a corresponding value, set returns a list of all of the legal choices for that property. For example, the command set(hndl,'LineStyle') will return a list of all legal line styles with the default choice in brackets:

```
» set(hndl,'LineStyle')
ans =
    '-'
    '--'
    ':'
    '-.'
    'none'
```

This function shows that the legal line styles are '-', '—', ':', '-.', and 'none', with the first choice as the default.

If the property does not have a fixed set of values, MATLAB returns an empty cell array:

```
» set(hndl,'LineWidth')
ans =
     {}
```

The function set(hndl) will return all of the possible choices for all of the properties of an object.

```
» xxx = set(hndl)
xxx =
                  Color: {}
              EraseMode: {4x1 cell}
              LineStyle: {5x1 cell}
              LineWidth: {}
                 Marker: {14x1 cell}
             MarkerSize: {}
        MarkerEdgeColor: {2x1 cell}
        MarkerFaceColor: {2x1 cell}
                  XData: {}
                  YData: {}
                  ZData: {}
          ButtonDownFcn: {}
               Children: {}
               Clipping: {2x1 cell}
              CreateFcn: {}
              DeleteFcn: {}
             BusyAction: {2x1 cell}
       HandleVisibility: {3x1 cell}
                HitTest: {2x1 cell}
          Interruptible: {2x1 cell}
```

```
               Selected: {2x1 cell}
     SelectionHighlight: {2x1 cell}
                    Tag: {}
          UIContextMenu: {}
               UserData: {}
                Visible: {2x1 cell}
                 Parent: {}
            DisplayName: {}
              XDataMode: {2x1 cell}
            XDataSource: {}
            YDataSource: {}
            ZDataSource: {}
```

Any of the items in this list can be expanded to see the available list of options.

```
» xxx.EraseMode
ans =
    'normal'
    'background'
    'xor'
    'none'
```

7.3.5 Finding Objects

Each new graphics object that is created has its own handle, and that handle is returned by the creating function. If you intend to modify the properties of an object that you create, then it is a good idea to save the handle for later use with `get` and `set`.

✱ Good Programming Practice

If you intend to modify the properties of an object that you create, save the handle of that object for later use with `get` and `set`.

However, sometimes we might not have access to the handle. Suppose that we lost a handle for some reason. How can we examine and modify the graphics objects? MATLAB provides four special functions to help find the handles of objects.

- `gcf`—Returns the handle of the current *figure*.
- `gca`—Returns the handle of the current *axes* in the current *figure*.
- `gco`—Returns the handle of the current *object*.
- `findobj`—Finds a graphics object with a specified property value.

The function `gcf` returns the handle of the current figure. If no figure exists, `gcf` *will create one* and return its handle. The function `gca` returns the handle of the current axes within the current figure. If no figure exists or if the current figure exists but contains no axes, gca *will create a set of axes* and return its handle. The function gco has the form

```
h_obj = gco;
h_obj = gco(h_fig);
```

where h_obj is the handle of the object and h_fig is the handle of a figure. The first form of this function returns the handle of the *current object in the current figure,* while the second form of the function returns the handle of the *current object in a specified figure.*

The current object is defined as the last object clicked on with the mouse. This object can be any graphics object except the root. There will not be a current object in a figure until a mouse click has occurred within that figure. Before the first mouse click, function gco will return an empty array []. Unlike gcf and gca, gco does not create an object if it does not exist.

Once the handle of an object is known, we can determine the type of the object by examining its 'Type' property. The 'Type' property will be a character string, such as 'figure', 'line', 'text', etc.

```
h_obj = gco;
type = get(h_obj,'Type')
```

The easiest way to find an arbitrary MATLAB object is with the findobj function. The basic form of this function is

```
hndls = findobj('PropertyName1',value1,...)
```

This command starts at the root object, and searches the entire tree for all objects that have the specified values for the specified properties. Note that multiple property/value pairs may be specified, and findobj will only returns the handles of objects that match *all* of them.

For example, suppose that we have created Figures 1 and 3. Then the function findobj('Type','figure') will return the results:

```
» h_fig = findobj('Type','figure')
h_fig =
      3
      1
```

This form of the findobj function is very useful, but it can be slow, since it must search through the entire object tree to locate any matches. If you must use an object multiple times, only make one call to findobj and save the handle for re-use.

Restricting the number of objects that must be searched can increase the execution speed of this function. This can be done with the following form of the function:

```
hndls = findobj(Srchhndls,'PropertyName1',value1,...)
```

Here, only the handles listed in array Srchhndls and their children will be searched to find the object. For example, suppose that you wanted to find all of the dashed lines in Figure 1. The command to do this would be:

```
hndls = findobj(1,'Type','line','LineStyle','--');
```

Good Programming Practice

If possible, restrict the scope of your searches with `findobj` to make them faster.

7.3.6 Selecting Objects with the Mouse

Function `gco` returns the handle of the current object, which is the last object clicked on by the mouse. Each object has a **selection region** associated with it, and any mouse click within that selection region is assumed to be a click on that object. This is very important for thin objects like lines or points—the selection region allows the user to be slightly sloppy in mouse position and still select the line. The width of and shape of the selection region varies for different types of objects. For instance, the selection region for a line is five pixels on either side of the line, while the selection region for a surface, patch, or text object is the smallest rectangle that can contain the object.

The selection region for an `axes` object is the area of the axes plus the area of the titles and labels. However, lines or other objects inside the axes have a higher priority, so to select the axes you must click on a point within the axes that is not near lines or text. Clicking on a figure outside of the axes region will select the figure itself.

What happens if a user clicks on a point that has two or more objects, such as the intersection of two lines? The answer depends on the **stacking order** of the objects. The stacking order is the order in which MATLAB selects objects. This order is specified by the order of the handles listed in the `'Children'` property of a figure. If a click is in the selection region of two or more objects, the one with the highest position in the `'Children'` list will be selected.

MATLAB includes a function called `waitforbuttonpress` that is sometimes used when selecting graphics objects. The form of this function is:

```
k = waitforbuttonpress
```

When this function is executed, it halts the program until either a key is pressed or a mouse button is clicked. The function returns 0 if it detects a mouse button click or 1 if it detects a key press.

The function can be used to pause a program until a mouse click occurs. After the mouse click occurs, the program can recover the handle of the selected object using the `gco` function.

Example 7.2—Selecting Graphics Objects

The program shown below explores the properties of graphics objects, and incidentally shows how to select objects using `waitforbuttonpress` and `gco`. The program allows objects to be selected repeatedly until a key press occurs.

```
% Script file: select_object.m
%
% Purpose:
%   This program illustrates the use of waitforbuttonpress
%   and gco to select graphics objects. It creates a plot
%   of sin(x) and cos(x), and then allows a user to select
%   any object and examine its properties. The program
%   terminates when a key press occurs.
%
% Record of revisions:
%     Date       Programmer      Description of change
%     ====       ==========      =====================
%   01/25/05    S. J. Chapman    Original code
%
% Define variables:
%   details      -- Object details
%   h1           -- handle of sine line
%   h2           -- handle of cosine line
%   handle       -- handle of current object
%   k            -- Result of waitforbuttonpress
%   type         -- Object type
%   x            -- Independent variable
%   y1           -- sin(x)
%   y2           -- cos(x)
%   yn           -- Yes/No

% Calculate sin(x) and cos(x)
x = -3*pi:pi/10:3*pi;
y1 = sin(x);
y2 = cos(x);

% Plot the functions.
h1 = plot(x,y1);
set(h1,'LineWidth',2);
hold on;
h2 = plot(x,y2);
set(h2,'LineWidth',2,'LineStyle',':','Color','r');
title('\bfPlot of sin \itx \rm\bf and cos \itx');
xlabel('\bf\itx');
ylabel('\bfsin \itx \rm\bf and cos \itx');
legend('sine','cosine');
hold off;

% Now set up a loop and wait for a mouse click.
k = waitforbuttonpress;
```

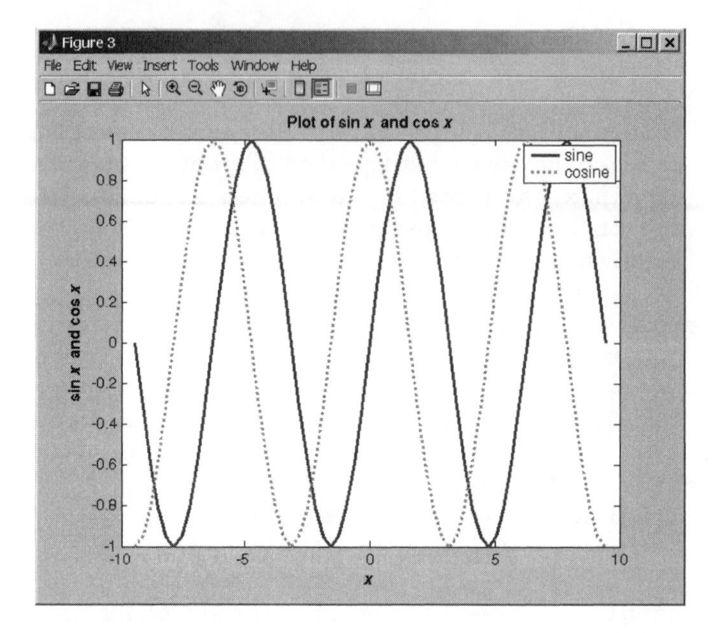

Figure 7.12 Plot of sin *x* and cos *x*.

```
while k == 0

   % Get the handle of the object
   handle = gco;

   % Get the type of this object.
   type = get(handle,'Type');

   % Display object type
   disp (['Object type = ' type '.']);

   % Do we display the details?
   yn = input('Do you want to display details? (y/n) ','s');

   if yn == 'y'
      details = get(handle);
      disp(details);
   end

   % Check for another mouse click
   k = waitforbuttonpress;
end
```

When this program is executed, it produces the plot shown in Figure 7.12. Experiment by clicking on various objects on the plot, and seeing the resulting characteristics.

◀

7.4 Position and Units

Many MATLAB objects have a 'position' property, which specifies the size and position of the object on the computer screen. This property differs slightly for different kinds of objects, as described below.

7.4.1 Positions of figure Objects

The 'position' property for a figure specifies the location of that figure on the computer screen using a four-element row vector. The values in this vector are [left bottom width height], where left is the leftmost edge of the figure, bottom is the bottom edge of the figure, width is the width of the figure, and height is the height of the figure. These position values are in the units specified in the 'Units' property for the object. For example, the position and units associated with a the current figure can be found as follows:

```
» get(gcf,'Position')
ans =
   176   204   672   504
» get(gcf,'Units')
ans =
pixels
```

This information specifies that the lower left corner of the current figure window is 176 pixels to the right and 204 pixels above the lower left corner of the screen, and the figure is 672 pixels wide by 504 pixels high. This is the drawable region of the figure, excluding borders, scrollbars, menus, and the figure title area.

The 'units' property of a figure defaults to pixels, but it can be inches, centimeters, points, characters, or normalized coordinates. Pixels are screen pixels, which are the smallest rectangular shape that can be drawn on a computer screen. Typical computer screens are at least 640 pixels wide $\times$ 480 pixels high, and screens can have more than 1,000 pixels in each direction. Since the number of pixels varies from computer screen to computer screen, the size of an object specified in pixels will also vary.

Normalized coordinates are coordinates in the range 0 to 1, where the lower left corner of the screen is at (0, 0) and the upper right corner of the screen is at (1, 1). If an object position is specified in normalized coordinates, it will appear in the same relative position on the screen regardless of screen resolution. For example, the following statements create a figure and place it into the upper left quadrant of the screen on any computer, regardless of screen size.[3]

```
h1 = figure(1)
set(h1,'units','normalized','position',[0 .5 .5 .45])
```

[3] The normalized height of this figure is reduced to 0.45 to allow room for the Figure title and menu bar, both of which are above the drawing area.

✱ Good Programming Practice

If you would like to place a window in a specific location, it is easier to place the window at the desired location using normalized coordinates, and the results will be the same regardless of the computer's screen resolution

7.4.2 Positions of `axes` Objects

The position of `axes` objects is also specified by a four-element vector, but the object position is specified relative to the lower-left hand-corner of the *figure* instead of the position of the screen. In general, the `'Position'` property of a child object is relative to the position of its parent.

By default, the positions of axes objects are specified in *normalized* units within a figure, with (0, 0) representing the lower left hand corner of the figure, and (1, 1) representing the upper right hand corner of the figure.

7.4.3 Positions of `text` Objects

Unlike other objects, `text` objects have a position property containing only two or three elements. These elements correspond to the x, y, and z values of the text object *within* an `axes` object. Note that these values are in the units being displayed on the axes themselves.

The position of the text object with respect to the specified point is controlled by the object's `HorizontalAlignment` and `VerticalAlignment` properties. The `HorizontalAlignment` can be `{Left}`, `Center`, or `Right`, and the `VerticalAlignment` can be `Top`, `Cap`, `{Middle}`, `Baseline`, or `Bottom`.

The size of `text` objects is determined by the font size and the number of characters being displayed, so there are no height and width values associated with them.

Example 7.3—Positioning Objects Within a Figure

As we mentioned earlier, axes positions are defined relative to the lower left hand corner of the frame that they are contained in, while text object positions are defined within axes in the data units being displayed on the axes.

To illustrate the positioning of graphics objects within a figure, we will write a program that creates two overlapping sets of axes within a single figure. The first set of axes will display sin x versus x, and have a text comment attached to the display line. The second set of axes will display cos x versus x, and have a text comment in the lower left hand corner.

A program to create the figure is shown below. Note that we are using the `figure` function to create an empty figure, and then two `axes` functions to create the two sets of axes within the figure. The position of the `axes` functions is specified in normalized units within the figure, so the first set of axes, which starts at (0.05, 0.05), is in the lower left hand corner of the figure, and the second set of axes, which starts at (0.45, 0.45), is in the upper right hand corner of the figure. Each set of axes has the appropriate function plotted on it.

The first `text` object is attached to the first set of axes at position $(-\pi, 0)$, which is a point on the curve. The `'HorizontalAlignment','right'` property is selected, so the *attachment point* $(-\pi, 0)$ is on the *right hand side* of the text string. As a result, the text appears to the *left* of the of the attachment point in the final figure. (This can be confusing for new programmers!)

The second `text` object is attached to the second set of axes at position $(-7.5, -0.9)$, which is near the lower left hand corner of the axes. This string uses the default horizontal alignment, which is `'left'`, so the attachment point $(-7.5, -0.9)$ is on the *left hand side* of the text string. As a result, the text appears to the right of the of the attachment point in the final figure.

```
% Script file: position_object.m
%
% Purpose:
%   This program illustrates the positioning of graphics
%   graphics objects. It creates a figure, and then places
%   two overlapping sets of axes on the figure. The first
%   set of axes is placed in the lower left hand corner of
%   the figure, and contains a plot of sin(x). The second
%   set of axes is placed in the upper right hand corner of
%   the figure, and contains a plot of cos(x). Then two
%   text strings are added to the axes, illustrating the
%   positioning of text within axes.
%
% Record of revisions:
%      Date        Programmer        Description of change
%      ====        ==========        =====================
%    01/25/05    S. J. Chapman       Original code
%
% Define variables:
%   h1              -- Handle of sine line
%   h2              -- Handle of cosine line
%   ha1             -- Handle of first axes
%   ha2             -- Handle of second axes
%   x               -- Independent variable
%   y1              -- sin(x)
%   y2              -- cos(x)
```

```
% Calculate sin(x) and cos(x)
x = -2*pi:pi/10:2*pi;
y1 = sin(x);
y2 = cos(x);

% Create a new figure
figure;

% Create the first set of axes and plot sin(x).
% Note that the position of the axes is expressed
% in normalized units.
ha1 = axes('Position',[.05 .05 .5 .5]);
h1 = plot(x,y1);
set(h1,'LineWidth',2);
title('\bfPlot of sin \itx');
xlabel('\bf\itx');
ylabel('\bfsin \itx');
axis([-8 8 -1 1]);

% Create the second set of axes and plot cos(x).
% Note that the position of the axes is expressed
% in normalized units.
ha2 = axes('Position',[.45 .45 .5 .5]);
h2 = plot(x,y1);
set(h2,'LineWidth',2,'Color','r','LineStyle','--');
title('\bfPlot of cos \itx');
xlabel('\bf\itx');
ylabel('\bfsin \itx');
axis([-8 8 -1 1]);

% Create a text string attached to the line on the first
% set of axes.
axes(ha1);
text(-pi,0.0,'sin(x)\rightarrow','HorizontalAlignment','right');

% Create a text string in the lower left hand corner
% of the second set of axes.
axes(ha2);
text(-7.5,-0.9,'Test string 2');
```

When this program is executed, it produces the plot shown in Figure 7.13. You should execute this program again on your computer, changing the size and/or location of the objects being plotted, and observing the results.

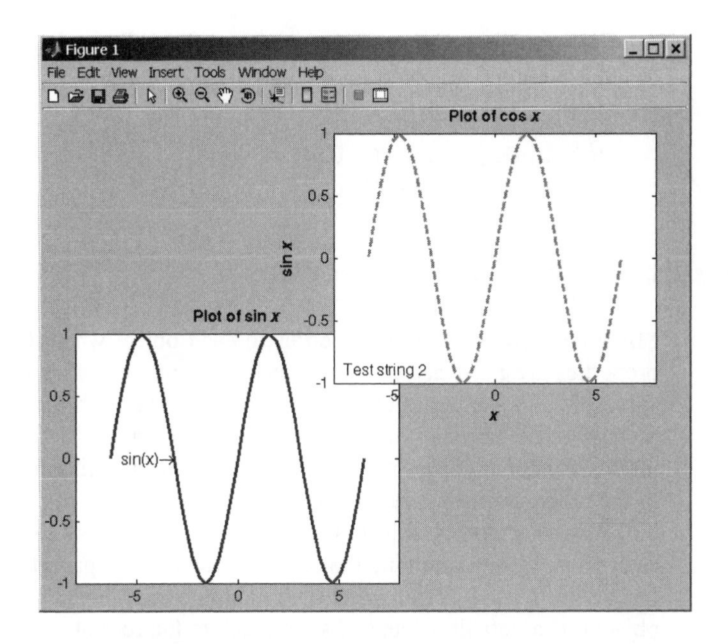

Figure 7.13 The output of program `position_object`.

7.5 Printer Positions

The `'Position'` and `'Units'` properties specify the location of a figure on the *computer screen*. There are also five other properties that specify the location of a figure on a sheet of paper *when it is printed*. These properties are summarized in Table 7.3 below.

Table 7.3 Printing-Related Figure Properties

Option	Description
`PaperUnits`	Units for paper measurements: `[ {inches} | centimeters | normalized | points ]`
`PaperOrientation`	`[ {portrait} | landscape ]`
`PaperPosition`	A position vector of the form `[left, bottom, width, height]`, where all units are as specified in `PaperUnits`.
`PaperSize`	A two-element vector containing the power size, for example `[8.5 11]`
`PaperType`	Sets paper type. Note that setting this property automatically updates the `PaperSize` property. `[ {usletter} | uslegal | A0 | A1 | A2 | A3 | A4 | A5 | B0 | B1 | B2 | B3 | B4 | B5 | arch-A | arch-B | arch-C | arch-D | arch-E | A | B | C | D | E | tabloid | <custom> ]`

For example, to set a plot to print out in landscape mode, on A4 paper, in normalized units, we could set the following properties:

```
set(hndl,'PaperType','A4')
set(hndl,'PaperOrientation','landscape')
set(hndl,'PaperUnits','normalized');
```

7.6 Default and Factory Properties

MATLAB assigns default properties to each object when it is created. If those properties are not what you want, then you must use `set` to select the desired values. If you wanted to change a property in every object that you create, this process could become very tedious. For those cases, MATLAB allows you to modify the default property itself, so that all objects will inherit the correct value of the property when they are created.

When a graphics object is created, MATLAB looks for a default value for each property by examining the object's parent. If the parent sets a default value, that value is used. If not, MATLAB examines the parent's parent to see if that object sets a default value, and so on back to the root object. MATLAB uses the *first* default value that it encounters when working back up the tree.

Default properties may be set at any point in the graphics object hierarchy that is *higher* than level at which the object is created. For example, a default `figure` color would be set in the `root` object, and then all figures created after that time would have the new default color. On the other hand, a default `axes` color could be set in either the `root` object or the `figure` object. If the default `axes` color is set in the `root` object, it will apply to all new axes in all figures. If the default `axes` color is set in the `figure` object, it will apply to all new axes in the current figure only.

Default values are set using a string consisting of `'Default'` followed by the object type and the property name. Thus the default figure color would be set with the property `'DefaultFigureColor'` and the default axes color would be set with the property `'DefaultAxesColor'`. Some examples of setting default values are shown below:

`set(0,'DefaultFigureColor','y')`	Yellow figure background—all new figures.
`set(0,'DefaultAxesColor','r')`	Red axes background—all new axes in all figures.
`set(gcf,'DefaultAxesColor','r')`	Red axes background—all new axes in current figure only.
`set(gca,'DefaultLineLineStyle',':')`	Set default line style to dashed, in current axes only.

If you are working with existing objects, it is always a good idea to restore them to their existing condition after they are used. *If you change the default properties of an object in a function, save the original values and restore them before exiting the function.* For example, suppose that we wish to create a series of figures in normalized units. We could save and restore the original units as follows:

```
saveunits = get(0,'DefaultFigureUnits');
set(0,'DefaultFigureUnits','normalized');
...
<MATLAB statements>
...
set(0,'DefaultFigureUnits',saveunits);
```

If you want to customize MATLAB to use different default values at all times, you should set the defaults in the `root` object every time that MATLAB starts up. The easiest way to do this is to place the default values into the `startup.m` file, which is automatically executed every time MATLAB starts. For example, suppose you always use A4 paper and you always want a grid displayed on your plots. Then you could set the following lines into `startup.m`:

```
set(0,'DefaultFigurePaperType','A4');
set(0,'DefaultFigurePaperUnits','centimeters');
set(0,'DefaultAxesXGrid','on');
set(0,'DefaultAxesYGrid','on');
set(0,'DefaultAxesZGrid','on');
```

There are three special value strings that are used with handle graphics: `'remove,'` `'factory,'` and `'default.'` If you have set a default value for a property, the `'remove'` value will remove the default that you set. For example, suppose that you set the default figure color to yellow:

```
set(0,'DefaultFigureColor','y');
```

The following function call will cancel this default setting and restore the previous default setting.

```
set(0,'DefaultFigureColor','remove');
```

The string `'factory'` allows a user to temporarily override a default value and use the original MATLAB default value instead. For example, the following figure is created with the factory default color despite a default color of yellow being previously defined.

```
set(0,'DefaultFigureColor','y');
figure('Color','factory')
```

The string `'default'` forces MATLAB to search up the object hierarchy until it finds a default value for the desired property. It uses the first default value that it finds. If it fails to find a default value, then it uses the factory default value for that property. This use is illustrated below:

```
% Set default values
set(0,'DefaultLineColor','k');    % root default = black
set(gcf,'DefaultLineColor','g'); % figure default = green

% Create a line on the current axes. This line is green.
hndl = plot(randn(1,10));
set(hndl,'Color','default');
pause(2);

% Now clear the figure's default and set the line color to the new
% default. The line is now black.
set(gcf,'DefaultLineColor','remove');
set(hndl,'Color','default');
```

7.7 Graphics Object Properties

There are hundreds of different graphic object properties, far too many to discuss in detail here. The best place to find a complete list of graphics object properties is in the Help Browser distributed with MATLAB.

We have mentioned a few of the most important properties for each type of graphic object as we have needed them (`'LineStyle'`, `'Color'`, and so forth). A complete set of properties is given in the MATLAB Help Browser documentation under the descriptions of each type of object.

7.8 Summary

Cell arrays are arrays whose elements are *cells,* containers that can hold other MATLAB arrays. Any sort of data may be stored in a cell, including structure arrays and other cell arrays. They are a very flexible way to store data, and are used in many internal MATLAB Graphical User Interface functions.

Structure arrays are a data type in which each individual element is given a name. The individual elements of a structure are known as fields, and each field in a structure may have a different type. The individual fields are addressed by combining the name of the structure with the name of the field, separated by a period. Structure arrays are useful for grouping together all of the data related to a particular person or thing into a single location.

Every element of a MATLAB plot is a graphics object. Each object is identified by a unique handle, and each object has many properties associated with it which affect the way the object is displayed.

MATLAB objects are arranged in a hierarchy with **parent objects** and **child objects**. When a child object is created, it inherits many of its properties from its parent.

The highest-level graphics object in MATLAB is the `root`, which can be thought of as the entire computer screen. Under the root there can be one or more Figure Windows. Each `figure` is a separate window on the computer screen that can display graphical data, and each figure has its own properties.

A figure can contain one or more sets of axes. Each set of axes can contain as many `lines`, `text` strings, `patches`, etc. as necessary to create the plot of interest.

The handles of the current figure, current axes, and current object may be recovered with the `gcf`, `gca`, and `gco` functions respectively. The properties of any object may be examined and modified using the `get` and `set` functions.

There are literally hundreds of properties associated with MATLAB graphics functions, and the best place to find the details of these of these functions is the MATLAB on-line documentation.

7.8.1 Summary of Good Programming Practice

The following guidelines should be adhered to:

1. Always preallocate all cell arrays before assigning values to the elements of the array. This practice greatly increases the execution speed of a program.
2. Use cell array arguments `varargin` and `varargout` to create functions that support varying numbers of input and output arguments.
3. If you intend to modify the properties of an object that you create, save the handle of that object for later use with `get` and `set`.
4. If possible, restrict the scope of your searches with `findobj` to make them faster.
5. If you would like to place a window in a specific location, it is easier to place the window at the desired location using normalized coordinates, and the results will be the same regardless of the computer's screen resolution.

7.8.2 MATLAB Summary

The following summary lists all of the MATLAB commands and functions described in this chapter, along with a brief description of each one.

`axes`	Creates a new axes/makes axes current.
`cell`	Pre-define a cell array structure.
`celldisp`	Display contents of a cell array.
`cellplot`	Plot structure of a cell array.

`cellstr`	Convert a two-dimensional character array to a cell array of strings.
`fieldnames`	Return a list of field names in a cell array of strings.
`figure`	Creates a new figure/makes figure current.
`findobj`	Finds an object based on one or more property values.
`gca`	Get handle of current axes.
`gcf`	Get handle of current figure.
`gco`	Get handle of current object.
`get`	Gets object properties.
`getfield`	Get current value from a field.
`rmfield`	Remove a field from a structure array.
`set`	Sets object properties.
`setfield`	Set new value into a field.
`waitforbuttonpress`	Pauses program, waiting for a mouse click or keyboard input.

7.9 Exercises

7.1 Write a MATLAB function that will accept a cell array of strings and sort them into ascending order according to the lexicographic order of the ASCII character set. (You may use function `c_strcmp` from Chapter 6 for the comparisons if you wish.)

7.2 Write a MATLAB function that will accept a cell array of strings and sort them into ascending order according to *alphabetical order*. (This implies that you must treat '`A`' and '`a`' as the same letter.)

7.3 Create a function that accepts any number of numeric input arguments, and sums up all of individual elements in the arguments. Test your function by passing it the four arguments `a` = 10, `b` = $\begin{bmatrix} 4 \\ -2 \\ 2 \end{bmatrix}$, `c` = $\begin{bmatrix} 1 & 0 & 3 \\ -5 & 1 & 2 \\ 1 & 2 & 0 \end{bmatrix}$, and `d` = [1 5 −2].

7.4 Modify the function of the previous exercise so that it can accept either ordinary numeric arrays or cell arrays containing numeric values. Test your function by passing it the two arguments `a` and `b`, where `a` = $\begin{bmatrix} 1 & 4 \\ -2 & 3 \end{bmatrix}$, `b{1}` = [1 5 2], and `b{2}` = $\begin{bmatrix} 1 & -2 \\ 2 & 1 \end{bmatrix}$.

7.5 Create a structure array containing all of the information needed to plot a data set. At a minimum, the structure array should have the following fields:

- `x_data` *x*-data (one or more data sets in separate cells)
- `y_data` *y*-data (one or more data sets in separate cells)
- `type` linear, semilogx, etc.
- `plot_title` plot title
- `x_label` *x*-axis label
- `y_label` *y*-axis label
- `x_range` *x*-axis range to plot
- `y_range` *y*-axis range to plot

You may add additional fields that would enhance your control of the final plot.

After this structure array is created, create a MATLAB function that accepts an array of this structure and produces one plot for each structure in the array. The function should apply intelligent defaults if some data fields are missing. For example, if the `plot_title` field is an empty matrix, then the function should not place a title on the graph. Think carefully about the proper defaults before starting to write your function!

To test your function, create a structure array containing the data for three plots of three different types, and pass that structure array to your function. The function should correctly plot all three data sets in three different figure windows.

7.6 Define a structure `point` containing two fields `x` and `y`. The `x` field will contain the *x*-position of the point, and the `y` field will contain the *y*-position of the point. Then write a function `dist3` that accepts two points, and returns the distance between the two points on the Cartesian plane. Be sure to check the number of input arguments in your function.

7.7 Write a function that will accept a structure as an argument, and return two cell arrays containing the names of the fields of that structure, and the data types of each field. Be sure to check that the input argument is a structure, and generate an error message if it is not.

7.8 Write a function that will accept a structure array of `student` as defined in this chapter, and calculate the final average of each one assuming that all exams have equal weighting. Add a new field to each array to contain the final average for that student, and return the updated structure to the calling program. Also, calculate and return the final class average.

7.9 Write a function that will accept two arguments, the first a structure array and the second a field name stored in a string. Check to make sure that these input arguments are valid. If they are not valid, print out an error message. If they are valid and the designated field is a string, concatenate all of the strings in the specified field of each element in the array, and return the resulting string to the calling program.

7.10 **Calculating Directory Sizes.** Function `dir` returns the contents of a specified directory. The `dir` command returns a structure array with four fields, as shown below:

```
» d = dir('chap7')
d =
36x1 struct array with fields:
    name
    date
    bytes
    isdir
```

The field `name` contains the names of each file, `date` contains the last modification date for the file, `bytes` contains the size of the file in bytes, and `isdir` is 0 for conventional files and 1 for directories. Write a function that accepts a directory name and path, and returns the total size of all files in the directory, in bytes.

7.11 **Recursion.** A function is said to be *recursive* if the function calls itself. Modify the function created in Problem 7.11 so that it calls itself when it finds a subdirectory, and sums up the size of all file in the current directory plus all subdirectories.

7.12 What is meant by the term "handle graphics"?

7.13 Use the MATLAB Help Browser to learn about the `Name` and `NumberTitle` properties of a `figure` object. Create a figure containing a plot of the function $y(x) = e^x$ for $-2 \leq x \leq 2$. Change the properties mentioned above to suppress the figure number and to add the title "Plot Window" to the figure.

7.14 Write a program that modifies the default figure color to orange and the default line width to 3.0 points. Then create a figure plotting the ellipse defined by the equations

$$\begin{aligned} x(t) &= 10 \cos t \\ y(t) &= 6 \sin t \end{aligned} \qquad (7\text{-}2)$$

from $t = 0$ to $t = 2\pi$. What color and width was the resulting line?

7.15 Use the MATLAB Help Browser to learn about the `CurrentPoint` property of an `axes` object. Use this property to create a program that creates an `axes` object, and plots a line connecting the locations of successive mouse clicks within the axes. Use the function `waitforbuttonpress` to wait for mouse clicks, and update the plot after each click. Terminate the plot when a keyboard press occurs.

7.16 Use the MATLAB Help Browser to learn about the `CurrentCharacter` property of a `figure` object. Modify the program created in Problem 7.15 by testing the `CurrentCharacter` property when a keyboard press occurs. If the character typed on the keyboard is a "c" or "C," change the color of the line being displayed. If the character typed on the keyboard is an "s" or "S," change the line style of the line being displayed. If the character typed on the keyboard is a "w," or "W," change the width of the line being displayed. If the character typed on the keyboard is an "x" or "X." terminate the plot. (Ignore all other input characters.)

7.17 Create a MATLAB program that plots the functions

$$\begin{aligned} x(t) &= \cos\frac{t}{\pi} \\ x(t) &= 2\sin\frac{t}{2\pi} \end{aligned} \tag{7-3}$$

for the range $-2 \leq t \leq 2$. The program should then wait for mouse clicks, and if the mouse has clicked on one of the two lines, the program should change the line's color randomly from a choice of red, green, blue, yellow, cyan, magenta, or black. Use the function `waitforbuttonpress` to wait for mouse clicks, and update the plot after each click. Use the function `gco` to determine the object clicked on, and use the `Type` property of the object to determine if the click was on a line.

7.18 The `plot` function plots a line and returns a handle to that line. This handle can be used to get or set the line's properties after it has been created. Two of a line's properties are `XData` and `YData`, which contain the x- and y-values currently plotted. Write a program that plots the function

$$x(t) = \cos(2\pi t - \theta) \tag{7-4}$$

between the limits $-1.0 \leq t \leq 1.0$, and saves the handle of the resulting line. The angle θ is initially 0 radians. Then, re-plot line over and over with $\theta = \pi/10$ rad, $\theta = 2\pi/10$ rad, $\theta = 3\pi/10$ rad, and so forth up to $\theta = 2\pi$ rad. To re-plot the line, use a `for` loop to calculate the new values of x and t, and update the line's `XData` and `YData` properties with `set` commands. Pause 0.5 seconds between each update, using MATLAB's `pause` command.

ASCII Character Set

MATLAB strings use the ASCII character set, which consists of the 127 characters shown in the table below. The results of MATLAB string comparison operations depend on the *relative lexicographic positions* of the characters being compared. For example, the character “a” in the ASCII character set is at position 97 in the table, while the character “A” is at position 65. Therefore, the relational operator `'a' > 'A'` will return a 1 (true), since 97 > 65.

Each MATLAB character is stored in a 16-bit field, which means that in the future MATLAB can support the entire Unicode character set.

The table shown below shows the ASCII character set, with the first two digits of the character number defined by the row and the third digit defined by the column. Thus, the letter `'R'` is on row 8 and column 2, so it is character 82 in the ASCII character set.

	0	**1**	**2**	**3**	**4**	**5**	**6**	**7**	**8**	**9**
0	nul	soh	stx	etx	eot	enq	ack	bel	bs	ht
1	nl	vt	ff	cr	so	si	dle	dc1	dc2	dc3
2	dc4	nak	syn	etb	can	em	sub	esc	fs	gs
3	rs	us	sp	!	"	#	$	%	&	'
4	(	)	*	+	,	-	.	/	0	1
5	2	3	4	5	6	7	8	9	:	;
6	<	=	>	?	@	A	B	C	D	E
7	F	G	H	I	J	K	L	M	N	O
8	P	Q	R	S	T	U	V	W	X	Y
9	Z	[	\	]	^	_	`	a	b	c
10	d	e	f	g	h	I	j	k	l	m
11	n	o	p	q	r	s	t	u	v	w
12	x	y	z	{	|	}	~	del		

APPENDIX 

MATLAB Input/Output Functions

In Chapter 2, we learned how to load and save MATLAB data using the `load` and `save` commands and how to write out formatted data using the `fprintf` function. This appendix includes additional details about MATLAB's input/output capabilities.

B.1 The `textread` Function

The `textread` function reads ASCII files that are formatted into columns of data, where each column can be of a different type, and stores the contents of each column in a separate output array. This function is *very* useful for importing tables of data printed out by other applications.

The form of the `textread` function is

```
[a,b,c,...] = textread(filename,format,n)
```

where `filename` is the name of the file to open, `format` is a string containing a description of the type of data in each column, and `n` is the number of lines to read. (If `n` is missing, the function reads to the end of the file.) The format string contains the same types of format descriptors as function `fprintf`. Note that the number of output arguments must match the number of columns that you are reading.

For example, suppose that file `test_input.dat` contains the following data:

```
James  Jones  O+  3.51  22  Yes
Sally  Smith  A+  3.28  23  No
```

This data could be read into a series of arrays with the following function:

```
[first,last,blood,gpa,age,answer] = ...
      textread('test_input.dat','%s %s %s %f %d %s')
```

When this command is executed, the results are:

```
» [first,last,blood,gpa,age,answer] = ...
           textread('test_input.dat','%s %s %s %f %d %s')

first =
   'James'
   'Sally'
last =
   'Jones'
   'Smith'
blood =
   'O+'
   'A+'
gpa =
   3.5100
   3.2800
age =
   42
   28
answer =
   'Yes'
   'No'
```

This function can also skip selected columns by adding an asterisk to the corresponding format descriptor (for example, `%*s`). The following statement reads only the first name, last name, and gpa from the file:

```
» [first,last,gpa] = ...
           textread('test_input.dat','%s %s %*s %f %*d %*s')

first =
   'James'
   'Sally'
last =
   'Jones'
   'Smith'
gpa =
   3.5100
   3.2800
```

Function `textread` is much more useful and flexible than the `load` command. The `load` command assumes that all of the data in the input file is of a single type—it cannot support different types of data in different columns. In addition, the load command stores all of the data into a single array. In contrast, the textread function allows each column to go into a separate variable, which is *much* more convenient when working with columns of mixed data.

Function textread has a number of additional options that increase its flexibility. Consult the MATLAB on-line documentation for details of these options.

B.2 MATLAB File Processing

To use files within a MATLAB program, we need some way to select the desired file and to read from or write to it. MATLAB has a series of C-like functions to read and write files, whether they are on disk, magnetic tape, or some other device attached to the computer. These functions open, read, write, and close files using a **file id** (sometimes known as **fid**). The file id is a number assigned to a file when it is opened, and used for all reading, writing, and control operations on that file. The file id is a positive integer. Two file ids are always open—file id 1 is the standard output device (stdout) and file id 2 is the standard error (stderr) device for the computer on which MATLAB is executing. Additional file ids are assigned as files are opened and released as files are closed.

Several MATLAB functions can be used to control disk file input and output. The file I/O functions are summarized in Table B.1. The file opening, closing, reading, and writing functions are described subsequently. For details of the positioning and status functions, see the MATLAB documentation.

Table B.1 MATLAB Input/Output Functions

Category	Function	Description
File opening and closing	fopen	Open file
	fclose	Close file.
Binary I/O	fread	Read binary data from file.
	fwrite	Write binary data to file.
Formatted I/O	fscanf	Read formatted data from file.
	fprintf	Write formatted data to file.
	fgetl	Read line from file, discard newline character.
	fgets	Read line from file, keep newline character.
File positioning, status, and miscellaneous	delete	Delete file.
	exist	Check for the existence of a file.
	ferror	Inquire file I/O error status.
	feof	Test for end-of-file.
	fseek	Set file position.
	ftell	Check file position.
	frewind	Rewind file.
Temporary files	tempdir	Get temporary directory name.
	tempname	Get temporary file name.

File ids are assigned to disk files or devices using the `fopen` statement and detached from them using the `fclose` statement. Once a file is attached to a file id using the `fopen` statement, we can read and write to that file using MATLAB file input and output statements. When we are through with the file, the `fclose` statement closes the file and makes the file id invalid. The `frewind` and `fseek` statements may be used to change the current reading or writing position in a file while it is open.

Data can be written to and read from files in two possible ways: as binary data or as formatted character data. Binary data consists of the actual bit patterns that are used to store the data in computer memory. Reading and writing binary data is very efficient, but a user cannot examine the data stored in the file. In contrast, data in formatted files is translated into characters that can be read directly by a user. However, formatted I/O operations are slower and less efficient than binary I/O operations. We discuss both types of I/O operations later in this appendix.

B.3 File Opening and Closing

The file opening and closing functions, `fopen` and `fclose`, are described in the following subsections.

B.3.1 The `fopen` Function

The `fopen` function opens a file and returns a file id number for use with the file. The basic forms of this statement are

```
fid = fopen(filename,permission)
[fid, message] = fopen(filename,permission)
[fid, message] = fopen(filename,permission,format)
```

where *`filename`* is a string specifying the name of the file to open, *`permission`* is a character string specifying the mode in which the file is opened, and *`format`* is an optional string specifying the numeric format of the data in the file. If the open is successful, `fid` will contain a positive integer after this statement is executed, and `message` will be an empty string. If the open fails, `fid` will contain a -1 after this statement is executed, and `message` will be a string explaining the error. If a file is opened for reading and it is not in the current directory, MATLAB will search for it along the MATLAB search path.

The possible permission strings are shown in Table B.2.

On some platforms such as Windows PCs, it is important to distinguish between text files and binary files. If a file is to be opened in text mode, then a `t` should be added to the permissions string (for example, `'rt'` or `'rt+'`). If a file is to be opened in binary mode, a `b` may be added to the permissions string (for example, `'rb'`), but this is not actually required, because files are opened in binary mode by default. This distinction between text and binary files does not exist on Macintosh, Unix, or Linux computers, so the `t` or `b` is never needed on those systems.

Table B.2 fopen File Permissions

File Permission	Meaning
`'r'`	Open an existing file for reading only (default).
`'r+'`	Open an existing file for reading and writing.
`'w'`	Delete the contents of an existing file (or create a new file) and open it for writing only.
`'w+'`	Delete the contents of an existing file (or create a new file) and open it for reading and writing.
`'a'`	Open an existing file (or create a new file) and open it for writing only, appending to the end of the file.
`'a+'`	Open an existing file (or create a new file) and open it for reading and writing, appending to the end of the file.
`'W'`	Write without automatic flushing (special command for tape drives).
`'A'`	Append without automatic flushing (special command for tape drives).

Table B.3 fopen Format Strings

File Permission	Meaning
`'native'` or `'n'`	Numeric format for the machine MATLAB is executing on (default)
`'ieee-le'` or `'l'`	IEEE floating point with little-endian byte ordering
`'ieee-be'` or `'b'`	IEEE floating point with big-endian byte ordering
`'ieee-le.l64'` or `'a'`	IEEE floating point with little-endian byte ordering and 64-bit long data type
`'ieee-le.b64'` or `'s'`	IEEE floating point with big-endian byte ordering and 64-bit long data type

The *format* string in the `fopen` function specifies the numeric format of the data stored in the file. This string is needed only when transferring files between computers with incompatible numeric data formats, so it is rarely used. A few of the possible numeric formats are shown in Table B.3; see the MATLAB Language Reference Manual for a complete list of possible numeric formats.

There are also two forms of this function that provide information rather than open files. The function

```
fids = fopen('all')
```

returns a row vector containing a list of all file ids for currently open files (except for `stdout` and `stderr`). The number of elements in this vector is equal to the number of open files. The function

```
[filename, permission, format] = fopen(fid)
```

returns the file name, permission string, and numeric format for an open file specified by file id.

Some examples of correct `fopen` functions are shown below.

Case 1: Opening a Binary File for Input

The function below opens a file named `example.dat` for binary input only.

```
fid = fopen('example.dat','r')
```

The permission string is `'r'`, indicating that the file is to be opened for reading only. The string could have been `'rb'`, but this is not required because binary access is the default case.

Case 2: Opening a File for Text Output

The functions below open a file named `outdat` for text output only.

```
fid = fopen('outdat','wt')
```

or

```
fid = fopen('outdat','at')
```

The `'wt'` permissions string specifies that the file is a new text file; if it already exists, the old file will be deleted and a new empty file will be opened for writing. This is the proper form of the `fopen` function for an *output file* if we want to replace preexisting data.

The `'at'` permissions string specifies that we want to append to an existing text file. If it already exists, it will be opened and new data will be appended to the currently existing information. This is the proper form of the `fopen` function for an *output file* if we don't want to replace preexisting data.

Case 3: Opening a Binary File for Read/Write Access

The function below opens a file named `junk` for binary input and output.

```
fid = fopen('junk','r+')
```

The function below also opens the file for binary input and output.

```
fid = fopen('junk','w+')
```

The difference between the first and the second statements is that the first statement required the file to exist before it is opened, while the second statement will delete any preexisting file.

B.3.2 The `fclose` Function

The `fclose` function closes a file. Its form is

```
status = fclose(fid)
status = fclose('all')
```

where `fid` is a file id and `status` is the result of the operation. If the operation is successful, `status` will be 0, and if it is unsuccessful, `status` will be –1.

The form `status = fclose('all')` closes all open files except for `stdout` (`fid` = 1) and `stderr` (`fid` = 2). It returns a status of 0 if all files close successfully and –1 otherwise.

B.4 Binary I/O Functions

The binary I/O functions, `fwrite` and `fread`, are described in the following subsections.

B.4.1 The `fwrite` Function

The `fwrite` function writes binary data in a user-specified format to a file. Its form is

```
count = fwrite(fid,array,precision)
count = fwrite(fid,array,precision,skip)
```

where `fid` is the file id of a file opened with the `fopen` function, `array` is the array of values to write out, and `count` is the number of values written to the file.

MATLAB writes out data in *column order,* which means that the entire first column is written out, followed by the entire second column, etc. For example, if `array` $= \begin{bmatrix} 1 & 2 \\ 3 & 4 \\ 5 & 6 \end{bmatrix}$, the data will be written out in the order 1, 3, 5, 2, 4, 6.

The optional *`precision`* string specifies the format in which the data will be output. MATLAB supports both platform-independent precision strings, which are the same for all computers that MATLAB runs on, and platform-dependent precision strings, which vary among different types of computers. *You should use only the platform-independent strings,* and those are the only forms presented in this book.

For convenience, MATLAB accepts some C and Fortran data-type equivalents for the MATLAB precision strings. If you are a C or Fortran programmer, you may find it more convenient to use the names of the data types in the language that you are most familiar with.

The possible platform-independent precisions are presented in Table B.4. All of these precisions work in units of bytes, except for `'bitN'` or `'ubitN'`, which work in units of bits.

The optional argument *`skip`* specifies the number of bytes to skip in the output file before each write. This option is useful for placing values at certain points in fixed-length records. Note that if *`precision`* is a bit format like `'bitN'` or `'ubitN'`, skip is specified in bits instead of bytes.

Table B.4 Selected MATLAB Precision Strings

MATLAB Precision String	C/Fortran Equivalent	Meaning
'char'	'char*1'	8-bit characters
'schar'	'signed char'	8-bit signed character
'uchar'	'unsigned char'	8-bit unsigned character
'int8'	'integer*1'	8-bit integer
'int16'	'integer*2'	16-bit integer
'int32'	'integer*4'	32-bit integer
'int64'	'integer*8'	64-bit integer
'uint8'	'integer*1'	8-bit unsigned integer
'uint16'	'integer*2'	16-bit unsigned integer
'uint32'	'integer*4'	32-bit unsigned integer
'uint64'	'integer*8'	64-bit unsigned integer
'float32'	'real*4'	32-bit floating point
'float64'	'real*8'	64-bit floating point
'bitN'		N-bit signed integer, $1 \leq N \leq 64$
'ubitN'		N-bit unsigned integer, $1 \leq N \leq 64$

B.4.2 The `fread` Function

The `fread` function reads binary data in a user-specified format from a file and returns the data in a (possibly different) user-specified format. Its form is

```
[array,count] = fread(fid,size,precision)
[array,count] = fread(fid,size,precision,skip)
```

where `fid` is the file id of a file opened with the `fopen` function, `size` is the number of values to read, `array` is the array to contain the data, and `count` is the number of values read from the file.

The optional argument *`size`* specifies the amount of data to be read from the file. There are three versions of this argument:

- `n`—Read exactly `n` values. After this statement, `array` will be a column vector containing `n` values read from the file.
- `Inf`—Read until the end of the file. After this statement, `array` will be a column vector containing all of the data until the end of the file.
- `[n m]`—Read exactly `n` × `m` values, and format the data as an `n` × `m` array.

If `fread` reaches the end of the file and the input stream does not contain enough bits to write out a complete array element of the specified precision, fread pads the last byte or element with zero bits until the full value is obtained. If an error occurs, reading is done up to the last full value.

The *precision* argument specifies both the format of the data on the disk and the format of the data array to be returned to the calling program. The general form of the precision string is

'*disk_precision* => *array_precision*'

where disk_precision and array_precision are both one of the precision strings found in Table B.4. The array_precision value can be defaulted. If it is missing, then the data is returned in a double array. There is also a shortcut form of this expression if the disk precision and the array precision are the same: '**disk_precision*'.

A few examples of precision strings are shown below:

'single'	Read data in single precision format from disk, and return it in a double array.
'single=>single'	Read data in single precision format from disk, and return it in a single array.
'*single'	Read data in single precision format from disk, and return it in a single array (a shorthand version of the previous string).
'double=>real*4'	Read data in double precision format from disk, and return it in a single array.

Example B.1—Writing and Reading Binary Data

The example script file shown below creates an array containing 10,000 random values, opens a user-specified file for writing only, writes the array to disk in 64-bit floating-point format, and closes the file. It then opens the file for reading and reads the data back into a 100 × 100 array. It illustrates the use of binary I/O operations.

```
% Script file: binary_io.m
%
% Purpose:
%   To illustrate the use of binary i/o functions.
%
% Record of revisions:
%   Date          Programmer        Description of change
%   ====          ==========        =====================
%   01/21/05     S. J. Chapman      Original code
%
% Define variables:
%   count          -- Number of values read / written
%   fid            -- File id
%   filename       -- File name
```

```
%   in_array     -- Input array
%   msg          -- Open error message
%   out_array    -- Output array
%   status       -- Operation status

% Prompt for file name
filename = input('Enter file name: ','s');

% Generate the data array
out_array = randn(1,10000);

% Open the output file for writing.
[fid,msg] = fopen(filename,'w');

% Was the open successful?
if fid > 0

    % Write the output data.
    count = fwrite(fid,out_array,'float64');

    % Tell user
    disp([int2str(count) ' values written...']);

    % Close the file
    status = fclose(fid);

else

    % Output file open failed. Display message.
    disp(msg);

end

% Now try to recover the data. Open the
% file for reading.
[fid,msg] = fopen(filename,'r');

% Was the open successful?
if fid > 0

   % Write the output data.
   [in_array, count] = fread(fid,[100 100],'float64');

   % Tell user
   disp([int2str(count) ' values read...']);

   % Close the file
   status = fclose(fid);

else

   % Input file open failed. Display message.
   disp(msg);

end
```

When this program is executed, the result are

```
» binary_io
Enter file name: testfile
10000 values written
10000 values read
```

An 80,000-byte file named `testfile` was created in the current directory. This file is 80,000 bytes long because it contains 10,000 64-bit values, and each value occupies 8 bytes.

◄

B.5 Formatted I/O Functions

The formatted I/O functions are described below.

B.5.1 The fprintf Function

The `fprintf` function writes formatted data in a user-specified format to a file. Its form is

```
count = fprintf(fid,format,val1,val2,...)
fprint(format,val1,val2,...)
```

where `fid` is the file id of a file to which the data will be written, and `format` is the format string controlling the appearance of the data. If `fid` is missing, the data is written to the standard output device (the Command Window). This is the form of `fprintf` that we have been using since Chapter 2.

The format string specifies the alignment, significant digits, field width, and other aspects of output format. It can contain ordinary alphanumeric characters along with special sequences of characters that specify the exact format in which the output data will be displayed. The structure of a typical format is shown in Figure B.1. A single `%` character always marks the beginning of a format—if an

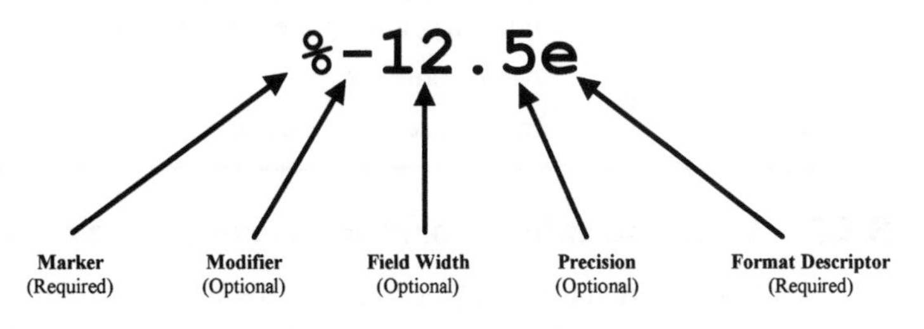

Figure B.1 The structure of a typical format specifier.

ordinary % sign is to be printed out, then it must appear in the format string as %%. After the % character, the format can have a flag, a field width and precision specifier, and a conversion specifier. The % character and the conversion specifier are always required in any format, while the field and field width and precision specifiers are optional.

The possible conversion specifiers are listed in Table B.5, and the possible flags are listed in Table B.6. If a field width and precision are specified in a format, then the number before the decimal point is the field width, which is the number of characters used to display the number. The number after the decimal point is the precision, which is the minimum number of significant digits to display after the decimal point.

In addition to ordinary characters and formats, certain special escape characters can be used in a format string. These special characters are listed in Table B.7.

Table B.5 Format Conversion Specifiers for `fprintf`

Specifier	Description
`%c`	Single character
`%d`	Decimal notation (signed)
`%e`	Exponential notation (using a lowercase `e` as in `3.1416e+00`)
`%E`	Exponential notation (using an uppercase `E` as in `3.1416E+00`)
`%f`	Fixed-point notation
`%g`	The more compact of `%e` or `%f`. Insignificant zeros do not print.
`%G`	Same as `%g`, but using an uppercase `E`
`%o`	Octal notation (unsigned)
`%s`	String of characters
`%u`	Decimal notation (unsigned)
`%x`	Hexadecimal notation (using lowercase letters `a–f`)
`%X`	Hexadecimal notation (using uppercase letters `A–F`)

Table B.6 Format Flags

Flag	Description
Minus sign (−)	Left-justifies the converted argument in its field (e.g., `%-5.2d`). If this flag is not present, the argument is right-justified.
+	Always print a + or - sign (e.g., `%+5.2d`).
0	Pad argument with leading zeros instead of blanks (e.g., `%05.2d`).

B.5.2 Understanding Format Conversion Specifiers

The best way to understand the wide variety of format conversion specifiers is by example, so we will now present several examples along with their results.

Table B.7 Escape Characters in Format Strings

Escape Sequences	Description
`\n`	New line
`\t`	Horizontal tab
`\b`	Backspace
`\r`	Carriage return
`\f`	Form feed
`\\`	Print an ordinary backslash (\) symbol.
`\''` or `''`	Print an apostrophe or single quote.
`%%`	Print an ordinary percent (`%`) symbol.

Case 1: Displaying Decimal Data

Decimal (integer) data is displayed with the `%d` format conversion specifier. The `d` may be preceded by a flag and a field width and precision specifier, if desired. If used, the precision specifier sets a minimum number of digits to display. If there are not enough digits, leading zeros will be added to the number.

Function	Result	Comment
`fprintf('%d\n',123)`	`----|----|` `123`	Display the number using as many characters as required. For the number 123, three characters are required.
`fprintf('%6d\n',123)`	`----|----|` `   123`	Display the number in a 6-character-wide field. By default the number is *right justified* in the field.
`fprintf('%6.4d\n',123)`	`----|----|` `  0123`	Display the number in a 6-character-wide field using a minimum of 4 characters. By default the number is *right justified* in the field.
`fprintf('%-6.4d\n',123)`	`----|----|` `0123`	Display the number in a 6-character-wide field using a minimum of 4 characters. The number is *left justified* in the field.
`fprintf('%+6.4d\n',123)`	`----|----|` ` +0123`	Display the number in a 6-character-wide field using a minimum of 4 characters plus a sign character. By default the number is *right justified* in the field.

If a nondecimal number is displayed with the `%d` conversion specifier, the specifier will be ignored and the number will be displayed in exponential format. For example,

```
fprintf('%6d\n',123.4)
```

produces the result `1.234000e+002`.

Case 2: Displaying Floating-Point Data

Floating-point data can be displayed with the `%e`, `%f`, or `%g` format conversion specifiers. They may be preceded by a flag and a field width and precision specifier, if desired. If the specified field with is too small to display the number, it is ignored. Otherwise, the specified field width is used.

Function	Result	Comment
`fprintf('%f\n',123.4)`	`----|----|` `123.400000`	Display the number using as many characters as required. The default case for `%f` is to display 6 digits after the decimal place.
`fprintf('%8.2f\n',123.4)`	`----|----|` `  123.40`	Display the number in an 8-character-wide field, with two places after the decimal point. The number is *right justified* in the field.
`fprintf('%4.2f\n',123.4)`	`----|----|` `123.40`	Display the number in a 6-character-wide field. The width specification was ignored because it was too small to display the number.
`fprintf('%10.2e\n',123.4)`	`----|----|` ` 1.23e+002`	Display the number in exponenial format in a 10-character-wide field using 2 decimal places. By default the number is *right justified* in the field.
`fprintf('%10.2E\n',123.4)`	`----|----|` ` 1.23E+002`	The same but with a capital `E` for the exponent.

Case 3: Displaying Character Data

Character data may be displayed with the `%c` or `%s` format conversion specifiers. They may be preceded by field width specifier, if desired. If the specified field width is too small to display the number, it is ignored. Otherwise, the specified field width is used.

Function	Result	Comment
`fprintf('%c\n','s')`	`----|----|` `s`	Display a single character.
`fprintf('%s\n','string')`	`----|----|` `string`	Display the character string.
`fprintf('%8s\n','string')`	`----|----|` `  string`	Display the character string in an 8-character-wide field. By default the string is *right justified* in the field.
`fprintf('%-8s\n','string')`	`----|----|` `string`	Display the character string in an 8-character-wide field. The string is *left justified* in the field.

B.5.3 The `fscanf` Function

The `fscanf` function reads formatted data in a user-specified format from a file. Its form is

```
array = fscanf(fid,format)
[array, count] = fscanf(fid,format,size)
```

where `fid` is the file id of a file from which the data will be read, `format` is the format string controlling how the data is read, and `array` is the array that receives the data. The output argument `count` returns the number of values read from the file.

The optional argument *`size`* specifies the amount of data to be read from the file. There are three versions of this argument:

- `n`—Read exactly `n` values. After this statement, `array` will be a column vector containing `n` values read from the file.
- `Inf`—Read until the end of the file. After this statement, `array` will be a column vector containing all of the data until the end of the file.
- `[n m]`—Read exactly `n × m` values and format the data as an `n × m` array.

The format string specifies the format of the data to be read. It can contain ordinary characters along with format conversion specifiers. The `fscanf` function compares the data in the file with the format conversion specifiers in the format string. As long as the two match, `fscanf` converts the value and stores it in the output array. This process continues until the end of the file or until the amount of data in *`size`* has been read, whichever comes first.

If the data in the file does not match the format conversion specifiers, the operation of `fscanf` stops immediately.

The format conversion specifiers for `fscanf` are basically the same as those for `fprintf`. The most common specifiers are shown in Table B.8.

Table B.8 Format Conversion Specifiers for `fscanf`

Specifier	Description
`%c`	Read a single character. This specifier reads any character including blanks, new lines, etc.
`%Nc`	Read *N* characters.
`%d`	Read a decimal number (ignores blanks).
`%e %f %g`	Read a floating-point number (ignores blanks).
`%i`	Read a signed integer (ignores blanks).
`%s`	Read a string of characters. The string is terminated by blanks or other special characters such as new lines.

To illustrate the use of `fscanf`, we will attempt to read a file called `x.dat` containing the following values on two lines:

```
10.00    20.00
30.00    40.00
```

1. If the file is read with the statement

```
[z, count] = fscanf(fid,'%f');
```

variable z will be the column vector $\begin{bmatrix} 10 \\ 20 \\ 30 \\ 40 \end{bmatrix}$ and `count` will be 4.

2. If the file is read with the statement

```
[z, count] = fscanf(fid,'%f',[2 2]);
```

variable z will be the array $\begin{bmatrix} 10 & 30 \\ 20 & 40 \end{bmatrix}$ and `count` will be 4.

3. Next, let's try to read this file as decimal values. If the file is read with the statement

```
[z, count] = fscanf(fid,'%d',Inf);
```

variable z will be the single value 10 and `count` will be 1. This happens because the decimal point in the 10.00 does not match the format conversion specifier and `fscanf` stops at the first mismatch.

4. If the file is read with the statement

```
[z, count] = fscanf(fid,'%d.%d',[1 Inf]);
```

variable z will be the row vector [10 0 20 0 30 0 40 0] and `count` will be 8. This happens because the decimal point is now matched in the format conversion specifier and the numbers on either side of the decimal point are interpreted as separate integers!

5. Now let's try to read the file as individual characters. If the file is read with the statement

```
[z, count] = fscanf(fid,'%c');
```

variable z will be a row vector containing every character in the file, including all spaces and newline characters! Variable `count` will be equal to the number of characters in the file.

6. Finally, let's try to read the file as a character string. If the file is read with the statement

```
[z, count] = fscanf(fid,'%s');
```

variable z will be a row vector containing the 20 characters 10.0020.0030.0040.00, and count will be 4. This happens because the string specifier ignores white space and the function found four separate strings in the file.

B.5.4 The fgetl Function

The fgetl function reads the next line *excluding the end-of-line characters* from a file as a character string. It form is

```
line = fgetl(fid)
```

where fid is the file id of a file from which the data will be read, and line is the character array that receives the data. If fgetl encounters the end of a file, the value of line is set to –1.

B.5.5 The fgets Function

The fgets function reads the next line *including the end-of-line characters* from a file as a character string. It form is

```
line = fgets(fid)
```

where fid is the file id of a file from which the data will be read and line is the character array that receives the data. If fgets encounters the end of a file, the value of line is set to –1.

B.6 The textscan Function

The textscan function reads ASCII files are formatted into columns of data, where each column can be of a different type and stores the contents into the columns of a cell array. This function is *very* useful for importing tables of data printed out by other applications. It is new in MATLAB 7.0. It is basically similar to textread, except that it is faster and more flexible.

The form of the textscan function is

```
a = textscan(fid, 'format')
a = textscan(fid, 'format', N)
a = textscan(fid, 'format', param, value, ...)
a = textscan(fid, 'format', N, param, value, ...)
```

where fid is the file id of a file that has already been opened with fopen, format is a string containing a description of the type of data in each column, and n is the number of times to use the format specifier. (If n is –1 or is missing, the function reads to the end of the file.) The format string contains the same types of format descriptors as function fprintf. Note that there is only one output argument, with all of the values returned in a cell array. The cell array will contain a number of elements equal to the number of format descriptors to read.

For example, suppose that file `test_input1.dat` contains the following data:

```
James    Jones    O+    3.51    22    Yes
Sally    Smith    A+    3.28    23    No
Hans     Carter   B-    2.84    19    Yes
Sam      Spade    A+    3.12    21    Yes
```

This data could be read into a cell array with the following function:

```
fid = fopen('test_input1.dat','rt');
a = textscan(fid,'%s %s %s %f %d %s',-1);
fclose(fid);
```

When this command is executed, the results are:

```
» fid = fopen('test_input1.dat','rt');
» a = textscan(fid,'%s %s %s %f %d %s',-1)
a =
  {4x1 cell}  {4x1 cell}  {4x1 cell}  [4x1 double]
  [4x1 int32]  {4x1 cell}
» a{1}
ans =
    'James'
    'Sally'
    'Hans'
    'Sam'
» a{2}
ans =
    'Jones'
    'Smith'
    'Carter'
    'Spade'
» a{3}
ans =
    'O+'
    'A+'
    'B-'
    'A+'
» a{4}
ans =
    3.5100
    3.2800
    2.8400
    3.1200
» fclose(fid);
```

This function can also skip selected columns by adding an asterisk to the corresponding format descriptor (for example, `%*s`). For example, the following statements read only the first name, last name, and `gpa` from the file:

```
fid = fopen('test_input1.dat','rt');
a = textscan(fid,'%s %s %*s %f %*d %*s',-1);
fclose(fid);
```

Function `textscan` is similar to function `textread`, but it is more flexible and faster. The advantages of `textscan` include:

1. The `textscan` function offers better performance than `textread`, making it a better choice when reading large files.
2. With `textscan`, you can start reading at any point in the file. When the file is opened with `fopen`, you can move to any position in the file with `fseek` and begin the `textscan` at that point. The `textread` function requires that you start reading from the beginning of the file.
3. Subsequent `textscan` operations start reading the file at a point where the last `textscan` left off. The `textread` function always begins at the start of the file, regardless of any prior `textread` operations.
4. Function `textscan` returns a single-cell array regardless of how many fields you read. With `textscan`, you don't need to match the number of output arguments with the number of fields being read, as you would with `textread`.
5. Function `textscan` offers more choices in how the data being read is converted.

Function `textscan` has a number of additional options that increase its flexibility. Consult the MATLAB on-line documentation for details of these options.

B.7 Function `uiimport`

Function `uiimport` is a GUI-based way to import data from a file or from the clipboard. This command takes the forms

```
uiimport
structure = uiimport;
```

In the first case, the imported data is inserted directly into the current MATLAB workspace. In the second case, the data is converted into a structure and saved in variable `structure`.

When the command `uiimport` is typed, the Import Wizard is displayed in a window (see Figure B.2 for the PC version of this window). The user can then select the file that he or she would like to import from and the specific data within that file. Many different formats are supported—a partial list is given in

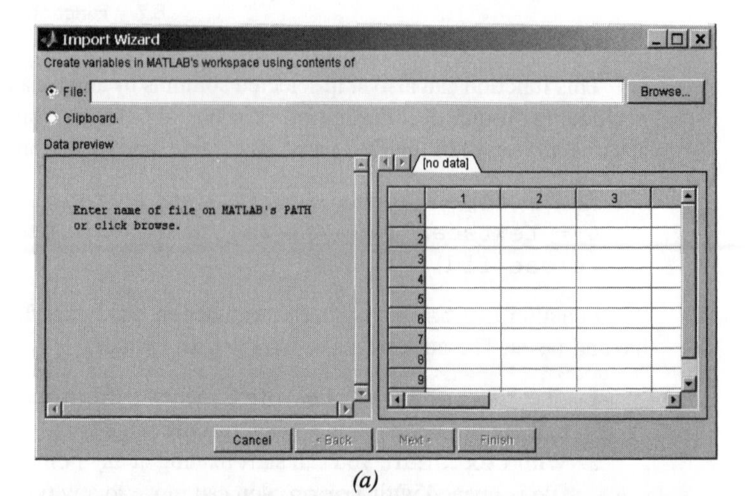

(a)

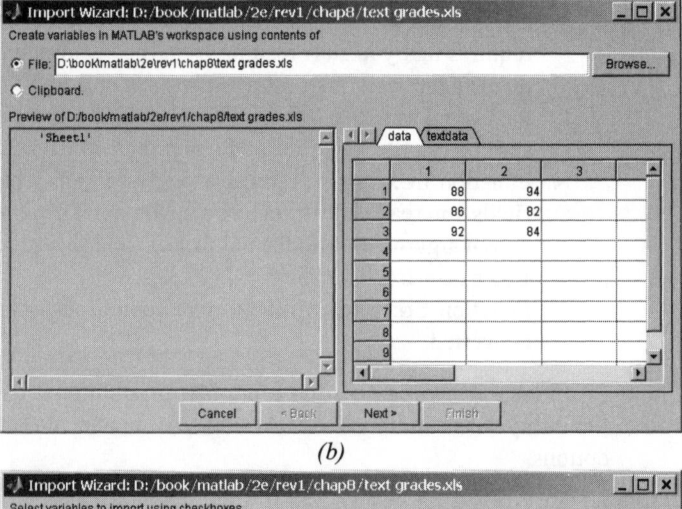

(b)

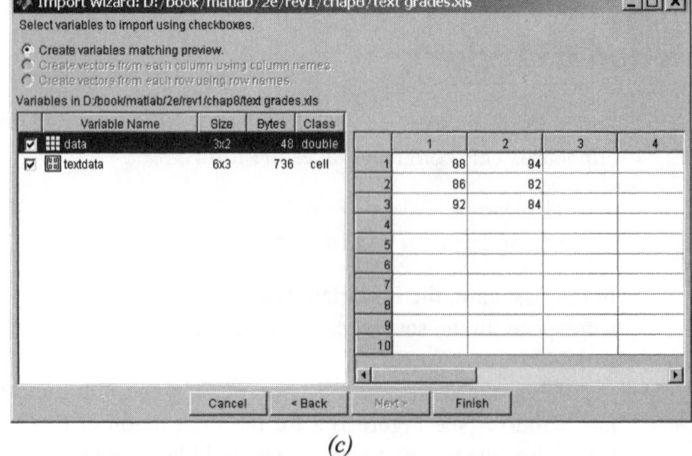

(c)

Figure B.2 Using `uiimport`: *(a)* The Import Wizard after it is started. *(b)* After a data file has been selected, one or more data arrays are created, and their contents can be examined. *(c)* Next, the user can select which of the data arrays will be imported into MATLAB.

Table B.9. In addition, data can be imported from almost *any* application by saving the data on the clipboard. This flexibility can be very useful when you are trying to get data into MATLAB for analysis.

Table B.9 Selected File Formats Supported by uiimport

File Extents	Meaning
*.gif	Image files
*.jpg	
*.jpeg	
*.ico	
*.png	
*.pcx	
*.tif	
*.tiff	
*.bmp	
*.cur	Cursor format
*.hdf	Hierarchical data format files
*.au	Sound files
*.snd	
*.wav	
*.avi	Movie files
*.csv	Spreadsheet files
*.xls	
*.wk1	
*.txt	Text files
*.dat	
*.dlm	
*.tab	

APPENDIX C

Answers to Quizzes

This appendix contains the answers to all of the quizzes in the book.

Quiz 1.1, page 18

1. The MATLAB Command Window is the window where a user enters commands. A user can enter interactive commands at the command prompt (») in the Command Window and they will be executed on the spot. The Command Window is also used to start M-files executing. The Edit/Debug Window is an editor used to create, modify, and debug M-files. The Figure Window is used to display MATLAB graphical output.
2. You can get help in MATLAB by:
   - Typing help <command_name> in the Command Window. This command will display information about a command or function in the Command Window.
   - Typing lookfor <keyword> in the Command Window. This command will display in the Command Window a list of all commands or functions containing the keyword in their first comment line.
   - Starting the Help Browser by typing helpwin or helpdesk in the Command Window, by selecting "Help" from the Start menu, or by clicking on the question mark icon (?) on the desktop. The Help Browser contains an extensive hypertext-based description of all of the features in MATLAB, plus a complete

copy of all manuals on-line in HTML and Adobe PDF formats. It is the most comprehensive source of help in MATLAB.

3. A workspace is the collection of all the variables and arrays that can be used by MATLAB when a particular command, M-file, or function is executing. All commands executed in the Command Window (and all script files executed from the Command Window) share a common workspace, so they can all share variables. The contents of the workspace can be examined with the `whos` command, or graphically with the Workspace Browser.
4. To clear the contents of a workspace, type `clear` or `clear variables` in the Command Window.
5. The commands to perform this calculation are:

```
» t = 5;
» x0 = 10;
» v0 = 15;
» a = -9.81;
» x = x0 + v0 * t + 1/2 * a * t^2
x =
   -37.6250
```

6. The commands to perform this calculation are:

```
» x = 3;
» y = 4;
» res = x^2 * y^3 / (x - y)^2
res =
     576
```

Questions 7 and 8 are intended to get you to explore the features of MATLAB. There is no single "right" answer for them.

Quiz 2.1, page 30

1. An array is a collection of data values organized into rows and columns and known by a single name. Individual data values within an array are accessed by including the name of the array followed by subscripts in parentheses that identify the row and column of the particular value. The term "vector" is usually used to describe an array with only one dimension, while the term "matrix" is usually used to describe an array with two or more dimensions.
2. (*a*) This is a 3×4 array; (*b*) `c(2,3)` $= -0.6$; (*c*) The array elements whose value is 0.6 are `c(1,4)`, `c(2,1)` and `c(3,2)`.
3. (*a*) 1×3; (*b*) 3×1; (*c*) 3×3; (*d*) 3×2; (*e*) 3×3; (*f*) 4×3; (*g*) 4×1.

4. `w(2, 1)` $= 2$
5. `x(2, 1)` $= -20i$
6. `y(2, 1)` $= 0$
7. `v(3)` $= 3$

Quiz 2.2, page 39

1. (*a*) `c(2,:)` $= [0.6 \quad 1.1 \quad -0.6 \quad 3.1]$

   (*b*) `c(:,end)` $= \begin{bmatrix} 0.6 \\ 3.1 \\ 0.0 \end{bmatrix}$

   (*c*) `c(1:2,2:end)` $= \begin{bmatrix} -3.2 & 3.4 & 0.6 \\ 1.1 & -0.6 & 3.1 \end{bmatrix}$

   (*d*) `c(6)` $= 0.6$

   (*e*) `c(4,end)` $= [-3.2 \quad 1.1 \quad 0.6 \quad 3.4 \quad -0.6 \quad 5.5 \quad 0.6 \quad 3.1 \quad 0.0]$

   (*f*) `c(1:2,2:end)` $= \begin{bmatrix} -3.2 & 3.4 & 0.6 \\ 1.1 & -0.6 & 3.1 \end{bmatrix}$

   (*g*) `c([1 3],2)` $= \begin{bmatrix} -3.2 \\ 0.6 \end{bmatrix}$

   (*h*) `c([2 2],[3 3])` $= \begin{bmatrix} -0.6 & -0.6 \\ -0.6 & -0.6 \end{bmatrix}$

2. (*a*) `a` $= \begin{bmatrix} 7 & 8 & 9 \\ 4 & 5 & 6 \\ 1 & 2 & 3 \end{bmatrix}$ (*b*) `a` $= \begin{bmatrix} 4 & 5 & 6 \\ 4 & 5 & 6 \\ 4 & 5 & 6 \end{bmatrix}$ (*c*) `a` $= \begin{bmatrix} 4 & 5 & 6 \\ 4 & 5 & 6 \end{bmatrix}$

3. (*a*) `a` $= \begin{bmatrix} 1 & 0 & 0 \\ 1 & 2 & 3 \\ 0 & 0 & 1 \end{bmatrix}$ (*b*) `a` $= \begin{bmatrix} 1 & 0 & 4 \\ 0 & 1 & 5 \\ 0 & 0 & 6 \end{bmatrix}$ (*c*) `a` $= \begin{bmatrix} 1 & 0 & 0 \\ 0 & 1 & 0 \\ 9 & 7 & 8 \end{bmatrix}$

Quiz 2.3, page 45

1. The required command is "`format long e`."
2. (*a*) These statements get the radius of a circle from the user and calculate and display the area of the circle. (*b*) These statements display the value of π as an integer, so they display the string: "`The value is 3!`."

3. The first statement outputs the value 12345.67 in exponential format, the second statement outputs the value in floating-point format, the third statement outputs the value in general format, and the fourth statement outputs the value in floating-point format in a field 12 characters wide, with four places after the decimal point. The results of these statements are:

```
value = 1.234567e+004
value = 12345.670000
value = 12345.7
value = 12345.6700
```

Quiz 2.4, page 52

1. (*a*) This operation is illegal. Array multiplication must be between arrays of the same shape, or between an array and a scalar. (*b*) Legal matrix multiplication: result $= \begin{bmatrix} 4 & 4 \\ 3 & 3 \end{bmatrix}$ (*c*) Legal array multiplication: result $= \begin{bmatrix} 2 & 1 \\ -2 & 4 \end{bmatrix}$ (*d*) This operation is illegal. The matrix multiplication `b * c` yields a 1×2 array, and `a` is a 2×2 array, so the addition is illegal. (*e*) This operation is illegal. The array multiplication `b .* c` is between two arrays of different sizes, so the multiplication is illegal.

2. This result can be found from the operation `x = A\B`: $x = \begin{bmatrix} -0.5 \\ 1.0 \\ -0.5 \end{bmatrix}$

Quiz 3.1, page 101

Expression	Result	Comment
1. `a > b`	1 (`logical` true)	
2. `b > d`	0 (`logical` false)	
3. `a > b && c > d`	0 (`logical` false)	
4. `a == b`	0 (`logical` false)	
5. `a & b > c`	0 (`logical` false)	

6. `~~b`	1 (`logical` true)	
7. `~(a > b)`	$\begin{bmatrix} 0 & 0 \\ 0 & 1 \end{bmatrix}$ (`logical` array)	
8. `a > c && b > c`	Illegal	The `&&` and `||` operators work only between *scalar* operands.
9. `c <= d`	Illegal	The `<=` operator must be between arrays of the same size or between an array and a scalar.
10. `logical(d)`	$\begin{bmatrix} 1 & 1 & 1 \\ 0 & 1 & 0 \end{bmatrix}$ (`logical` array)	
11. `a * b > c`	$\begin{bmatrix} 1 & 0 \\ 0 & 1 \end{bmatrix}$ (`logical` array)	The expression `a * b` is evaluated first, producing the `double` array $\begin{bmatrix} 2 & -4 \\ 0 & 20 \end{bmatrix}$, and the `logical` operation is evaluated second, producing the final answer.
12. `a * (b > c)`	$\begin{bmatrix} 2 & 0 \\ 0 & 2 \end{bmatrix}$ (`double` array)	The expression `b > c` produced the `logical` array $\begin{bmatrix} 1 & 0 \\ 0 & 1 \end{bmatrix}$, and multiplying that `logical` array by 2 converted the results back into a `double` array.
13. `a*b^2 > a*c`	0 (`logical` false)	
14. `d || b > a`	1 (`logical` true)	
15. `(d | b) > a`	0 (`logical` false)	

16. `isinf(a/b)`	0 (`logical` false)	
17. `isinf(a/c)`	1 (`logical` true)	
18. `a > b && ischar(d)`	1 (`logical` true)	
19. `isempty(c)`	0 (`logical` false)	
20. `(~a) & b`	0 (`logical` false)	
21. `(~a) + b`	−2 (`double` value)	`~a` is a logical 0. When added to `b`, the result is converted back to a `double` value.

Quiz 3.2, page 116

1.
```
if x >= 0
   sqrt_x = sqrt(x);
else
   disp('ERROR: x < 0');
   sqrt_x = 0;
end
```

2.
```
if abs(denominator) < 1.0E-300
   disp('Divide by 0 error.');
else
   fun = numerator / denominator;
   disp(fun);
end
```

3.
```
if distance <= 100
   cost = 0.50 * distance;
elseif distance <= 300
   cost = 50 + 0.30 * (distance - 100);
else
   cost = 110 + 0.20 * (distance - 300);
end
```

4. These statements are incorrect. For this structure to work, the second `if` statements would need to be an `elseif` statement.
5. These statements are legal. They will display the message `"Prepare to stop."`

6. These statements will execute, but they will not do what the programmer intended. If the `temperature` is 150, these statements will print out `"Human body temperature exceeded."` instead of `"Boiling point of water exceeded."`, because the `if` structure executes the *first* `true` condition and skips the rest. To get proper behavior, the order of these tests should be reversed.

Quiz 3.3, page 134

1.
```
x = 0:pi/10:2*pi;
x1 = cos(2*x);
y1 = sin(x);
plot(x1,y1,'-ro','LineWidth',2.0,'MarkerSize',6,...
     'MarkerEdgeColor','b','MarkerFaceColor','b')
```
3. `'\itf\rm(\itx\rm) = sin \theta cos 2\phi'`
4. `'\bfPlot of \Sigma \itx\rm\bf^{2} versus \itx'`
5. This string creates the characters τ_m
6. This string creates the characters $\boldsymbol{x_1^2 + x_2^2}$ (units: $\mathbf{m^2}$)
7. The backslash character is displayed using a double backslash (`'\\'`).

Quiz 4.1, page 174

1. 4 times
2. 0 times
3. 1 time
4. 2 times
5. 2 times
6. `ires` = 10
7. `ires` = 55
8. `ires` = 25;
9. `ires` = 49;
10. With loops and branches:
```
for ii = -6*pi:pi/10:6*pi
   if sin(ii) > 0
      res(ii) = sin(ii);
   else
      res(ii) = 0;
end
```

With vectorized code:

```
arr1 = sin(-6*pi:pi/10:6*pi);
res = zeros(size(arr1));
res(arr1>0) = arr1(arr1>0);
```

Quiz 5.1, page 222

1. Script files are collections of MATLAB statements that are stored in a file. Script files share the Command Window's workspace, so any variables that were defined before the script file starts are visible to the script file, and any variables created by the script file remain in the workspace after the script file finishes executing. A script file has no input arguments and returns no results, but script files can communicate with other script files through the data left behind in the workspace. In contrast, each MATLAB function runs in its own independent workspace. It receives input data through an input argument list, and returns results to the caller through an output argument list.

2. The `help` command displays all of the comment lines in a function until either the first blank line or the first executable statement is reached.

3. The H1 comment line is the first comment line in the file. This line is searched by and displayed by the `lookfor` command. It should always contain a one-line summary of the purpose of a function.

4. In the pass-by-value scheme, a *copy* of each input argument is passed from a caller to a function, instead of the original argument itself. This practice contributes to good program design because the input arguments may be freely modified in the function without causing unintended side effects in the caller.

5. A MATLAB function can have any number of arguments, and not all arguments need to be present each time the function is called. Function `nargin` is used to determine the number of input arguments actually present when a function is called, and function `nargout` is used to determine the number of output arguments actually present when a function is called.

6. This function call is incorrect. Function `test1` must be called with two input arguments. In this case, variable `y` will be undefined in function `test1`, and the function will abort.

7. This function call is correct. The function can be called with either one or two arguments.

Quiz 6.1, page 283

1. (*a*) `result` = 1 (true), because the comparion is made between the real parts of the numbers (*b*) `result` = 0 (false), because the absolute values of the two numbers are identical (*c*) `result` = 25
2. The function `plot(array)` plots the imaginary part of the array versus the real part of the array, with the real part on the x axis and the imaginary part on the y axis.
3. The vector can be converted using the `double` function.
4. These statements concatenate the two lines together, and variable `res` contains the string `'This is a test!This line, too.'`.
5. These statements are illegal—there is no function `strcati`.
6. These statements are illegal—the two strings must have the same number of columns, and these strings are of different lengths.
7. These statements are legal, producing the result `res` = $\begin{bmatrix}\texttt{This is another test!}\\ \texttt{This line, too.}\end{bmatrix}$. Note that each line is now 21 characters long, with line 2 padded out to that length.
8. These statements are legal, and the result `res` = 1, since the two strings are identical in their first five characters.
9. These statements are legal, and the result is `res` = `[4 7 13]`, since the letter "s" is at those locations in the string.
10. These statements are legal. Each space in the original string is replaced by an `'x'`, and the final string is `'Thisxisxaxtest!xx'`.
11. These statements are legal. The function isstrprop returns a `1 (true)` for alphanumeric characters, and a `0 (false)` for other characters. The result is

```
res =

       1   1   1   1   0   1   1   1   1   0   0   0
```

12. These statements are legal, with the result `res` = `'ThiS IS a test!'`.
13. These statements are legal. The results are `11` = 9, `12` = 9, `13` = 18, `14` = 6, and `15` = 12.
14. These statements are illegal—you must specify the number of characters to compare in the two strings when using function `strncmp`.

Quiz 7.1, page 331

1. A cell array is an array of "pointers," each element of which can point to any type of MATLAB data. It differs from an ordinary array in that each element of a cell array can point to a different type of data, such as a numeric array, a string, another cell array, or a structure. Also, cell arrays use braces `{}` instead of parentheses `()` for selecting and displaying the contents of cells.

2. *Content indexing* involves placing braces `{}` around the cell subscripts, together with cell contents in ordinary notation. This type of indexing defines the contents of the data structure contained in a cell. *Cell indexing* involves placing braces `{}` around the data to be stored in a cell, together with cell subscripts in ordinary subscript notation. This type of indexing creates a data structure containing the specified data and then assigns that data structure to a cell.

3. A structure is a data type in which each individual element is given a name. The individual elements of a structure are known as fields, and each field in a structure may have a different type. The individual fields are addressed by combining the name of the structure with the name of the field, separated by a period. Structures differ from ordinary arrays and cell arrays in that ordinary arrays and cell array elements are addressed by subscript, while structure elements are addressed by name.

4. Function `varargin` appears as the last item in an input argument list, and it returns a cell array containing all of the actual arguments specified when the function is called, each in an individual element of a cell array. This function allows a MATLAB function to support any number of input arguments.

5. (*a*) `a(1,1) = [3x3 double]`. The contents of cell array element `a(1,1)` is a 3×3 double array, and this data structure is displayed.

   (*b*) `a{1, 1}` $= \begin{bmatrix} 1 & 2 & 3 \\ 4 & 5 & 6 \\ 7 & 8 & 9 \end{bmatrix}$. This statement displays the *value* of the data structure stored in element `a(1,1)`.

   (*c*) These statements are illegal, since you can not multiply a data structure by a value.

   (*d*) These statements are legal, since you *can* multiply the contents of the data structure by a value. The result is $\begin{bmatrix} 2 & 4 & 6 \\ 8 & 10 & 12 \\ 14 & 16 & 18 \end{bmatrix}$.

(*e*) `a{2, 2} =` $\begin{bmatrix} -4 & -3 & -2 \\ -1 & 0 & 1 \\ 2 & 3 & 4 \end{bmatrix}$.

(*f*) This statement is legal. It initializes cell array element `a(2,3)` to be a 2×1 double array containing the values $\begin{bmatrix} -17 \\ 17 \end{bmatrix}$.

(*g*) `a{2, 2}(2, 2) = 0`.

6. (*a*) `b(1).a - b(2).a =` $\begin{bmatrix} -3 & 1 & -1 \\ -2 & 0 & -2 \\ -3 & 3 & 5 \end{bmatrix}$.

(*b*) `strncmp(b(1).b, b(2).b, 6) = 1`, since the two structure elements contain character strings that are identical in their first six characters.

(*c*) `mean(b(1).c) = 2`

(*d*) This statement is illegal, since you cannot treat individual elements of a structure array as though it were an array itself.

(*e*)
```
b = 1x2 struct array with fields:
        a
        b
        c
```

(*f*) `b(1).('b') = 'Element 1'`

(*g*)
```
b(1) =
    a:  [3x3 double]
    b:  'Element 1'
    c:  [1 2 3]
```

Index

Note: **Boldface** numbers indicate illustrations or tables.